Applied Crime Analysis

Applied Crime Analysis

A Social Science Approach to Understanding Crime, Criminals, and Victims

Wayne Petherick

AMSTERDAM • BOSTON • HEIDELBERG • LONDON
NEW YORK • OXFORD • PARIS • SAN DIEGO
SAN FRANCISCO • SINGAPORE • SYDNEY • TOKYO

Anderson Publishing is an imprint of Elsevier

Acquiring Editor: Sara Scott
Editorial Project Manager: Marisa LaFleur
Project Manager: Punithavathy Govindaradjane
Designer: Russell Purdy

Anderson Publishing is an imprint of Elsevier
225 Wyman Street, Waltham, MA 02451, USA

Library of Congress Cataloging-in-Publication Data
Petherick, Wayne.
    Applied crime analysis : a social science approach to understanding crime, criminals, and vicitms / Wayne Petherick.
        pages cm
    ISBN 978-0-323-29460-7
1. Crime analysis. 2. Criminal psychology. 3. Victims of crimes. 4. Crime–Sociological aspects. I. Title.
HV7936.C88P466 2015
363.25–dc23   2014006137

British Library Cataloguing in Publication Data
A catalogue record for this book is available from the British Library

ISBN: 978-0-323-29460-7

For information on all Anderson publications visit our website at http://store.elsevier.com

This book has been manufactured using Print On Demand technology. Each copy is produced to order and is limited to black ink. The online version of this book will show color figures where appropriate.

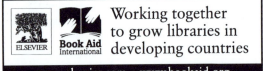

ELSEVIER Book Aid International Working together to grow libraries in developing countries

www.elsevier.com • www.bookaid.org

Contents

Acknowledgments

I was once told at a meeting that it takes, on average, 3 to 5 years to write a book. There were no specifics as to the type of book: a discipline based dictionary or encyclopaedia; a single authored text; an edited text; any other type of work? The message was also not clear as to how many chapters, or how long each chapter was. Still, I remember thinking at the time: *that seems awfully long*.

As the reader will be aware (at least those who have read the Foreword will know), I never believed I was up to the task of writing a text. *Who I am to write a book, the kid who spent half his school year in detention?* I scoffed at the very idea, but a plain and simple challenge, the throwing down of the proverbial gauntlet, got me to make a plan, seek out a publisher, and execute it over the course of about 18 months from conception to printing. When you don't believe you can do something, the best thing to do is to prove that you can.

Three to five years? Surely not.

It is at this point that I should toot my own horn, ring my own bell, blow my own whistle, and shout from the rooftops what an amazing writer I am. How the words flow from my fingers, and magically appear on the screen in front of me by the thousands. How the copy editors mail me with their praise and claims that they can now go home early because so little work needed to be done on some of the most amazing writing they have ever seen.

And then the bubble bursts, the coconut-borne drink in my hand disappears in a cloud of Looney Tunes-style dust, the sandy beach I am reclining on turns back into my desk, and my reality comes crashing back to my not-too-comfortable chair. The breeze in my hair isn't a Pacific breeze carried on trade winds, and the rustling isn't the gentle back and forth sway of palm fronds. It is the gale of the fan, and the rustle of the hundreds of pieces of paper that line my desk at any given moment, ready for the transition from one chapter to the next.

In this reality, my Mac and I engage in daily staring competitions (we are currently 64, computer's favor). An important journal article it took me forever to find is blown off the table by my fan where it is promptly chewed by the dog, or returned as a colorful drawing by my well meaning nine-year-old, or as a list of friends my now teenager has drawn up for her weekend plans. And there is always just enough pen or pencil over that actual part that I wanted the article for in the first place as to render it useless. Oh, and this reality involves far too much coffee and cursing.

This fantasy carries me from work to work, but the reality is somewhat different. While I can look back proudly at what has come from these efforts, and they are an effort, I cannot claim grand success as a writer or the Bradley Cooper-esque drug-enhanced-speed-writing authorship of the movie *Limitless*. If I am honest with myself, writing is something that I have always struggled with. The words don't always come easy, and they don't always flow. Books don't magically appear on my computer, and copy editors don't email me praise.

But I think writing shouldn't be easy. I think if it was, we would be missing something. And even when the words do seem to come (such as the report writing chapter, compiled in less than 8 hours or so), it is because they are the result of hundreds of hours of reading, research, thinking, distilling, culminating, critiquing, applying, testing, falsifying, reworking, reconsidering, teaching, and abandoning. Even a chapter that takes less than 8 hours to complete is a labor of hundreds, if not thousands, of hours of work that has led you to that point.

So while we may be flippant with the time it took to actually write, the effort is far more than the time it took where the end result is words on a page.

Returning now to the time it takes to write. I will admit I have never taken 3 years to do any book project and would likely lose interest if it did. Approximately 90% of this book was finished in a period of six to seven weeks. So this then begs the question of how these projects are possible, given the usual timeframe of several years, and the more than occasional struggle with writing?

Well, by accident or design, the answer is relatively simple: you surround yourself with good and competent people and you rely on and trust them to do the job you have set for them. If they can't, you have chosen the wrong people. But when you find them, and you will not mistake when you do, each and every project becomes easier and easier as you know each other's strengths, weaknesses, style, and approach. They will know what you want before you ask for it, and they will each have their part finished before it is due. It is working with the right people that allows these projects to be finished well before the 3- or 5- year point. They will cut down the total amount of work required, and render any editorial work obsolete beyond the most basic of keystroke error. They will make every effort an absolute joy, and they will guarantee that you will look back on every single aspect of every single project with a smile and a sense of satisfaction.

This I can tell you from experience.

So, with that in mind, I offer up the following heartfelt thanks, in no particular order. Without those below, this work would have taken me 3 years (more realistically 10).

Well there is only one Claire Ferguson, and some would say thank God. I recall the very first class Claire took as a Postgraduate student and remember thinking *you'll go far*. And she did. Not only did she top pretty much every class, but she became my first PhD student as a supervisor. When her first draft arrived, I read it and thought I had failed her: I couldn't find anything wrong and assumed it was my error as a critical supervisor. As it turns out, the three examiners it went to couldn't find anything wrong either, and it just so happens it was a quality work with a unanimous three *pass no revisions* verdict returned. It was just that good (I'd like to take more credit but I can't). Since then, we have worked together on many projects, chapter, journal, and text, and it has always seemed less like work. We taught together for four and a half years, and they were among the best work years of my life, such that it never seemed like actual work. But something conspired against us and another institution now enjoys the fruits of her labor (her current boss refers to her as their Black Caviar- a reference to an Australian thoroughbred racehorse undefeated in 25 races). Despite this, we will always write together, and she may not have as much choice in this as she thinks. Love working with you Brosephine!

Those who know Grant Sinnamon would describe him as a cross between Thor and the Energiser Bunny. He is like Hawking without the wheelchair, and by 2015, he will have more degrees than a thermometer. From his knowledge of the history of the world through to his understanding of functional neuroanatomy, every conversation is like a lesson wrapped in a mental gym session. Grant proves the if-he-doesn't-know-it-it's-not-worth-knowing rule. Beyond this he is a humanitarian and a damned fine person and his family are equally great. While we had a gap in the middle, I consider him one of my oldest and best friends.

Andy Rowan and I met many moons ago (I won't count, it makes me feel old), and since then our paths have crossed a number of times. I run into Andy now and then at an Australian and New Zealand Forensic Science Society meeting, or should he be able to teach the odd subject here and there. It's always a pleasure to see him and if I'm honest I'd say I'd like to catch up for a few more coffees.

Despite my best efforts, I lost contact over the chapter project with Andy, and despite this he didn't reject my 11th hour pleading and came on board and this chapter turned out to be one of my favorites in the book. In short, you're a bloody legend!

Jess Gormley took a subject with me whilst doing her Master of Psychology, and she was a great student, and she has become a great practitioner. She answered the call and provided a great overview on risk assessment including problems with the practice, and a number of other areas. She has extensive experience in the field, and has since returned to her home town but I am sure our paths will cross repeatedly while she is doing her PhD. At least that is my hope. Thanks for your input Jess, your contribution to the book was greatly appreciated, as was your insight. Keep your computer handy, your name is now on my list!

I first met Jim Cawood in Las Vegas in 2000. I was familiar with Jim's work in *The Psychology of Stalking*, and can say he came across as a genuinely great guy, and very down to earth. This is rare for someone of Jim's celebrity (I know he probably wouldn't be comfortable with my use of that term!), and we had a great conversation over a beer at the bar of the Holiday Inn on the strip. He asked me what areas I was interested in and when I mentioned stalking, he asked "have you ever been stalked?" When I replied in the negative, he said "stick with this area long enough, and you will be". Over the years our paths crossed from time to time, and he has given me great advice when I needed it. As a result, when I decided to include a chapter on threat management, I could think of no one better. Thanks for your efforts and your input Jim, both were greatly appreciated and this work is so much the better for it.

And of course who could forget Natasha Petroff? I get many Facebook friend requests from people I don't know and this is one I am glad I accepted. Within a brief period of time, I could tell she had the drive necessary to succeed and she took on editorial duties with the appropriate amount of humility, gusto, and vigor. I piled enough work on her to floor a pack mule, and she rose to the challenge without so much as a sniffle, whine or "but my back…" She cracked the whip when needed, listened to my numerous complaints without complaint, and provided support and encouragement as and when required. I couldn't have asked for any more, you were part of the team Tash, and you never disappointed. I'd tell you to stock up on red pens for the next project, but you already have more stationery than Officeworks, so I'll just say grab your favorite.

Last but by no means least is Yolande Huntingdon. Yolande is one of my many students-turned-colleague and she has given me great ideas and has been inspirational on many fronts. When asked to compile a chapter on bullying and harassment for *Profiling and Serial Crime* Yolande was the first to come to mind and she didn't disappoint. Her efforts could only be described as professional and tireless. When she commits to a job, it gets done to the best of her abilities (which are considerable) and usually well within the timeframe she is given. If not for the fact that I know she has her own time consuming research projects, I would have asked her to take more of an active role in this book. When asked if she would like some editing work, she took to the task in her typical way - efficient, professional, and in a timely manner. Thanks Yolande, your efforts are greatly appreciated as usual!

There will undoubtedly be new projects, and hopefully new editions, and these will be added to and revised as the need arises. What some of these may be is a mystery at this point, and will be dictated by new developments and industry demands. The variables are too numerous to account for as we stare into the looking glass, but one variable you can count on will be those names that appear above.

About the Authors

Wayne Petherick, PhD

Wayne is associate professor of criminology in the Faculty of Society and Design at Bond University on Queensland's Gold Coast and is author or editor of *Profiling and Serial Crime* now in its third edition, and *Forensic Criminology*, and has published journal articles in the areas of criminal profiling and false reports in stalking. He teaches criminal profiling, criminal motivations, crime and deviance, forensic criminology, applied crime analysis, and profiling and crime analysis at the undergraduate and postgraduate levels. Wayne also supervises higher degree by research students with supervision ranging from the development of risk assessment tools to an examination of the emotional and psychological factors that lead to victimization. Wayne is currently conducting research on the use of heuristics in case linkage, and risky online behaviors for children and adolescents.

In addition, Wayne has lectured on criminal profiling, stalking, and bullying in Australia and the United States and he has consulted on a variety of cases including numerous homicides, sexual assaults, stalking, and risk assessments in Australia and overseas. He is a frequent media commentator, through radio, television, and print, on stalking, homicide, risk and threat, and other matters of criminological importance.

Wayne can be contacted on wpetheri@staff.bond.edu.au or wpetheri@me.com.

Claire Ferguson, PhD

Claire holds a Bachelor of Arts (Honors) in Psychology from the University of Western Ontario in Canada, a Masters degree in Criminology, and a PhD in Criminology, both from Bond University in Australia. Dr Ferguson has worked for St. Leonard's Society, as well as interned with Queensland Fire and Rescue Services (Fire Investigation Unit). In 2008, she undertook an internship with Forensic Solutions, and in 2012 and 2013 Claire worked as a project manager and senior research fellow at The Australian Centre for Arson Research and Treatment, where she conducted research on homicides involving fire in Australia. She is currently a lecturer, researcher, and consultant in forensic criminology at the University of New England in New South Wales, Australia.

Dr Ferguson's doctorate (2011) was an analysis of staged crime scenes in homicide cases internationally. This project was the first systematic review of these types of cases to ever be conducted. Dr Ferguson has been nominated for the "Excellence in Teaching and Learning" award and the "UniJobs Lecturer of the Year" award by her students. She has published in the areas of victimology, crime scene staging, false allegations, profiling, and forensic criminology.

Jessica Gormley

Jessica holds a Bachelor of Applied Science (Psychology; Honors) from Royal Melbourne Institute of Technology in Melbourne, Victoria, and a Master of Psychology (Forensic) from Bond University. Her Masters dissertation focused on validating and expanding the existing typologies of sexual offenders.

Jessica has worked with a diverse range of offenders across correctional settings and has also tutored in research methods, forensic psychology, and psychological assessment at Bond University. She is currently working as a psychologist and is also a PhD Candidate in the Faculty of Society and Design at Bond University, with her research focusing on risk of recidivism for sexual offenders. Jessica's area of interests include psychological assessment, sexual and violent offenders, psychopathy, drug related crime, and personality disorders.

Grant Sinnamon, BPsych (Hons), MCouns, PhD

Grant has a background in adult education, psychology, counseling and medicine (psychiatry and psychiatric neuroscience). Grant's interest in criminology stems from his work in the area of early-life adversity and stress and their impact on neurobiological developmental trajectories, personality formation and consequent neuropsychological, behavioral and physiological functioning. Grant divides his time between research, teaching, private counseling practice, consulting, and facilitating professional development workshops for mental health, family law, education, and social welfare professionals. Grant's passions have a dual focus. The first is childhood onset autoimmune disease and his work in understanding the psychoneuroimmunology of Type 1 diabetes. Grant's second passion is the field of the developmental neurosciences and in particular understanding and treating the neurological deficits that can occur as a result of early-life developmental interference such as neglect, abuse, trauma, and illness as well as those associated with developmental disorders such as autism spectrum disorder, sensory processing disorder, attention deficit disorders, and other conditions that impact brain development and function.

Grant is the developer of the REPAIR™ Model of Neurofunctional Intervention: a direct neuroplasticity-based six-step intervention model for children and adults with neurofunctional deficits, and is the cofounder of the HART-BEAT Research Group (Health professionals And Researchers Together—Blood, Endothelium, And Tissue) at James Cook University in Australia—a multidiscipled research initiative that brings clinicians and research scientists together to work on clinical health problems. Grant is married to Natalie and together they have four children.

James S. Cawood

Jim is president of Factor One, which is a California-based corporation specializing in violence risk assessment and management, threat assessment, behavioral analysis, security consulting, and investigations. Mr. Cawood has worked in the area of threat and violence risk assessment and management, behavioral analysis, violence prevention, security analysis, and incident resolution for more than 25 years. He has successfully assessed and managed over 4000 violence-related cases for federal and state government agencies, universities and colleges, public and private corporations, and other business entities throughout North America. Mr. Cawood has also consulted and trained these types of organizations on the design and implementation of threat assessment, violence risk assessment, and incident response protocols. This included participation in the development of the P.O.S.T. (Peace Officers Standards and Training) telecourse for Workplace Violence for California and Arizona law enforcement, California P.O.S.T. development of a Fear and Anger Control telecourse, Association of Threat Assessment Professional's Risk Assessment Guideline Elements for Violence (RAGE-V), ASIS International's Workplace Violence Prevention and Response guideline, and participation in the development of a threat assessment protocol by the California District Attorney's Association for use in training judges, District Attorneys, and others in the California justice system. He also has served as an expert witness in dozens of cases involving questions concerning investigative and security issues, including threat assessment and violence in the workplace.

Mr. Cawood serves on the Editorial Board of the *Journal of Threat Assessment and Management* (APA) and is the former President of the Association of Threat Assessment Professionals (ATAP). He has also served as Association 2nd Vice President and the President of the Northern California Chapter of ATAP; on the ASIS International Foundation Board and was Secretary of the Board; as the Chairman of the Board of the California Association of Licensed Investigators and has served as Chairman of their Legislative Committee; as a Board member for the Association of Workplace Investigators; and as Law Liaison to the Student Section of the American Psychology-Law Society (Division 41) of the American Psychological Association.

Mr. Cawood is a graduate of U.C. Berkeley and holds a Masters in Forensic Psychology from Argosy University. He has served on the faculties of Golden Gate University, in their Security Management degree program and the University of California, Santa Cruz extension, teaching threat management. He is a Certified Protection Professional, Professional Certified Investigator, Physical Security Professional, Certified Fraud Examiner, Certified Security Professional, Certified Professional Investigator, and a Certified International Investigator. He has written articles and book chapters for various professional publications including *Security Management* and a coauthored chapter in *The Psychology of Stalking: Clinical and Forensic Perspectives*, published by Academic Press in 1998. He has also coauthored the book, *Violence Assessment and Intervention: The Practitioner's Handbook* with CRC Press in 2003, and contributed to *Ending Campus Violence: New Approaches to Prevention*, by Brian Van Brunt, Routledge, in 2012.

Andrew (Andy) Rowan

Andrew Rowan holds a B. App. Sci. (Built Environment) from Queensland University of Technology and an M. Sci (Forensic Science) from Griffith University. He is a Senior Sergeant of Police with the Queensland Police Service and commenced his forensic career as a Scenes of Crime Officer in 1994 attending volume and major crime scenes, eventually progressing to a management role in charge of a large crime scene unit. In 2002 he moved to a forensic scientist position performing major crime scene examinations. Andrew currently holds the position of Senior Forensic Scientist leading a small specialist unit that performs origin and cause determination of fire and explosion scenes.

Foreword

I have been working in tertiary education for over 15 years now. Previously, while serving in the Australian Regular Army as a Driver/Signaller in an M113 Armoured Personnel Carrier in the Royal Australian Armoured Corp, I decided to pursue studies in law in the hope of becoming a police officer. While studying law, I encountered a subject titled Human Behaviour Perspectives for Police, and found this side of human behavior and psychology to be far more interesting than any aspect of law I had studied. I changed the direction of my studies and eventually found myself enrolled in a Bachelor of Social Science (Psychology) at the Queensland University of Technology.

While studying, I had the opportunity to meet with and learn from an amazing lecturer, Dr Tricia Fox. Tricia's teaching style and the ability to build rapport with her students was amazing, and I found myself having many talks outside of the class with her about the area, my studies, and my future prospects. I learned that she had worked in the prison system, and found this aspect of her past employment fascinating. I started to direct my studies toward the broader criminal justice system, and urged on by Tricia, wrote all the essays I would on criminal behavior. I became "that" undergraduate who was never happy with the stock essay questions provided by the lecturers, and became frustrated when Tricia would let me follow my interests but others would not. Still, I wrote what I could on criminal behavior, and with no shortage of nagging, other teaching staff eventually caved, allowing me to follow my interests.

Over the last year of my study, I had begun to doubt—for a variety of reasons—my imagined career trajectory. I blazed through and applied for a number of fourth year programs, and was accepted into three for the beginning of my advanced studies in psychology. However, looking at the path ahead and considering the rigid and relatively inflexible pathways characteristic of the discipline, I became increasingly uncertain that this was my right choice.

Flustered, confused, and uncertain, I once again sought Tricia's counsel. She told me that she had been waiting for this conversation and that she had known for some time where my real interests lay. It was her opinion that psychology was not for me, and that I should pursue my interests in criminal behavior in a way that would allow me the freedom to pursue those areas in which I had most interest. She recommended I call Bond University as they had a criminology program and may be able to guide me.

I called the university and was invited down. From the second I walked on campus, I knew this is where I had to study. Located in a residential area of the suburb of Robina on Queensland's Gold Coast (one of the biggest tourist destinations in Australia), the sandstone fascia of the buildings and the sprawl around a purpose built keyhole lake were nothing but stunning. A light breeze seemed to blow nonstop through campus taking the sting out of the summer heat. I met with staff from the criminology program, filled out the paperwork, and left for the day knowing I would be back.

I went home and organized my finances, hearing back shortly after that my application had been accepted and that I was due to start the following semester. I could not have been happier.

I went to my first class around the middle of January in 1998. Still a little dazed and confused I attended my first lecture and tried to find my student colleagues. It was exciting but daunting at the same time. It became obvious that many people here already knew one another, some better than others.

I was to later learn that some of those enrolled in the Master of Criminology program were tutoring in some areas. I found this prospect fascinating.

I made an appointment to meet with the criminology staff to introduce myself and discuss my studies. I was to find that the department was very new, and that they were looking to expand their subject offerings. I proposed the subject criminal profiling, as it was a topic very much en vogue within law enforcement agencies around the world. My proposal was accepted, and I took the project on as a special study in criminology in the second semester of my enrollment. I compiled readings, sought out other resources, put together some cursory lectures slides, and planned applied tutorial activities. At the end of semester, I submitted my assessment materials and found that I had gone above and beyond the expectations of the study. My supervisor felt that this warranted a place in the subject beyond compiling materials and asked me to give two lectures. I had made the right decision in coming here—and I teach this same subject to this day.

Over time I got to know the other teaching staff well, and was asked to step in for a tutor who was sick one day to run a tutorial. I found that I did pretty well, and some of the students wrote in their evaluations that I should be given more work. Around this time, it was common for the postgraduate students to meet up for a drink at the campus tavern. The teaching staff would often join us. Some of the current tutors had taken on a lot of work and were becoming less available for criminology tutoring. Luckily, this was about the same time as a boom in student numbers, meaning there was an increased demand for tutoring work and a decrease in the number of tutors available. I said that I would be happy to help as a stop-gap measure in between my search for employment in the industry.

I started tutoring in September of 1998.

The numbers grew and grew and so did my hours. I found myself tutoring in a diverse array of subjects from crime and deviance to criminal profiling to crime prevention and even in the core subject communications skills. I had a voracious appetite for knowledge and read all that I could get my hands on. By this time I had a good grasp on criminal profiling and had started to network with experts in Australia and overseas. I had submitted papers to [then] The Crimelibrary on profiling, victimology and risk assessment, which were accepted and put online. I was also asked to be a research assistant; a position I happily took on.

In addition to my tutoring work and the writing I was doing, my newfound position as a research assistant added a whole new dimension to what I was doing. Working with established professionals exposed me to a variety of things I would not usually have access to as a student. I was doing research in criminology and forensic psychology. I was helping do research on the impact of the media on crime perceptions, the role of victim's groups, security in the banking industry and how this impacts on armed robberies, and even on homicide cases.

My supervisor received a telephone call about a stalking case, admitting that he had no experience in these matters, and asked if I would be interested in looking at it. I had developed an interest in this area knowing it was a prevalent crime, and knowing that profiling in serial murders in Australia was a largely misplaced goal. I was introduced to two private security professionals and together we worked on a cyberstalking case, being able to identify some key features, including the likely offender, and were able to work with the family to successfully resolve the case. This was my first foray into the world of what I now know as *Applied Crime Analysis*.

The proverbial finger was out of the dyke now, and beyond advancing my own studies further, I continued to do research work and assist others in the preparation of their reports. My own casework expanded over time, and based on networks I had established I was asked to prepare a chapter for a second edition work to be published early in the new millennium.

Around this same time I was asked to look into a homicide for the University of Melbourne Innocence Project. RB had been imprisoned for the murder of a family friend, with the victim's wife listed as a coaccused. Looking through the materials, I was to find that there were a number of investigative and legal shortcomings that needed to be addressed before "beyond reasonable doubt" became reasonable, despite the fact RB was prosecuted and jailed. I had begun to understand that the legal system provided different levels of protection depending on what you could afford and the degree to which odds were stacked against you, by the prosecutor and a host of other factors.

Following this I was asked to look at a multiple homicide case and was able to further refine my analytical skills by not only learning what needed to be done and how to do it, but also from working with others where information was needed from areas outside my own. I worked with forensic psychologists and forensic scientists, at home and abroad, and had the opportunity to take training in many areas including advanced studies in criminal profiling, victimology, psychopathology, crime reconstruction, and arson. I did as much as I could whenever I could, and found as a result I was asked to work on more cases.

Noting a number of deficiencies in the profiling books available, I had discussed my issue with this with a colleague. He was to suggest that if I was not happy, that I should do something about it. I recall thinking during my undergraduate studies that a textbook must be a mammoth effort based on decades of research, study, and experience. At this point I had no confidence that I could put together anything meaningful. I was told, "look around you and if you think you could do better, then you are the right person for the job". I found a publisher, put together a plan, and started to write. Having taught the profiling subject for a number of years at this point, I used the subject outline as a plan. From here, the first edition of *Serial Crime* was born and published in 2004, now in its third edition as *Profiling and Serial Crime* with Andersen, an imprint of Elsevier Science. Serial Crime followed hot on the heels of the publication of *Criminal Profiling* published by The Book of the Month Club and Reader's Digest in Australia, the UK, and the USA.

Throughout this time, the offerings at Bond expanded exponentially. Largely as a result of the cases I was working, I proposed new subjects in the areas of alcohol, drugs, and crime, and criminal motivations. It was around this time that we also introduced a new advanced profiling subject.

While I found the profiling useful for understanding many aspects of a case, I found an increasing reluctance on the part of the criminal justice and legal communities toward the use of profiling. This was likely owing to the publication of a number of failures in addition to the failure of the community to come together on key aspects of theory. Around this time I started to believe something more was needed. Something more holistic that could address a far greater range of criminal behaviors than profiling alone would allow; something that could provide more direction based on the unknowns as well as the knowns.

Over time, I started to depart from traditional profiling practices, and consider crime analysis as a distinct practice that resided in its own universe. While it was good to rely on the physical evidence as a relatively independent arbiter of the facts, physical evidence can be and is often misinterpreted, misrepresented, or simply not collected. Even when it is collected, it is not always presented, and so aspects of a case may remain unknown. In addition, current "best practice" leant toward what is known as *Structure Professional Judgement*. This means that while the analyst (or clinician) is free to make judgments in a given case based on their experience and training, these judgments must be anchored or weighted within the research studies carried out in respective areas. This approach goes a long way to reducing the human errors inherent in many endeavors.

My experience with cases and the reading I was doing was to restructure my teaching in a variety of ways. Most notably, the way I approached profiling had changed subtly, taking a step back from this as a single approach and incorporating it into a larger practice, if and where appropriate and relevant. I also started to use my cases as teaching material, in case-based reasoning. My students had access to the same material I had, they were asked the same questions I was asked, and they produced the reports that I produced.

I found that in many (approaching all) of the cases that I had been working as a social scientist, there were significant shortcomings in a variety of domains, including false or questionable alibis from key parties, information that had been treated as a fact even though it had not been established, suspect victim or witness statements, and the lack of examination of physical evidence in cases where it was pivotal to the final verdict. I am not suggesting at all that these cases involved corruption or incompetence, but all aspects of the criminal justice system are under considerable strain, either for time or resources, and things may be missed or overlooked as a result. Having said that, corruption and incompetence no doubt play a role in some cases.

Changing from an advanced profiling subject to *Applied Crime Analysis*, I found there were no texts in the area that adequately covered the subject areas we were addressing. Books of readings were provided that highlighted aspects of the case the students were being assessed on, but even these could not meet every demand. Early in 2013, I was approached by a student asking for a publication that outlined, in detail, the writing style and structure required of the report for assessment. Knowing there was not anything that suited this request, I sat down and compiled a "how to" guide for this very purpose.

Upon completion, I looked at what had become a lengthy guide, and felt that it would make a good book chapter. In as much as this chapter addressed those issues which were its focus—those elements of crime analysis, such as content and layout—it seemed to stand alone. What of those issues relating to victimology? Risk assessment? Threat management? Or physical evidence? Where then, was the rest of the work?

And in that moment, on that overcast Saturday, sitting on my couch, watching a movie, this book was born.

The general area known as crime analysis is a truly multidisciplinary area, relying on the input and knowledge of a number of different individuals, all of whom bring something different to the table. There is the forensic scientist whose role is to examine physical evidence, the victimologist who speaks to issues relating to the victim and their role within the offense, and the psychologist or other practitioner who speaks to risk and psychopathology, among others. This field combines aspects of practicality, such as what one does; with theories on which assessments are often based within the guidelines of research and best practice; in addition to the academic practitioner who can offer the best of both worlds.

As such, no one discipline can lay claim to the totality of the area of applied crime analysis. We need to acknowledge that work product, and not celebrity, is the key to answering difficult questions about criminal behavior and the context in which it occurs. This poses a challenge to the applied crime analyst who may, dependent on jurisdiction, be surrounded by practitioners who rely more on who they are than what they do. These challenges may not be so easily overcome.

To compound this, the contribution of certain disciplines may be downplayed, dismissed, or overlooked in an attempt to provide a quick and easy answer. Social scientific inquiry takes time, an enduring course of investigation that brings to bear the best of what methods are available and the manner in which these are employed. The scientific method is a time consuming and resource intensive endeavor that, by established standards, provides no fast answer regardless of the demand of consumers and clients. Quick and easy are grist for the media and occasionally for lawyers that demand simple answers, but no such things exist in the complex realm of human—and by extension—criminal behavior.

As such, the best answers take time. They involve scrutiny, skepticism, and disbelief. They require that when all others have abandoned hope, that someone in the room is still asking *who, what, when, where, how,* and *why.* The social scientist is identified by the requirement that they demand more proof than supposition or accusation. That falsification not assumption is their core skill. And they must be able to demonstrate why this is not only important from a theoretical but also a practical point of view. And perhaps most importantly they must understand complexity. That even the minutiae of behavior can involve complexity that must be established, accounted for, and explained. And they will do this to the point that they are unpopular, so long as their premises are sound and rational, based on fact and good science. For it is them who demand that others do their job, and do it properly.

Within this, complexity is not to be taken lightly. For it is in complexity that we distinguish between tertiary education and many other fields of education. A university education, comprised of the proper elements, will have within it enough scope to consider the diverse and multidetermined nature of the topics encompassed within a degree spanning several years. Indeed, it is as much this which distinguishes such an educational tract as distinct from a short course or vocational model, for within these approaches there is little time and scope in which to canvas the depth of the subjects and their many facets. This is not, however, a pathway to discourage such endeavors. Rather, it is intended as a backdrop to the extensive course of study one must undertake if they wish to truly comprehend the nature of criminal behavior, account for its various elements, and to communicate this in a meaningful way to a variety of different audiences. This is no small task, and the input that culminates in a meaningful output cannot, nor should not, be downplayed.

It is at this juncture that the forensic practitioner, a broad mantle under which I would include the applied crime analyst, stands. They must be prepared to defend a position, again within scientific boundaries, at which their opinion lay. This is in no small part contributed to by their education, training, and experience.

But this too places a great demand on them. In the immortal words of John Thornton, *it is not unreasonable, when the life or liberty of an individual hangs in the balance, that the forensic practitioners know what they are doing.*

It is incumbent, therefore, on the applied crime analyst that they know what they are doing. That their analysis is the result of careful and thoughtful deliberation of the evidence. That they bring the scientific method, critical thinking and logic to bear in what may appear on the surface to be even the most basic of cases. For it is here that their training will be most useful. They will uncover what others have not, identify areas where others have failed, and provide answers where others have nothing but questions.

But this is not the end. Most importantly: Hold yourself as ethical when all others would not; sleep well when others do not; live with yourself when others cannot live with themselves. In the long run you will hold yourself as others would not, sleep when others do not, and live your life as others cannot.

As a final word on the subject, it is critical to note that the job of the crime analyst is not an easy one. There is a good chance that someone has died or suffered harm or loss to provide you with an opportunity to do your job. You will see some of the worst behaviors, encounter some disordered personalities, and witness the most horrible things that we humans can do to one another. You will learn coping mechanisms that you will need to do your job to the best of your abilities. You will laugh at things you see, not because they are funny, but because they are not. Surround yourself with good and competent people that allow you to focus on the task at hand without distraction, and to analyze and interpret that which you may not be able to. They will be your support and sometimes your crutch. They will help you provide the advice, understanding, and interpretation that form the body of your work product. They will challenge you to do and be better.

But most of all they make things easier, and I know because some of the best in the business helped me put this book together.

Wayne Petherick, PhD
Associate Professor of Criminology
Bond University

An Introduction to Applied Crime Analysis

1

Wayne Petherick

Bond University, Gold Coast, QLD, Australia

Key Terms

Crime analysis
Research analysis
Strategic analysis
Tactical analysis
Operational analysis
Intelligence analysis
Criminal profiling
Applied crime analysis
Premises liability
Crime prevention
Risk assessment
Threat management
Psychopathology

INTRODUCTION

Crime analysis is a general term for a relatively amorphous group of analytic procedures relating to crime and criminal behavior. Although the term conjures typical notions of the police analyst sitting behind a computer extracting crime trends and links from a database, it is a large area involving many different types of analysis. Despite this rather broad inclusionary criteria, this chapter—indeed this book—does not discuss every single area of potential analysis, but looks first at those analytical techniques more commonly referred to as crime analysis. From this, a specific form of crime analysis, dubbed Applied Crime Analysis, is discussed and is the focus of the remainder of this work.

This chapter demonstrates the distinct but often related areas of crime analysis; intelligence analysis; geospatial analysis; criminal profiling (a specific type of crime analysis); and finally, the applied form, relying on case-based reasoning. The following have been adapted from Burns and Gebhardt (2005), although not all types are discussed, especially those that relate more to management practices than to the analysis of crime or criminal behavior specifically.

1

1.1 Types of crime analysis

Crime analysis has existed since before the time of Sherlock Holmes, when the public was introduced to the analysis of facts and information to solve the crimes of the time (Dixon & Schaub, 2004). In reality, since there has been crime, there have been those who attempt to understand causes and contexts.

Because crime analysis is a very broad term, it encompasses a variety of different practices with different focuses and outcomes. Each type has particular characteristics that subtly change or alter the purpose of analysis. This is discussed by Boba (2013, p. 3):

> *The central focus of crime analysis is the study of crime (e.g., rape, robbery, and burglary) and disorder (e.g., noise complaints, burglary alarms, and suspicious activity) problems and information related to the nature of incidents, offenders, and victims or targets (i.e., inanimate objects, such as buildings or property) of these problems. Crime analysts also study other police-related operational issues, such as staffing needs and areas of police service. Thus, even though this discipline is crime analysis, in practice it includes much more than just the examination of crime.*

In considering the general purpose of crime analysis, Ribaux, Girod, Walsh, Margot, Mikzrahi, and Clivaz (2003, p. 54) state that "crime analysis aims at revealing problems, analyzing their potential causes and trying to foresee their development in order to determine where to best target law enforcement resources." Analysis of the applied type may also help to direct limited resources, but is less about foreseeability, typically focusing on events that have already occurred and trying to understand them in their totality. Any or all types may be used by an agency, even in the examination of single incidents of crime.

1.1.1 Research analysis

Research analysts (RAs) occupy the more traditional academic role whose function is to be familiar with the latest academic research, carry out longitudinal studies, and serve as the agency librarians (Burns & Gebhardt, 2005). While keeping up to date with the latest research reflecting good practice, the task of RAs may be to inform the department that certain policies or procedures are lagging or that current theory has surpassed current practice. Unfortunately, although timely advice may be provided about best practice, even this may be rapidly outdated with the slow turn of bureaucratic wheels.

1.1.2 Strategic analysis

It is the task of strategic analysts (SAs) to examine crime trends, and they may be found in agencies that have a data-driven approach to directing resources (Burns & Gebhardt, 2005). These analysts may be involved in examining long-term trends within data with a view to providing long-term outcomes. These analysts provide *nomothetic* data that relates to groups rather than individuals.

The primary purposes of SAs are to assist in the identification and analysis of long-term problems such as drug activity or vehicle theft and to study and evaluate responses and procedures (Boba, 2001). The types of procedures examined include "deployment and staffing, redistricting of beats or precincts, data entry and integrity, and the reporting process" (Boba, 2001, p. 14). The output of strategic crime analysis may end in a "round table" presentation of the findings to commanding officers, key stakeholders, and in some cases the public. Examples of this include the New York City Police Department's (PD's) "compstat" and Boston PD's crime analysis meetings (Manning, 2004).

1.1.3 **Tactical analysis**

Tactical analysts (TAs) are more involved in interpretation than the previous two types of analysts. Their task is to associate people or suspects with events, people with people, and events with events (Burns & Gebhardt, 2005). Two of the specific roles of TAs are to provide case linkage capabilities (discussed in Chapter 6) and by using complex statistical models to determine where a serial criminal may strike next. Burns et al. note that these analysts must possess a wealth of knowledge and a capacious memory because of the amount of case materials that must be read and processed.

Boba (2001) goes into more depth, stating that the role of the TA is to provide specific information about criminal events such as the method and point of entry, suspect actions, type of victim, type of weapon used, and further particulars of the crime itself such as the day, date, time, and location. Once compiled, this information is then disseminated to patrol officers and detective branches.

1.1.4 **Operational analysis**

It is suggested that operational analysts (OAs) are more akin to investigators and that they must know the suspect, follow up on leads, and interface with other units within the same agency and also with external agencies (Burns & Gebhardt, 2005). It should be noted that OAs may be civilians in some jurisdictions, and as such their investigative capacities may be limited. In fact, some jurisdictions may not use OAs at all. Others still may be relying more on this type of crime analyst in an effort to diffuse roles and to free up other investigators for duties requiring sworn personnel, such as the arrest and questioning of suspects.

1.1.5 **Intelligence analysis**

Intelligence analysts (IAs) can be found in both law enforcement and military agencies, as well as other "alphabet agencies" that are typically government run. Such agencies would include the Central Intelligence Agency, Australian Security Intelligence Organisation, the National Institute of Justice, and a host of others. According to Burns and Gebhardt (2005), IAs are involved in relationship and association diagramming (also called *Social Network Analysis*, see van der Hurst, 2009), financial analysis in money laundering and other areas, criminal enterprise analysis, telephone toll analysis, and analysis related to terrorist activity.

According to Boba (2001), a related goal of intelligence analysis is to link and prioritize information and to identify relationships and point out further areas of investigation. This information may come from the public, but much of it is generated by law enforcement or the agencies themselves and is collected using surveillance, informants and informant networks, and participant observation.

1.1.6 **Crime analysis**

Depending on the jurisdiction, a *crime analyst* (CA) may be a general term for any individual, sworn or civilian, that examines case materials, crime data, crime trends, or any other aspect of the crime with a view to addressing specific issues. In some jurisdictions, a CA would include both someone who scours computer databases, such as the Violent Crime Linkage and Analysis System, looking for matches and someone who compiles information on threats to public officials in an effort to determine the veracity of those threats. The reader should be aware that the title and the distinction for various types of analysts is dictated not only by their roles and responsibilities but also by the jurisdiction in which they operate. Burns and Gebhardt (2005, p. 61) close with a similar refrain:

> *Paramount to understanding the realm of crime analysis is the notion of an analyst wearing many hats. It is rare that an analyst will strictly perform a single type of analysis. More likely, the analyst will perform two or three analyses in their job duties. One day, an analyst may be compiling statistics for a trend analysis and the next day they are working on operational analysis for a known burglar.*

The next type of CA to consider is the criminal profiler, made popular by movies such as *Silence of the Lambs*, and by television shows such as *Criminal Minds*.

1.1.7 Criminal profiling

Criminal profiling has received much attention from law enforcement, academic, and media audiences. As a subtype of applied crime analysis, criminal profiling needs to be described and the crime analysis-profiling nexus explained. Given the similarities, there is necessary overlap between the two practices.

Chapter 2 provides a detailed overview of logic and reasoning within crime analysis, and so these areas are not explicitly detailed here. Suffice to say, there are two predominant types of thinking that dictate the way the evidence is examined, the type of conclusions drawn, and the specificity of those conclusions to the current case.

The first type of thinking is induction. Induction relies on statistical or correlational reasoning (Petherick, 2014), where the conclusion is a matter of probability. Analysts using this reasoning determine the general type of crime or behavior they are examining, and then seek out research or literature on this crime type or behavior. The results of this search are then provided as a "profile" or aggregate picture of what the offender may be like. As a rule, these reports may be abstract, in that they exist in thought but may not be representative of any one physical entity because they are a statistical composite of possibilities.

The second type of thinking is deduction. Deduction is based on the notion that, if we can determine the factual accuracy of the evidence or information, then certain conclusions can be drawn that will be true providing certain conditions are met. Inductive reasoning is used within a deductive process whereby theories or hypotheses arise that are subsequently tested back against the available evidence in the extant case. Because of this, deductive processes rely heavily on the scientific method, a systematic process for obtaining knowledge through hypothesis testing (see Turvey & Petherick, 2010).

Within each type of reasoning there exists distinction between (usually) competing theoretical paradigms. There are a great many such approaches, but these could be grouped using the most base of distinctions as *Criminal Investigative Analysis*, *Investigative Psychology*, *Behavioral Evidence Analysis*, *Geographic Profiling*, and *Diagnostic Evaluations*.

1.1.8 Criminal investigative analysis

Criminal Investigative Analysis (CIA) is the method of profiling devised by the U.S. Federal Bureau of Investigation (FBI). It is largely inductive, and frequent reference is made in reports to research and the experience of the Behavioral Analysis Unit. This approach has changed significantly over time, starting as a largely evidentiary approach with Howard Teten and Pat Mullany. Later, John Douglas and Robert Ressler (both mostly famous for their popular true crime memoirs) took over the profiling responsibilities, making significant changes to the method, including the language used to convey profile characteristics. As stated in Ressler and Shachtman (1992), the language had to be more police friendly, using the organized classification to explain offenders who planned their crimes, cleaning up evidence or the crime scene, and were fairly orderly in other aspects. These offenders were also referred to as psychopathic. The classification disorganized was used to describe an offender who was the opposite; leaving evidence behind with little to no planning, and who was fairly disordered in other aspects. These offenders were also referred to as psychotic.

Although this method could be described as prevalent among law enforcement groups, the certification program offered by the International Criminal Investigative Analysis Fellowship (ICIAF) does not comprise the largest group of profilers around the world. Although actual figures are hard to ascertain, their website (http://www.iciaf.org) states they have 37 understudies around the world. While it had been the case in the past, one no longer needs to be sworn law enforcement personnel to join, with the ICIAF taking a number of civilians in the last decade or so. Perhaps as a function of its almost celebrity status, the method has been widely critiqued (see Canter & Wentink, 2004; Petherick, 2014).

1.1.9 Investigative psychology

Concerned about the degree to which the FBI relied on personal experience and intuition, David Canter, a British psychologist, developed a method that is largely inductive and rather than drawing on personal experience relies on published research in the determination of offender characteristics. According to the program's website (http://www.i-psy.com, 2013), investigative psychology (IP) involves "the application of psychological principles to all aspects of the analysis, investigation and prosecution of crime." Despite this broad mandate, much of the published work product revolves around profiling. Indeed, IP has established its own journal in the *Journal of Investigative Psychology and Offender Profiling*.

Drawing from the discipline of psychology, the theory base of IP is exceptionally broad and often involves the statistical testing of theories and assumptions used in the profiling process. This base includes an analysis of the different types of aggression (see Salfati, 2000); the homology assumption (see Doan & Snook, 2008; Mokros & Alison, 2002); the validity of the FBI's organized and disorganized offender typology (see Canter & Wentink, 2004); and the geographic behavior of offenders (see Canter & Larkin, 1993).

1.1.10 Behavioral evidence analysis

Behavioral evidence analysis (BEA) espouses a deductive orientation, although an analysis of individual reports still reveals a degree of reasoning to the best explanation. This method proposes that the physical evidence be the cornerstone of all examinations, which underpins all other levels of analysis. It could be said that this method closely resembles that originally practiced by the FBI when Howard Teten and Pat Mullany were the mainstays of the profiling program (referred to as applied criminology in the early 1970s).

BEA uses a four-stage process (see Turvey, 2014), the first of which is the *Forensic Analysis*. The purpose of this step is to ensure that a full and adequate collection of the available physical materials has been undertaken along with a full reconstruction of the crime event. In *Victimology*, a thorough and detailed study of the victim is conducted, including their lifestyle, hobbies, habits, and routines. The ultimate aim of this stage is to determine where, exactly, do the sources of risk come from in this victim's life. The third stage, *Crime Scene Characteristics*, involves determining the exact physical characteristics of the crime scene. This includes whether it is indoors or outdoors, where the victim and offender met (referred to as the point of contact), where most of the physical evidence can be found (the primary crime scene), or if it is simply where the body came to rest (the dump or disposal site). The last stage, *Offender Characteristics*, is interpretive and includes the compilation of all of the necessary materials where themes or patterns are established that suggest aspects of the offender's personality or association with the criminal event. Although many methods are expansive in terms of what they offer, BEA suggests only four features are investigatively relevant: knowledge of the crime scene, knowledge of the victims, knowledge of methods and materials, and criminal skill.

1.1.11 Geographic profiling

Geographic profiling can be considered a form of profiling, only insofar as it meets a core aspect of definition. Profiling involves attempts to determine features of the criminal from the crime, and geographic profiling involves an attempt to discern just one feature: physical location. To do so, a complex mathematical algorithm interprets all relevant aspects of the crime, such as the time and date of the offense; the physical location; the order of the crime in the series; and depending on the program, the hunting and attack styles of the offender. Many different computer-based programs exist for this purpose.

Once all of this information is entered into the computer, the result is usually a range restriction around crime sites that are deemed to be more important within the algorithm (usually dictated by a circle or some other geometric shape defining search parameters), and may be accompanied by probability estimates, with the suggested range indicating the offender's home base. Other programs (Rigel developed by Rossmo, for example), provide a color-coded map referred to as a *Jeopardy Surface* (see Rossmo, 2000) that is overlaid onto a topographical map to provide search priorities. Typically, hotter colors, such as red, indicate a higher priority, and cooler colors, such as blue, suggest areas of lesser importance.

Although popular in television dramas, questions have been raised about the utility of geographic profiling, most notably in the Washington Sniper case where its application was brought into question with the offenders having no fixed address.

1.1.12 Diagnostic evaluations

A diagnostic evaluation (DE) is a little more difficult to describe in terms of a unified or common approach. This is because it is the general term used to describe the often ad hoc analysis of offender characteristics based on principles and theories within the general fields of psychology and psychiatry (Wilson, Lincoln & Kocsis, 1997). Individual analyses will be heavily influenced by the theoretical orientation of the individual, being largely driven by general personality and learning theories, humanist/cognitive perspectives, and alternate/Eastern philosophies (Boon, 1997). Although historically more prevalent, psychoanalytic perspectives may not be so common today (Woodworth & Porter, 2001).

1.1.13 Applied crime analysis

As discussed above, a criminal profile is a specific form of crime analysis aimed at inferring characteristics of an offender from the physical evidence left behind. Although *Applied Crime Analysis* may certainly delve into a series of purported characteristics in much the same way, it is also a term used to describe a particular type of analysis aimed at giving a specific set of information a certain meaning. This could include the following:

- An examination of behavior to determine investigative and legal shortcomings at trial
- An examination of behavior to provide legal counsel with a set of behaviorally relevant questions to raise during an appeals process
- The compilation of a timeline to examine the veracity of a series of witness statements to an event
- An examination of a crime or crime series to determine whether there is any risk present to the victim or victims, and the source of this risk
- An examination of the behavior of a victim or victim series to determine intervention practices to reduce or eliminate the behavior (usually referred to as *Threat Management*)
- An examination of a potential series of crimes to determine which, if any, are the work of the same offender(s)
- An examination of a crime or crime series to determine the motivation of the offender(s), or the escalation or deescalation of the offender(s) motive over time

Although this is not an exhaustive list, it certainly gives enough of an idea as to the range of the analyses that may be undertaken by the applied CA in their duties, regardless of the organization that uses them. But what is it that separates the applied CA from all of the other types discussed previously, if anything?

Applied Crime Analysis refers to the process of identifying, collating, interpreting, and reporting the results of the ideographic study of a single case for the purpose of providing a holistic understanding of a crime or crime series. Given the scope of their potential involvement as discussed, it could be said that applied CAs are an amalgam of the types discussed at the outset of this chapter.

They must possess the academic knowledge of the RA, and act as librarian for those who seek their services. They must know about crime trends and what constitutes a spike versus a crime wave. They must be able to, as the TA can, associate various people with places and with other people within any criminal event, regardless of crime type. The investigations of the OA must be second nature, and the nature and utility of intelligence, regardless of type, must be recognized and incorporated into the analytic process.

The applied CA should have training in several different disciplines and a variety of skills that can be brought to bear in a variety of situations. They must be intimately familiar with their role and responsibilities and know where their duty ends and that of another begins. They must speak a variety of different technical languages because they are required to interface and communicate with many different experts from many different fields. They are required to interpret and understand autopsy reports, legal reports, and investigative reports and to distill these reports into a common language for dissemination to any number of individuals or groups. This can range from friends and family of victims or offenders, lawyers and other legal experts, police, or administrators, among others. As such, they must possess good oral and written communication skills.

This is perhaps closest to what Vellani (2006, p. 74) refers to as *crime specific analysis,* noting the following:

> *Crime-specific analysis focuses not only on the type of crimes committed at a facility by enumerating the amount of crimes such as robberies and assaults, but also on whether the victim was a business or an individual. Further specificity aids management in knowing the specific type of problem, to what degree it exists, and indirectly what specific crime prevention measures can be used to reduce the opportunity for those problems, if not eradicate them completely. Another benefit of this type of*

analysis is that a breakdown by crime will help to indicate whether the asset targeted was a person or property, whether the crime was violent or not, the resulting loss or damage to that particular target, and the implications of that loss or damage.

Although the summary above refers to burglary, any crime could be substituted here, and the same considerations would apply.

1.2 Applications

The number of areas of application are as vast and as broad as the potential number of questions that any invested party can have about crime and criminal behavior. The issues that come across the desk of the CA may be specific in nature (What was the degree of planning in this crime? Was there any crime scene staging in this crime? What is the motive in this case?) or they can be broad and all inclusive (What happened here? What are the main places, times, and events?). Even specific questions may demand a complete analysis of the situation to provide a big picture backdrop for the resolution of case specific inquiries.

Although some may fit within the purview of the applied crime analyst, they are not discussed herein. Certain crimes will not be of sufficient prevalence to demand a CA's involvement and nor would their scale warrant the financial or logistical cost. Low-value burglaries are but one example. Other areas, such as white collar crime, although important areas of criminological interest, also are not discussed further. Owing to the nature of this offense, any examination of this type of activity will likely be undertaken by forensic accountants or a variety of business analysts. It is the author's experience that applied crime analysis is more suited to violent and interpersonal crime replete with information from many sources such as the victim, witnesses, documentary and physical evidence, and a host of others. As such, the coming discussion is limited to these areas, with a few exceptions.

Some of these areas are discussed at length in chapters of this volume, and some are touched upon only briefly below. Note, there is overlap between many areas such that the analysis of a criminal matter may come to be used in civil matter involving crime prevention, risk assessment, or both. There are no bright yellow lines between the following areas then.

1.2.1 Criminal cases

A crime is generally defined as any act or omission that constitutes a violation of law that is punishable. The range of possible offenses are considerable and will differ partly by time and location. For example, the phrase *a rule of thumb* is a reference to the permissible beating of a wife by her husband, providing that the switch (stick or other like object) was no thicker than his thumb. Although this may have been allowable hundreds of years ago, most Western and developed nations would consider this a form of domestic violence today and the offense would be punishable by law. Before 1990, stalking as a crime was not recognized, although "anti-stalking" laws spread rapidly in the United States, Australia, and many other countries after this time.

As a result, what is allowable today may be an offense tomorrow, and the landscape will change with the passage of time. In general, certain offenses, such as murder and sexual assault, will be against the law. It should be noted however that there are even exceptions to this general rule. For example, in many Middle

Eastern countries, honor killings are still allowed where male perpetrators kill female family members or spouses for indiscretions, often as minor as wearing the wrong type of clothing. Under Jordanian law, Article 340 exempts a man from penalty if he kills his wife or female relative after the discovery of adultery, whereas Article 98 allows for a reduced penalty if he commits the crime in a fury providing the victim was engaged in unlawful or dangerous acts (http://www.change.org have an international petition against these laws). In Iran, Article 630 allows a husband to kill his wife should he suspect her of adultery.

In addition, what is a crime in one context or situation may be excusable under the criminal law in another context. Homicide, the killing of one person by another, may be murder with intent and without lawful excuse, or it may be self-defense, providing the response is proportionate to the threat and the individual feared for their safety, or the safety of another.

The point here is simple. Laws change with time and place, and the subject of analysis will likely also change over time and between places. However, a few examples of crimes that may be analyzed include stalking (e.g., threat management, risk assessment, profiling, victimology), homicide (e.g., intent, planning, profiling, risk assessment, self-defense and provocation), and sexual assault (e.g., risk assessment, threat management, victimology, profiling).

1.2.2 Civil cases

Civil cases revolve around private relationships such as debts, unpaid fines, damages and loss, discrimination, tenancy, divorce, custody, and various workplace relations issues. Two such examples from the authors' files involve bank tellers suing their respective financial institutions for psychological trauma suffered as a result of being repeat victims in armed holdups. After each event, there was a failure to adequately debrief the staff and subsequently to increase physical security to thwart or deter future events. These cases involved a crime prevention assessment of the location, an interview with the clients that incorporated a separate psychological assessment indicating psychological trauma, and an examination of the literature to prevent bank robberies (including, e.g., training, staff response during events, the physical layout of the environment, employment of security guards).

In another type of civil assessment, the analyst determines the degree to which a property licensee, manager, or owner is responsible for damages occurring to clients or others using their facilities. This is called premises liability and is discussed in its own section below, but is not expanded upon in other parts of this work.

1.2.3 Premises liability

Premises liability is the civil responsibility incurred by property owners who have a special relationship to their clients, lessees, and customers where they fail to provide reasonable and adequate security (Petherick & Turvey, 2010). These may involve criminal matters, where the civil suit follows from prosecution for harm or death and may, in turn, be used to leverage the civil claim. They may also be used as a remedy for victims where the unfavorable outcome of a criminal matter may not satisfy their need for justice.

There has been an increase in the use of private security to control instances of crime and incivility as a result of the failure of state law enforcement to effect the same outcome (Meadows, 1991), thereby providing frontline enforcement of antisocial behavior, especially in areas regularly used by the public such as shopping malls, entertainment and nightclub districts, and in some locations, schools and universities. There is a subsequent increase in liability claims by victims for harm caused as a result of criminal acts (Voigt & Thornton, 1996).

Given the applied analysts' focus on crime and criminality, it makes sense that their role may broaden to include the loss and damage that results from criminal behavior, and as such they may see themselves reporting on, or testifying to, the harm arising from criminal activity. Generally, there are two main issues involved in assessing premises liability. The first issue is foreseeability; foreseeability relates to the ability to anticipate the likelihood that something will occur. Distinct from concrete prediction, foreseeability is actually about determining whether an event was more likely than not (Petherick & Turvey, 2010). The second issue is deterability. If an offense was foreseeable, the second issue to arise revolves around whether this event would have been deterrable had the appropriate mechanisms been used. There are two specific components of deterability. The first component is general deterrence and seeks to establish whether offenders of this type of crime (or any type of crime) would be deterred if measures were implemented. Private security guards are an example of a general deterrent. The second component is specific deterrence, or whether this particular offender in this particular environment would have been deterred from committing this particular crime.

Specific deterrence warrants a detailed examination of the case and all of its peculiarities. This involves an assessment of psychopathology, victimology, offender motivation, the physical environment, the nature of the area (e.g., commercial, industrial, residential), and the types of crime that are common in that area. This is critical as specific deterrence cannot be automatically assumed from general deterrence. For example, a static security guard is a general deterrent for bank robberies because this effectively hardens the target and makes the offense higher risk (Australian Institute of Criminology, 1989; Barnes, 2004–2005). However, one static guard who is unarmed may be seen as a minor inconvenience to a group of heavily armed robbers who are well prepared and have knowledge that a particular institution has a large cash reserve at the time of the intended hit. Specific deterrence in this instance may not be achievable if the only measure was lone private security.

1.2.4 Crime prevention

Crime prevention, like crime analysis, is a catch-all term and includes many different approaches to the prevention and management of a wide array of crimes and criminal behaviors. Situational crime prevention aims to undermine the opportunity structure for crime by increasing risks and reducing rewards (Clarke, 1995); developmental crime prevention addresses the crime problem by identifying behaviors and attitudes learned early in life that may predispose someone to crime and other antisocial patterns (Tremblay & Craig, 1995); and community crime prevention seeks to change social conditions that contribute to criminality in residential communities (Hope, 1995).

Any of the approaches to crime prevention can be achieved by applying the research and theory relevant to particular crime problems of an area, applications that may or may not involve a detailed examination of that area, including socioeconomic factors, ethnicity, demographic factors such as age, and specific types of crime. Such an examination may seem an obvious necessity so that one knows which research to rely on, but this may not always be the case. Budgetary and time constraints, or the reliance on census data (which may be outdated), are but a few examples where an in-depth study of particular communities may not be undertaken.

The type of analysis requested may also dictate the suitability of an applied crime analysis. Because of its idiographic focus, applied analysis may be more suited to a crime prevention assessment of a financial institution that has been repeatedly subjected to armed robberies than it would be to an examination of a community and its crime problems. Threat management, aimed at reducing or eliminating stalking behaviors (see Cawood, this volume), incorporates elements of situational crime prevention, regardless of whether the intervener has a crime prevention background, or it may be based entirely on situational crime prevention theories and practice.

1.2.5 Case linkage

Case linkage is discussed at great length in Chapter 6 of this volume and is only covered briefly here. When investigators are faced with a potential series of crimes and no indication of the identity of the offender, or whether the series is the work of one or more offenders acting in concert, attempts may be made to link the crimes using the available behavior and evidence. This is a difficult practice, depending on the availability of physical evidence, and the nature and the strength of links used to assess similarity.

1.2.6 Risk assessment

Risk assessment has historically been practiced by forensic psychologists and psychiatrists, although this is changing more and more. Today, police officers, drug and alcohol counselors, educators, taxation investigators, postal service workers, criminologists, and a host of others are conducting risk assessments. Risk is a nebulous term, and so current practices have moved from general risk conceptualizations to more targeted assessments of violence incorporating management strategies.

Risk assessment is discussed in Chapter 10.

Because applied analysts should also be voracious readers of literature and research, they will have a substantial understanding of what factors contribute to crime and specific instances of criminality—so called risk factors—and are thus well placed to determine the level of risk that may be present in any given situation. Because an in-depth examination of the case will have been made, they will be intimately familiar with those things that signal danger as they exist in this case, and they will also know those things that may contribute toward safety or mitigate risk. These things are known as protective factors and may include having stable employment, a good social network with strong support, insight, resilience, and good personal security.

1.2.7 Assessment of psychopathology

For obvious reasons, any identification of psychopathology should be limited to those with training in psychology and other mental health areas. However, if a case has not been previously assessed and the applied analyst is undertaking the first serious examination, or if evidence of psychopathology has been previously overlooked or missed, the analyst may be best placed to identify not only the evidence of dysfunction but also what this may indicate. If identified, the information can be passed along for further and more complete analysis and diagnosis by the appropriately qualified individual.

This process is discussed in Chapter 12 of this volume.

Evidence of psychopathology may provide pivotal information on which the outcome of a case may turn. This psychopathology includes the full range of psychotic disorders, organic as well as drug induced; the personality disorders, particularly Axis II, Cluster B; and any other psychological condition that may impact behavior and cognition. Where possible, this assessment should be undertaken on the offender(s) and victim(s) to provide as complete a picture as possible. In one of the author's cases, the individual who ultimately became the victim of homicide had previously been diagnosed with antisocial personality disorder and borderline personality disorder, with histrionic personality features and drug-induced psychosis. These factors combined and, over a long time with an extensive history of conflict with interpersonal partners, resulted in an explosive homicide event with her current partner in 2009, which resulted in her demise.

Conclusions

A crime analyst is a general term used to describe a variety of individuals involved in various and different stages of the investigation of crime and crime-related problems. Crime analysis is a catch-all term used to describe various types of analysis, including geospatial crime mapping, intelligence analysis, research analysis, and strategic analysis. Applied crime analysis is a particular type of analysis that involves the process of identifying, collating, interpreting, and reporting the results of the study of a single case to answer forensic or investigative questions. It is an amalgam of many other types of crime analysis, but rather than considering crime or crime types as a general social problem, the focus is on a single case to develop an understanding of the interactions between offender, victim, and place.

REVIEW QUESTIONS

1. Which type of analyst occupies a more traditional academic role and carries out research?
   a. Tactical analyst
   b. Research analyst
   c. Intelligence analyst
   d. Strategic analyst
   e. None of the above

2. Crime analysis is a general term for individuals, sworn or civilian, that examines case materials, crime data, and other aspects of the crime. True or False?

3. Criminal profiling is a particular type of crime analysis. True or False?

4. Premises liability revolves around two main issues: _____ and _____.

5. Name and briefly define two types of criminal profiling.

6. What are some of the areas in which applied crime analysis may be useful?

References

Australian Institute of Criminology. (1989). *Armed robbery from an offender's perspective: implications for prevention.* Available from http://aic.gov.au/documents/C/4/F/%7BC4F63B1B-7712-4134-BF2B-AA2417BF32 EB%7Darobbery.pdf. Accessed 17.12.14.

Barnes, G. (2004–2005). Bank robbery: the FBI perspective. *Western Independent Bankers. Western Banking Magazine.* Available from http://www.wib.org/wb_articles/crime_dec04/fbi_dec04.htm. Accessed 05.12.14.

Boba, R. (2001, November). Introductory guide to crime analysis and mapping. *Community Oriented Policing Services (COPS).* U. S. Department of Justice.

Boba, R. (2013). *Crime analysis and crime mapping* (3rd ed.). Thousand Oaks: Sage.

Boon, J. (1997). The contribution of personality theories to psychological profiling. In J. J. Jackson & D. Bekerian (Eds.), *Offender Profiling: Theory, Research and Practice*. Chichester: Wiley.

Burns, S., & Gebhardt, C. (2005). Understanding the crime analysis paradigm. *Law and Order, 53*(3), 56–58, 60–61.

Canter, D., & Larkin, P. (1993). The environmental range of serial rapists. *Journal of Environmental Psychology, 13*, 63–69.

Canter, D. V., Alison, L. J., Alison, E., & Wentink, N. (2004). The organised/disorganised typology of serial murder. Myth or model? *Psychology, Public Policy, and Law, 10*(3), 293–320.

Clarke, R. (1995). Situational crime prevention. *Crime and Justice, 19*, 21–89.

Dixon, D., & Schaub, C. M. (2004). Modern crime analysis: a "pulse check" of crime analysis in California. *Journal of California Law Enforcement, 38*, 22–30.

Doan, B., & Snook, B. (2008). A failure to find empirical support for the homology assumption in criminal profiling. *Journal of Police and Criminal Psychology, 23*, 61–70.

Hope, T. (1995). Community crime prevention. *Crime and Justice, 19*, 21–89.

Manning, P. (2004). Police technology: crime analysis. *Criminal Justice Matters, 58*(1), 26–27.

Meadows, R. J. (1991). Premises liability and negligent security: issues and applications. *The Journal of Contemporary Criminal Justice, 7*, 167–178.

Mokros, A., & Alison, L. J. (2002). Is offender profiling possible? Testing the predicted homology of crime scene actions and background characteristics in a sample of rapists. *Legal and Criminological Psychology, 7*, 25–43.

Petherick, W. A. (2014). Induction and deduction in criminal profiling. In W. A. Petherick (Eds.), *Profiling and Serial Crime: Theoretical and Practical Issues* (3rd ed.). Boston: Andersen Publishing.

Petherick, W. A., & Turvey, B. E. (2010). Premises liability. In W. A. Petherick, B. E. Turvey, & C. E. Ferguson (Eds.), *Forensic Criminology*. Boston: Elsevier Science.

Ressler, R. H., & Shachtman, T. (1992). *Whoever fights monsters: my 20 years tracking serial killers for the FBI*. New York: St. Martin's Paperbacks.

Ribaux, O., Girod, A., Walsh, S. J., Margot, P., Mikzrahi, S., & Clivaz, V. (2003). Forensic intelligence and crime analysis. *Law, Probability and Risk, 2*, 47–60.

Rossmo, D. K. (2000). *Geographic Profiling*. Boca Raton: CRC Press.

Salfati, G. (2000). The nature of expressiveness and instrumentality in homicide: implications for offender profiling. *Homicide Studies, 4*(3), 265–293.

Tremblay, R. E., & Craig, W. M. (1995). Developmental crime prevention. *Crime and Justice, 19*, 21–89.

Turvey, B. E. (2014). *Criminal Profiling: An Introduction to Behavioral Evidence Analysis* (4th ed.). San Diego: Elsevier Science.

Turvey, B. E., & Petherick, W. A. (2010). An introduction to forensic criminology. In W. A. Petherick, B. E. Turvey, & C. E. Ferguson (Eds.), *Forensic Criminology*. Boston: Elsevier Science.

van der Hurst, R. (2009). Introduction to social network analysis as an investigative tool. *Trends in Organized Crime, 12*, 101–121.

Vellani, K. (2006). *Strategic security management: a risk assessment guide for decision makers*. Burlington: Butterworth-Heinemann.

Voigt, L., & Thornton, W. (1996). Sociology and negligent security: premises liability and crime prevention. In P. J. Jenkins, & S. Kroll-Smith (Eds.), *Witnessing for Sociology: Sociologists in Court*. Westport: Praeger.

Wilson, P., Lincoln, R., & Kocsis, R. (1997). Validity, utility and ethics of profiling for serial violent and sexual offenders. *Psychiatry, Psychology and the Law, 4*(1), 1–11.

Woodworth, M., & Porter, S. (2001). Historical foundations and current applications of criminal profiling in violent crime investigations. *Expert Evidence, 7*, 241–264.

Logic and Reasoning in Crime Analysis

Wayne Petherick

Bond University, Gold Coast, QLD, Australia

Key Terms

Induction
Deduction
Heuristics
Analysis of Competing Hypotheses
Preconceived theories
Metacognition
Complexity
Structured Professional Judgment

INTRODUCTION

Despite the protest of some, common sense is not a form of logic or reasoning. Although common sense may have a role in the analytic process (it would not be a common sense or logical argument to suggest that the offender has no fingers because there are no fingerprints), it should be replaced at every stage possible with sound logic and reasoning that is based on solid theory and practice, of events that can be observed, with results that can be reproduced and that are open to testing. These are, after all, many of the core characteristics of science on which aspects of applied crime analysis are based. Logic itself is a science, and at the very least this will help to base the analysis on a solid foundation.

This chapter begins by defining logic and reasoning and provides a background to cognitive errors, followed by guides to thinking known as heuristics. From there, induction and deduction are explained. Induction is perhaps the most common type of reasoning, and follows that if the information on which an argument is based is accurate, the conclusion drawn from this information will *likely* be accurate. In deduction, when the evidence on which an argument is based is accurate, then the conclusion is a certainty. Structured professional judgment (SPJ) is then discussed; and finally, the work of Heuer (1999) on resolving conflicts between Analysis of Competing Hypotheses (ACHs) is addressed.

2.1 Logic and reasoning

An understanding of logic in applied crime analysis cannot be overstated. A working knowledge of the fundamentals of logic and cognitive errors coupled with an appreciation of the scientific method can literally be the difference between sound conclusions that are easily defensible and confusing nonsense unable to withstand even the most cursory review. At the conclusion of a case, any person completely unfamiliar with the nuances of the incident should be able to read the report and understand what is being argued and why. Lines of reasoning should be clearly stated, and the information on which conclusions are based should also be outlined. Furthermore, it is vital that the difference between assumptions, inferences, and facts be provided.

An assumption is defined as taking something for granted, or supposing a thing without proof (Merriam-Webster, 2011). An assumption therefore is essentially a conclusion drawn in the absence of supporting information or evidence. Given their potential to introduce error into the analysis, assumptions should be absent or relied upon only in the most exceptional cases, such as those cases where public safety is an issue. For example, where a serial offender is in operation and only limited information is known about elements such as modus operandi or victim selection criteria. In almost all other circumstances, assumptions should be omitted. At a Coroner's Court hearing in the Australian state of New South Wales, Kristina Illingsworth (Coroner's Court, 2003), a state police crime analyst/profiler was giving evidence as to the circumstances surrounding the disappearance of 16-year-old Gordana Kotevski. Several assumptions are clearly identified and stated:

Q. Now any opinion you express, however, needs to be qualified insofar as you've acted on the assumption that the Toyota Hilux motor vehicle in question is a vehicle which is owned by the offender himself, is that correct?

A. Yes.

Q. And not merely a borrowed vehicle or a leased or hired vehicle?

A. That's right.

Q. i.e., Put very simply, it is his vehicle.

A. Yes.

Q. And you were also acting on the assumption in any opinion that you express that the vehicle is owned by the driver?

A. Yes.

Q. i.e., The offender?

A. Yes.

Q. And not by any co-offenders who may be in the motor vehicle?

A. That's correct.

Q. And there is perhaps a further important assumption and that is that Gordana's abduction was by the person who was firstly the owner of the vehicle and he was involved in the abduction of Gordana?

A. Yes that's right.

As the analyst then goes on to base the conclusions about the offender's personality on vehicle ownership, these assumptions are important for the analysis that follows. Put another way, without these assumptions there is not enough information available to offer up possible characteristics. In this case, the vehicle ownership was used to infer machismo:

Q. Now what about the image that one would expect the offender to project. You referred in your report to him projecting a macho image. Why do you say that?

A. Yes. There's not enough behavior in the actual abduction phase to draw that inference but when you look at the vehicle on the assumption that he is the owner of that vehicle. The vehicle was very macho and sporty in itself. It was a 4-wheel drive that had immaculate condition to it and spotlights et cetera. This would be what we see as a reflection of the driver's personality. So in turn we expect his personality to permeate throughout his life. This means that we think he would be sporty, macho, and fit in himself with his clothes, his possessions and vehicle, any other vehicles that he might own in the future I would also expect to be well maintained, immaculate internally and externally, and in turn when you put those factors together as well, as well as the intelligence level, he's probably someone of a—at least a middle socioeconomic range.

Despite this assertion, there is another logical possibility from the evidence: that the so-called macho driver had in fact commanded the nondominant partner to drive.

Q. You assume that—to state the obvious someone had to drive away and someone had to subdue Gordana. Who would you expect to be the driver and who would you expect to be the subduer?

A. I don't know if it is possible to say.[a] I guess it'd be more likely that the owner of the vehicle, the driver, would also drive away, but if he's the dominant one then he could well be controlling Gordana and the other person could have to drive away. It's not really possible to say.

Q. But I thought you said you would expect the owner of the vehicle to be the macho one?

A. Yes.

Q. So you do not equate macho with dominance or?

A. Well driving around in that vehicle and arriving at that point of the abduction, that would be the macho one and at that point during the abduction when things start to get a little bit out of control where you have a victim who is struggling and resisting, it is not possible for me to say who might be holding her and who might be driving away.

An inference, in contrast, is defined as "the act or inferring by deduction or induction; that which is inferred; a truth or proposition drawn from another which is admitted or supposed to be true" (Merriam-Webster, 2011). Inference therefore is a type of conclusion that is suggested by the evidence or information available, also known as reasoning to the best explanation. These conclusions are a more valid form of argumentation than assumption for reasons that should be apparent. Consider the following example from the author's files:

> The offender spent considerable time cleaning up the crime scene, placing the victims in the spa, and removing other evidence. Nothing of value was taken (beyond a camera and some jewelry, please refer to the discussion on staging) though one drawer in the upstairs lounge and several drawers in Neelma's bedroom were opened, and some jewellery boxes were overturned. While this activity alone was not necessarily time consuming, it would have added to the overall duration of the offence (it is my opinion this constitutes part of the staging also, discussed subsequently). To have engaged in the number of precautionary acts, staging including staging evident, the offender felt no fear they would be discovered at the scene, allowing them to take time and effort to conceal their crime. In short, the offender's behaviour suggests they knew they would not be interrupted during the extensive clean up.

The inference, based on the available evidence, suggests that the offender knew the rest of the family would not be returning on the night the victims were killed because of the amount of time spent at

---

[a] Recall that the previous conclusion was based on the vehicle being owned by the driver to make the determination of his macho personality.

the crime scene after the murders. Although there is no proof as to time spent at the crime scene, these efforts were considerable and would have taken some time. They included filling a mop and bucket with bleach before mopping an entire tiled section downstairs, and then placing these items back in the laundry; moving all victims from the location where they were originally killed to an upstairs bathroom; moving bed covering and other materials into the bathroom and covering the victims; searching through cupboards and drawers; and moving items as part of general staging efforts.

Logic is a science that revolves around the act of valid thought or the objects of valid thinking (Bhattacharyya, 1958). Others have defined logic as a "unified discipline which investigates the structure and validity of ordered knowledge" (Farber, 1942, p. 41). As stated, a healthy grasp of logic is a requisite to carrying out a thorough analysis and preparing defensible reports.

Logic and reasoning are necessary components of attempts to gather knowledge and to answer questions put forth during any analysis. As such, they relate to the area of philosophy known as epistemology that relates to the nature of knowledge and its acquisition. Popper (1960, pp. 68–69) provides the following excellent discussion on epistemology and the sources of knowledge. The applied crime analyst would do well to bear these points in mind, the lessons they teach in regard to how we know what we know, and how to ground our thinking during the process:

1. There are no ultimate sources of knowledge. Every source, every suggestion, is welcome; and every source, every suggestion, is open to critical examination. Except in history, we usually examine the facts themselves rather than the sources of our information.
2. The proper epistemological question is not one about sources; rather, we ask whether the assertion made is true, that is to say, whether it agrees with the facts (that we can work, without getting involved in antinomies, with a notion of objective truth in the sense of correspondence to the facts, has been shown by the work of Alfred Tarski). And we try to find this out, as well as we can, by examining or testing the assertion itself; either in a direct way, or by examining or testing its consequences.
3. In connection with this examination, all kinds of arguments may be relative. A typical procedure is to examine whether our theories are consistent with our observations. But we may also examine, for example, whether our historical sources are internally consistent.
4. Quantitatively and qualitatively by far the most important source of our knowledge, apart from inborn knowledge, is tradition. Most things we know we have learned by example, by being told, by reading books, and by learning how to criticize, how to take and accept criticism, and how to respect truth.
5. The fact that most of the sources of our knowledge are traditional condemns antitraditionalism as futile. But this fact must not be held to support a traditionalist attitude: every bit of our traditional knowledge (and even our inborn knowledge) is open to critical examination and may be overthrown. Nevertheless, without tradition, knowledge would be impossible.
6. Knowledge cannot start from nothing—from a *tablua rasa*—nor yet from observation. The advance of knowledge consists, mainly, in the modification of earlier knowledge. Although we may sometimes, for example in archaeology, advance through a chance observation, the significance of the discovery will usually depend upon its powers to modify our earlier theories.
7. Pessimistic and optimistic epistemologies are about equally mistaken. The pessimistic cave story of Plato is the true story, and not his optimistic story of anamnesis (even though we should admit that all men, like all other animals, and even all plants, possess inborn knowledge). But although the world of appearances is indeed a world of mere shadows on the walls of our cave, we all

constantly reach beyond it, and although, as Democritus said, the truth is hidden deep, we can probe into the deep. There is no criterion of truth at our disposal, and this fact supports pessimism. But we do possess criteria that, if we are lucky, often allow us to recognize error and falsity. Clarity and distinctiveness are not criteria of truth, but such things as obscurity and confusion indicate error. Similarly, coherence does not establish truth, but incoherence and inconsistency establish falsehood. And, when they are recognized, our own errors provide the dim red lights that help us in feeling our way out of the dark of our cave.

8. Neither observation nor reason is an authority. Intellectual intuition and imagination are most important, but they are not reliable: they may show things very clearly, and yet they may mislead us. They are indispensable as the main sources of our theories; but most of our theories are false anyway. The most important function of observation and reasoning, and even of intuition and imagination, is to help us in the critical examination of those bold conjectures that are the means by which we probe into the unknown.

9. Every solution of a problem raises new unsolved problems; the more so the deeper the original problem and the bolder its solution. The more we learn about the world, and the deeper our learning, the more conscious, specific, and articulate will be our knowledge of our ignorance. For this, indeed, is the true source of our ignorance—the fact that our knowledge can only be finite, whereas our ignorance must necessarily be infinite.

Understanding the sources of our knowledge, and how they can help us or hinder us, is a powerful remedy to making errors of thought and reasoning. There are a variety of cognitive tools we can use to assist in the process, and there are a variety of cognitive errors that we must understand and avoid.

2.2 Cognitive tools

Although there are many sources of error within the analytic process, there are also many tools that can be used to ameliorate the intrusive effects of bias and distortion. Properly used, these can aid in overcoming cognitive errors and other deficiencies, and they can also help the analyst make complex decisions about the landscape before them that is comprised of hundreds or thousands of items of seemingly disconnected information.

These cognitive tools operate regardless of the type of logic used, and they are tempered by education, training, and experience. This of course means that the tools are sufficient to help make informed decisions and that they do not simply represent time over which numerous mistakes have been made, misidentified, uncorrected, and repeated.

2.2.1 Heuristics

A heuristic is an experiential guide to problem solving that may otherwise be referred to as a mental shortcut. According to Aickelen and Clark (2011), heuristics operate whereby on the basis of experience or judgment they may be more reliable in producing a good solution, although there is no guarantee that the solution will be optimum. These heuristics can take many forms, and their actual utility will be dictated by the situation at hand, the questions being asked of that situation (Is the crime staged? Are these cases linked?), and the type of heuristic used. In this last instance, the quality of the output through use of the heuristic may be heavily influenced by a variety of factors, such as experience with the type of case currently being analyzed. Problems encountered may not be the fault of the heuristic per se, but rather the divergence of the correct statistical approach where serious errors of inference are made (Nisbett, Krantz, Jepson, & Kunda, 1983).

Following are some of the more common heuristics.

2.2.2 Anchoring and adjustment

Anchoring and adjustment refers to the tendency to exaggerate the importance of the first piece of evidence revealed to us. Or as Tversky and Kahnemann (1974, p. 14) state "differing starting points yield different estimates, which are biased toward the initial values." This can be particularly problematic when that first piece of information tends to direct our cognitions of a particular event in a certain direction. In one case, the first piece of information provided was that the murder weapon used to kill the victim was owned by the accused. That itself is a damning piece of evidence that potentially skews initial thoughts one way or the other. However, after preliminary analysis of the material, it was discovered that many others knew of the weapon's location, including how to access it from outside of the property. Furthermore, ballistics tests failed to confirm this was in fact the murder weapon. Relying too much on the initial information will not only be misleading but also may initiate cognitive biases that are difficult to overcome.

The impact these theories have was noted by Hans Gross (1924, p. 10), an Austrian jurist, who referred to these as preconceived theories, noting that they are "the most deadly enemy of all inquiries":

> *Preconceived theories are so much the more dangerous as it is precisely the most zealous Investigating Officer, the officer most interested in his work, who is most exposed to them. The indifferent investigator who makes a routine of his work has as a rule no opinion at all and leaves the case to develop itself. When one delves into the case with enthusiasm one can easily find a point to rely on; but one may interpret it badly or attach an exaggerated importance to it. An opinion is formed that cannot be gotten rid of. In carefully examining our own minds (we can scarcely observe phenomena of a purely psychical character in others), we shall have many opportunities of studying how preconceived theories take root: we shall be astonished to see how accidental statements of almost no significance and often purely hypothetical have been able to give birth to a theory of which we can no longer rid ourselves without difficulty.*
>
> *Nothing can be known if nothing has happened; and yet, while still awaiting the discovery of the criminal, while yet only on the way to the locality of the crime, one comes unconsciously to formulate a theory, doubtless not quite void of foundation but only having superficial connection with the reality; one has already heard a similar story, or perhaps formerly seen an analogous case; one has had an idea for a long time that things would turn out in such and such a way. This is enough: the details of the case are no longer studied with the freedom of mind. Or a chance suggestion thrown out by another, a countenance which strikes one, a thousand other fortuitous incidents, above all losing sight of the association of ideas, end in a preconceived theory, which neither rests upon juridical reasoning nor is justified by actual facts.*

In short, the first information we encounter may be used to form an initial opinion that becomes difficult to unhinge, even in the face of future contradictory information. This may arise from a first meeting with a client where background information is provided; the report of another previously involved in the case; media coverage of the victim, suspect, or crime; and any other number of potential sources.

2.2.3 Availability heuristic

The availability heuristic refers to predictions that occur based on how easily past similar situations come to mind (see Tversky & Kahneman, 1974). This heuristic assumes that people will judge event frequency by the mental availability of relative instances to which they have been exposed (Pachur, Hertwig, & Steinmann, 2012).

As such, this heuristic will be heavily influenced by the education, training, and experience of the analyst in applying past experience to current problems. For example, one may be more inclined to state that cases are linked because this particular constellation of behavior may never have been seen before. However, if the analyst has limited experience with this crime type, the availability of similar situations will be limited. The analyst may be more inclined then to assume uniqueness.

The negative impact of the availability heuristic can be mitigated by reading widely, attending professional development courses, expanding educational horizons, and inviting peer review.

2.2.4 Representativeness heuristic

When there is an element of uncertainty about an event, the representativeness heuristic may play a role. This heuristic is the best studied and probably the most important (Nisbett et al., 1983), and it refers to the degree to which an event or outcome is similar to the general population to which it belongs. When this heuristic is used, error is likely because the analyst may confuse representation with likelihood. Just because something contains the quintessential qualities of its parent population does not mean it is more or less likely.

This heuristic is at play in risk judgments where larger reference groups are used as a point for risk estimates about specific individuals of that group. This problem is voiced by Meloy (1998, p. 8):

> *Nomothetic (group) studies on threats and their relationship to behaviour are not necessarily helpful in idiographic (single case) research or risk management, beyond the making of risk probability statements if the stalker fits closely into the reference group. Such studies may overshadow the common-sense premise that threats have one of three relationships to subsequent violence in single stalking cases: they inhibit violence, they disinhibit violence, or they have no relationship to the individual's violence. Careful scrutiny of the subject's threat/violence history should be the investigative focus when this relationship is analysed in an individual stalking case; and the importance, or weight, of threats in a risk management situation should be determined by searching for other factors (Monahan and Steadman, 1994).*

To effectively use this heuristic, the analyst must make every effort to ensure that the case under investigation is actually a "good fit" compared with the standard or reference group to which it belongs. A failure to do so may result in a reliance on false behavioral or evidentiary markers used to describe the features of the current case, thereby over- or underestimating risk.

2.2.5 Affect heuristic

Emotion is strongly linked to memory formation and recall. The affect heuristic is therefore the degree to which a risk or event is predicted on the emotional content of the risk or event. Put another way, people use affect as a cue to a hazard when estimating the risk presented by the hazard (Keller, Siegrist, & Gutscher, 2006). This heuristic may play a role in particularly emotive events, such as crimes involving children when the analyst is a parent, or when the analyst encounters a situation that is emotionally jarring. Thus, the analyst may be more likely to recall similar events (the *availability heuristic*) because these events have greater emotional connectivity.

Finucane, Alhakami, Slovic, and Johnson (2000) conducted research on the degree to which affect impacted judgments of risk and benefit. The impetus for this research (p. 3):

> *Is that images, marked by positive and negative affective feelings, guide judgement and decision making. Specifically, we propose that people use an affect heuristic to make judgements. That is, representation of objects and events in people's minds are tagged to varying degrees with affect. People consult or refer to and 'affective pool' (containing all the positive and negative tags associated with the representations consciously or unconsciously) in the process of making judgements. Just as imaginability, memorability, and similarity serve as cues for probability judgements (e.g. the availability and representativeness heuristics), affect may serve as a cue for many important judgements.*

Because this heuristic is widely recognized as playing a role in risk-taking relative to rewards, it may not only play a role in decisions that the analyst makes about events with emotional salience (such as the type of crime being analyzed, or the perceived innocence of the victim) but also impact upon risk and threat assessments. This may occur when the analyst imparts moral and value judgments on others (such as "I wouldn't have done that, so this offender must be particularly brazen"). Insight into one's emotional state and the physiological cues to arousal can be effective moderators in these situations.

2.3 Cognitive errors

Where biases and other cognitive problems arise, these errors must be acknowledged and controlled. Some, such as faulty heuristics and cognitive distortions, may be reduced, but some simply cannot put their biases aside no matter how hard they try. If an analyst knows there is a bias and that it is pervasive, it would be considered unethical to accept work in the case. Identifying any bias and attending to it during the analysis will be paramount in not allowing disgust or moral outrage to creep unnoticed into the assessment.

2.3.1 Metacognition

In broad terms, metacognition could be described as the process of thinking about thinking, although this does the field of considerable research depth little justice. Metacognition also includes introspection and reflection on one's thinking processes. Indeed, it could be said that these latter two processes are what give thinking about thinking its actual meaning; if you are not aware of when you are in error and why, you cannot make the appropriate cognitive corrections to reduce the potential for erroneous thought. The weight of such maladaptive rumination can be seen in the following citation from Kruger and Dunning (1999, p. 1121):

> *In 1995, McArthur Wheeler walked into two Pittsburgh banks and robbed them in broad daylight, with no visible attempt at disguise. He was arrested later that night, less than an hour after videotapes of him taken from surveillance cameras were broadcast on the 11 o'clock news. When police later showed him the surveillance tapes, Mr. Wheeler stared in incredulity. "But I wore the juice," he mumbled. Apparently, Mr. Wheeler was under the impression that rubbing one face with lemon juice rendered it invisible to video cameras (Fuocco, 1996).*

We bring up the unfortunate affairs of Mr. Wheeler to make three points…First, in many domains in life, success and satisfaction depend on knowledge, wisdom, or savvy in knowing which rules to follow and which strategies to pursue…Second, people differ widely in the knowledge and strategies they apply…with varying levels of success. Some of the knowledge and theories that people apply to their actions are sound and meet favourable results. Others, like the lemon juice hypothesis of McArthur Wheeler, are imperfect at best and wrong-headed, incompetent, or dysfunctional at worst.

Perhaps more controversial is the third point, the one that is the focus of this article. We argue that when people are incompetent in the strategies they adopt to achieve success and satisfaction, they suffer a dual burden. Not only do they reach erroneous conclusions and make unfortunate choices, but their incompetence robs them of the ability to realise it. Instead, like Mr. Wheeler, they are left with the mistaken impression that they are doing just fine.

The metacognitively impaired crime analyst is one who examines case materials (probably incorrectly), assembles unrelated facts and treats them as part of a homogenous whole, and interprets and reports findings either in a way that is inconsistent with the actual evidence, or in a way that makes sense to their befuddled thinking, but to few others. These reports may contain not only basic faults such as factual inaccuracies, but they may also contain outright fabrications, or be fit to burst with logical fallacies and other stereotypes that have no bearing on the case in question.

Although these analyses and subsequent reports may be easily identifiable, convincing their author of the flaws within may be a less than easy, and a far less than enviable, task. In short, if they did not have a problem with the way they did things in the first place, they are equally as likely to fail to understand the problem when it is laid out before them. This time and effort may be moot anyway: if they accept they are wrong, they are not likely to know why, or how to address the problem in the future.

Overconfidence bias plays a role here. Gigerenzer (1991) states that this occurs when we judge the probability of an event as more likely than it is in reality. This occurs because we have an overinflated view of our own judgment and possibly do not like to accept that we may not be as accurate in our assessments of possibility. This may occur as a result of metacognition that is linked to the better-than-average effect (Kruger & Dunning, 1999). These authors further explain (p. 1122):

We focus on the metacognitive skills of the incompetent to explain, in part, the fact that people seem to be so imperfect in appraising themselves and their abilities. Perhaps the best illustration of this is the "above average effect", or the tendency of the average person to believe he or she is above average…We believe focusing on the metacognitive deficits of the unskilled may help explain this overall tendency towards inflated self-appraisals…The failure to recognise that one has performed poorly will instead leave one to assume that one has performed well. As a result, the incompetent will tend to grossly overestimate their skills and abilities.

This tendency toward self-belief (based on metacognitive deficiencies) may explain overconfidence bias.

2.3.2 **Denying complexity**

Some situations are what they appear to be on their surface. But when we consider criminal behavior and the variety of factors that exist within any crime and its vast array of influences, it would be safe to say that things may rarely be as they appear on the face of it. For example, police enter the home where a disturbance has been reported. Initial communication simply identifies raised voices, followed by loud noises, and sirens soon thereafter. The scene appears to speak for itself: the male occupant of the house is standing, knife in hand, leaning over the bloodied body of his wife. But how straightforward is this scenario? Has the husband in fact killed his wife, or has he simply discovered a homicide at the hands of another? A jilted lover? Or could this be a robbery gone wrong? Or a suicide? Has the knife simply been picked up out of curiosity or confusion?

Although the law allows for situations in which the liability is self-evident (*res ipsur loquitur* or the thing speaks for itself), such instances of criminal behavior may be so rare that they infrequently pass over the desk of the crime analyst. Because of this, cases will be comprised of complexity, to varying degrees, and being able to identify the complexity and understand it will be pivotal to being effective.

In The Logic of Failure, cognitive psychologist Dietrich Dörner (1989, pp. 38–39) provides the following excellent commentary on complexity:

> *Complexity is the label we will give to the existence of many interdependent variables in a given system. The more variables and the greater their interdependence, the greater the system's complexity. Great complexity places high demands on a planner's capacities to gather information, integrate findings, and design effective actions. The links between the variables oblige us to attend to a great many features simultaneously, and that, concomitantly, makes it impossible for us to undertake only one action in a complex system…*
>
> *A system of variables is "interrelated" if an action that affects or is meant to affect one part of the system will always affect other parts of it. Interrelatedness guarantees that an action aimed at one variable will have side effects and long-term repercussions. A large number of variables will make it easy to overlook them.*
>
> *We might think that complexity could be regarded as an objective attribute of a system. We might even think we could assign a numerical value to it, making it, for instance, the product of the number of features times the number of interrelationships. If a system had ten variables and five links between them, then its "complexity quotient", measured in this way, would be fifty. Such attempts to measure the complexity of a system have in fact been made. But it is difficult to arrive at a satisfactory measure of complexity because the measurement should take into account not only the links themselves but also their nature. And, in any case, it is misleading (at least for our purposes here) to postulate complexity as a single concept.*
>
> *Complexity is not an objective factor but a subjective one. Take, for example, the everyday activity of driving a car. For a beginner, this is a complex business. He must attend to many variables at once, and that makes driving in a busy city a hair-raising experience for him. For an experienced driver, on the other hand, this situation poses no problem at all. The main difference between these two individuals is that the experienced driver reacts to many "supersignals". For her, a traffic situation is not made up of a multitude of elements that must be interpreted individually. It is a "gestalt", just as the face of an acquaintance, instead of being a multitude of contours, surfaces, and colour variations, is a "face".*
>
> *Supersignals reduce complexity, collapsing a number of features into one. Consequently, complexity must be understood in terms of a specific individual and his or her supply of supersignals. We learn super signals from experience, and our supply can differ greatly from another individual's. Therefore there can be no objective measure of complexity.*

This passage has several important lessons for the crime analyst, which will be restated for posterity. The lessons of these words, in fact the whole of Dörner's work, is of vital importance and the reader should make every effort to read it in its whole.

Crime is a constellation of victim, offender, and environmental factors. There is the state of mind, impact of alcohol or other drugs, emotional disturbance and psychopathology, opportunity, risk taking and risk managing, planning and preparation, and the possibility of any given number of X-factors (those situations that arise that are unplanned and unintentional). As a result, crime is by definition complex. This complexity "places high demands" (to use Dörner's words) on the analyst. It requires the ability to understand what one is looking at, what it means, and how each piece relates to each other piece and the whole. It requires the systematic deconstruction and reconstruction of events for an audience that may be totally unfamiliar with the case or the theory on which the analyst's version of events are based.

The second point is that of Dörner's objective attributes of a system and how this relates to applied analysis. Although it would be helpful, and in some instances favorable, to use such a factor-analytic approach (if a system had 10 variables and five links between them …), the reality is that such reductive strategies are not available to the analyst for the majority of their endeavors. The output of most analyses are still opinion, formed in the fire of intensive labor and tempered by countless hours of study, thought, and preparation.

The final point relates to what are called "supersignals," signals that relate to the degree to which a task becomes easier (ergo, its complexity is reduced) when one is practiced at it. This is the result of having done something repeatedly to the point where it becomes second nature: knowing what one is looking at; knowing what questions to ask to successfully acquire the correct materials; knowing what questions to ask once the material is acquired; knowing the theory on which any opinion will be based; and knowing how to successfully communicate that theory to any audience, professional or lay. But that is not to say that all practice is good practice. As stated previously, if one is unaware of their deficits, fails to identify when they are in error, and does not understand what steps are required to correct their course, experience is more hindrance than help.

2.4 Inductive logic

Induction is a form of reasoning that relies on probabilities and possibilities based on the supporting evidence (Petherick, 2014). A good inductive argument should clearly state the information on which it is based, in addition to providing the source of this information. Providing this information is not simply an academic exercise; it allows the reader to determine the veracity of the source and the level of confidence that they can have about the degree to which that evidence relates to the case. For example, a conclusion that is stated to occur in 20% of cases may not instill as much confidence as one stated to occur 80% of the time. An anecdote may not instill as much confidence as robust empirical research.

According to van den Brink-Budgen (2010), inductive arguments represent most of these made in critical thinking. In one of the clearest explanations of inductive logic and the nature of subjective interpretation, he goes on to state (p. 26–27):

> *In such arguments, the relationship between what's in the reason-side makes the conclusion at best probably rather than certainly true (or, we might prefer, that the conclusion 'is the case'). The process of inductive inference is one of extracting significance from the reason-side claim(s).*
>
> *Claim > it means this > so this follows.*

> *When we use the words 'extracting significance', we mean that the arguer has given a signifi-*
> *cance to the claim(s) and taken significance to give the inference. It would be more accurate there-*
> *fore to describe inductive inference as 'the process of extracting imputed significance from a claim'.*
> *In other words, the person doing the inference takes the claim to have significance (even though,*
> *crucially, others may not).*

Heuristics are one way we can extract significance.

Because inductive arguments are based on probabilities, they are easy to identify by statistical probabilities and qualifiers. This, however, relies on the analyst clearly stating not only their conclusion, but also what it is based on and accurately communicating its uncertain nature. Unfortunately, this is not always the case. Statements such as "the offender might be known to the victim," "it is possible that the victim was under the influence of drugs," or "it is likely that the victim was surveilled" are all strong indications that inductive logic has been used to arrive at the stated conclusions. The following is an example of an opening caveat typically given in inductive reports:

> *It [this analysis] is not a substitute for a thorough, well planned investigation and should not be*
> *considered all inclusive. Any information provided is based on reviewing, analysing, and researching*
> *criminal cases similar to the case submitted by your agency. The final analysis is based upon prob-*
> *abilities, noting however, that no two criminal acts or criminal personalities are exactly alike and,*
> *therefore, the offender at times may not always fit the analysis in every category.*

Statistical arguments can be drawn from a variety of sources, such as government crime data. In the United States, this would include the Federal Bureau of Investigation Uniform Crime Reports, the British Crime Survey in England and Wales, and Australian Crime Facts and Figures compiled by the Australian Institute of Criminology. Published literature and research can also provide statistics on which to base inductive inferences. Although these are too numerous to cite in full, a few examples would include risk factors in stalking (McEwan, Mullen, & Purcell, 2007), psychological and behavioral characteristics of gang members (Alleyne & Wood, 2010), and the spatial patterns of sex offenders (Beauregard, Proulx, & Rossmo, 2005). An example of a set of statistical arguments along with qualifiers can be found in Alison, Goodwill, and Alison (2011).

2.5 Current research on abduction murder

The following section describes the figures on prevalence of this sort of offence, the circumstances surrounding the offences (and the extent to which they are in line with other abductions); possible characteristics of the offender(s), and the probable sequence of events. We are assuming that this is a non-family abduction murder. However, we would remind the enquiry team that individuals familiar (where familiarity can mean a casual acquaintance or friend) to the child are more often responsible for child murders than strangers (Boudreaux, Lord & Jarvis, 2001).

Characteristics of the Offense

(unless otherwise stated, figures are from Hanflanf, Keppel, and Weis (1997); U.S. study)

- Abduction of juveniles (under 16 years of age is rare; 2% of violent crime against juveniles).
- Most victims are girls (76%), average age 11 years old (supported by Boudreaux, Lord and Dutra, 1999).
- 58% reported as "missing child" with, typically, a 2 h delay in making the initial report from the time of abduction.

- Children were "…not particularly vulnerable or high risk victims" (p. 21).
- Thus, there is nothing especially unusual about any of the victims in these regards, although Amelie is a little older than the average age. However, we have been informed that she looks physically younger.

Victim Selection, Procurement, and Disposal

- 80% of cases initial contact between the killer and the victim is within 0.4 km (0.25 mile) of the victim's residence.
- 57% of victim selection is opportunistic.
- 15% selected on the basis of a prior relationship with victim.
- Two-thirds use a "blitz" attack; the majority and not subtle, clever predators using deceptive means to abduct.
- Nonfamily abduction is often associated with other offences such as robbery or sexual assault, as a means of isolating the child (Finkelhor and Ormorod, 2000).
- Family involvement is less likely than stranger offenders for victims in this age group (Bourdreaux, Lord and Dutra, 1999).
- The majority (74%) of the abducted children who are murdered are dead within 3 h of the abduction.
- 52% conceal the body.

Another way to identify inductive arguments is to look for comparisons or similarities drawn between the current case and other cases that are similar. This approach can be seen in the report on the homicide of Roberta Ann Butler in California (Prodan, 1995):

> Based upon similar cases encountered at the NCAVC we would conclude such actions would indicate an association or "closeness" between the assailant and the victim.

This can also be seen in the report on the homicide of Shawna Michelle Edgar (Safarik, 2000), also in California:

> The NCAVC has observed in many cases, offenders interact with their homicide victims in a manner described as "Undoing". Undoing behavior are actions engaged in by an offender who has a close association with the victim and tries symbolically to undue [sic] a homicide…The NCAVC would describe the offenders interaction with the victim's body in terms of both her movement and placement as undoing.

Although there is a push from many communities to provide statistical or research evidence when providing opinions, and although it may make perfect sense to do so, relying on inductive reasoning alone can be perilous. Potential problems with relying on research alone, without any examination of the actual features of the case, are numerous. These include, but are not limited to the following:

> *Cross-cultural validity and reliability*: There is no guarantee that studies done in one country will apply to crime problems in other countries. Jurisdictional differences in the classification of crimes also means that certain behaviors relevant in one place may not even be recorded in another. This will impact on the research done and the degree to which it represents knowledge about crimes under examination in other jurisdictions.
> *Sample size issues*: When dealing with violent and/or aberrant criminal behavior, the frequency of certain behaviors may be so small as to render study into them pointless. At best, these may

constitute an anecdotal case-series. When confronted with these statistically infrequent behaviors, there may be no research to reference or rely on to provide advice or guidance.

Poor research methodology: When we rely on peer review, we rely on the reviewers being discriminating enough to find and identify flaws in research. These reviewers are human, however, and may miss flaws in data or methodology, allowing poor studies to be published in the professional literature. This, of course, relies on peer review being done at all.

Patch protection: Those who work as editors for peer-reviewed publications are often themselves industry or profession subject matter experts. As such, they may be especially protective of the subjects in which they teach, research, or publish. This may prompt them to reject an otherwise good research article because it represents competition (or worse, it may bring into question some of their own research findings or methodologies).

Crime classification differences: The classification of a criminal event can be somewhat arbitrary, based on the amount of available evidence, or what crime(s) can actually be proven. Furthermore, prosecution may only be undertaken with the worst crime that an offender has committed, and/or the most easy crime to prove within the confines of the evidence or the boundaries of the law. This means that the crime data relied on for some research may not be a true and accurate picture of the crime problem.

Intracrime classifications: Although there are obvious differences between crime types (sometimes considerable, as between arson and stalking, or they can be subtle, as between murder and manslaughter), there are also differences in subtypes of crime within the same classification. For example, in murder, there are domestic homicides, stranger homicides, serial homicides, spree homicides, and mass homicides. Each will have subtle differences such that the general data on homicide may not accurately reflect the nuances of a stranger homicide, and data on mass homicide may not reflect the nuances of a domestic homicide. Inappropriately grouping all research together may not provide a useful or accurate picture of a murder.

2.5.1 **Conditional probability**

As a final statement on induction, the problem of conditional probability must be considered. Over the years, the author has encountered numerous people who are of the opinion that having two conditions, where one relies on the other for truth, strengthens the argument as a type of statistical piling-on. This is incorrect.

Where the truth of one argument is dependent (or a condition of) another argument, the statistical probability reduces as a function of the original probabilities. The best example to use to illustrate this problem is the toss of a coin. Given that a coin has two sides, the probability of either outcome (heads/tails) is 50% or 0.5. So, if after one toss, the result is heads and we want to determine the probability of tossing a second heads in a row, we must multiply the individual probability of each event. In this case, 0.5×0.5 would result in a conditional probability of 25% or 0.25. Consider now that we are attempting to determine from studies the chance that a threatener may harm a victim, where research shows that threateners are past intimates approximately 75% of the time and that these past intimates harm their target another 75% of the time (these figures are illustrative only and do not reflect actual research findings). Because each individual outcome is known, and the outcome of the second is dependent on the first, we see that a conditional outcome is only approximately 56% (0.75×0.75). Although each individual outcome is promising (three-quarters of the time), the combined result is less so (just over one-half of the time).

2.6 Deductive logic

Deduction is not necessarily the antithesis to induction, but it could be considered the next point along the logical continuum. Deduction is the result of testing each and every theory (inductive possibility) against the known evidence. In a deductive argument, the conclusion follows from the premises (individual items of evidence or information), such that if the premises are true, the conclusion must also be true (Bevel & Gardiner, 2008). This type of logic was made popular by the Sherlock Holmes stories of Arthur Conan Doyle, although in reality Holmes' style of reasoning is best characterized as hypothetico-deduction, where missing premises are hypotheses (Musgrave, 2009).

In a deductive argument, syllogisms consist of two or more reasons that together lead to the conclusion put forth (van den Brink-Budgen, 2010). Because deductive arguments are structured so that the conclusion is implied from the premises (Petherick, 2014), it is critical that each and every premise is examined and the accuracy of facts and evidence established. If not, the conclusion has a much greater chance of being inaccurate resulting in a non sequitur argument (where the conclusion does not follow logically from the premises), also called a false deduction.

A good deductive argument, one following sound reasoning that is based on good information, will take us from truth to truth. Alexandra, Matthews, and Miller (2002, p. 65) state the following:

- It is not logically possible for its conclusions to be false if the premises are true.
- Its conclusions must be true, if its premises are true.
- It would be contradictory to assert its premises yet deny its conclusions.

The following example from Petherick (2014, p. 31) illustrates how deductive logic can be applied to crime scene behavior:

> *During an anal sexual assault, an offender approaches the victim from the front, allowing her to get a good look at him. He then wraps a belt around her neck, pulls down his pants before pulling her shirt up over her head, thereby revealing her breasts...After completing the sodomy, the offender then leaves the victim with her shirt in place, which covers her face.*

The basic information contained within the above example presents a number of possible scenarios with regards to the component behaviors. This includes but are not limited to the following possibilities:

- The offender is a male.
- The offender pulled the shirt up over the victim's head to obscure identification.
- The offender pulled the shirt up to provide access to the victim's breasts.

Although basic, the following example illustrates how to deductively establish the offender's sex based on the information above, by using an if-then structure:

> If the victim had not had sex before the attack;
> If only males have a penis;
> If only males ejaculate semen;
> If there is semen present;
> If the victim reports seeing a penis;
> Then the offender is a male.

Any of the behaviors under consideration must be considered as a theoretical possibility and this process then undertaken on each to determine its veracity. Ideally, when the total number of possibilities is compiled, falsifying them will leave the analyst with only one logical conclusion (a true deductive result), or several possible conclusions (when reasoning to the best conclusion is the only possible outcome with the evidence that is available). It is important to bear in mind that, despite deduction being the most ideal possible outcome, not every conclusion can be deductive. This will be especially true where evidence is scarce or weak in nature, or where little is known otherwise.

It is necessary at this point to note that although the above-mentioned structure would be the logical structure, actual deductive arguments are not always put forth in this way, and in this much detail. That is because, depending on the nature of the analysis and the amount of evidence available, such detail would make reports cumbersome and lengthy. As such, although care and thought may have gone into the logic of the arguments put forth, they may not at face value appear deductive.

When a premise that is needed to make a sound deductive argument is excluded or not stated, this is referred to as a suppressed premise, or what Moore and Parker (2009) call an unstated premise. Suppressed premises may be the result of laziness or ineptitude, oversight, or the desire for brevity. Although it may not be vital to include them in a report, one should be able to account for them if questioned over conclusions. This is to ensure that so-called deductions are not inductive hypotheses in disguise.

To provide a quick and useful summary of the differences between induction and deduction, "the premise of a good deductive argument ... proves the conclusion. This brings us to the second kind of argument, the inductive argument. The premises of inductive arguments do not prove the conclusions; they support them" (Moore & Parker, 2009, p. 45).

To enable us to move from evidentiary considerations to concrete inference, deduction relies heavily on the scientific method, which is discussed next.

2.6.1 **The scientific method**

At every stage of analysis, it is critical to think like a scientist and bring all the tools of science to bear so as to ensure the most rational, logical, and practical assessment of any evidentiary materials. Among the most important is the scientific method, a method for obtaining knowledge through testing and falsification. Far from being an inaccessible approach granted to only an elite few, the scientific method is relatively commonplace and is applied in a general way by scientists and nonscientists alike in situations involving everyday life. Bevel and Gardiner (2008, pp. 89–90) add:

> *Often in attacking an analysis, opposing experts and lawyers imply that only the "scientist" can apply the scientific method; mere mortal such as analysts, investigators, and police are incapable of understanding the mysterious tenets of this thing called the "scientific method". In such arguments, one can be sure that the counselor will use great effort and intonation each time he uses the word "science". More often than not, these same individuals will have no clue of what the scientific method is or exactly what it entails.*

The scientific method is a process used during investigation of the unknown and is comprised of several steps or stages. The number depends on the source, although four to six steps are not uncommon. Typically, these steps include the following:

- Observation of a problem or event.
- Collection of information or evidence.

- Examine for patterns and associations.
- Hypothesize or theorize (induction).
- Test or attempt to disprove the hypothesis (the cornerstone of the scientific method).
- Conclude (deduction).

Consider this rather mundane problem: your desk lamp will not turn on. This problem can be solved by using the scientific method to determine both the cause of the problem and its solution, by using an if-then or problem-solution approach. When applied to the lamp problem, we can see how thorough and powerful this process is in attempting to problem solve:

Observation: The lamp is not turning on.
Collect information: The lamp has a bulb and is plugged into the outlet. The outlet is switched on. All other devices plugged into the outlet are working properly.
Patterns and associations: The lamp has previously worked. The last time the lamp failed, the bulb had blown and replacing the bulb fixed the problem.
Hypothesis: If the lamp is not working, then the problem is one of many possible things: the outlet is faulty or not switched on, there is no power, the lamp is not plugged into the socket, or the bulb is blown.
Test or disprove: If there is power to the other devices, then current is not the problem. If the lamp is plugged into another outlet and still does not work, then the outlet is not faulty. If the lamp is plugged into the outlet, then this is not the issue. If the bulb is blown, then replacing the faulty bulb will result in the lamp turning on.
Conclusion: The lamp works after having the bulb replaced; therefore, the faulty bulb was the problem.

It should be obvious that this method is useful when applied to complex problems, but it can be time-consuming. This is not the criticism that it would appear, however, when the liberty of an individual hangs on the findings of the applied crime analyst; there is a reasonable expectation that the analyst will have taken as much care and exercised as much diligence as possible.

2.7 Structured professional judgment

SPJ involves drawing conclusions about the evidence while "weighting" this professional judgment with empirical research and literature in the domain of interest. This can help overcome the over- and underestimation of certain values and probabilities that may occur when left to their own mental devices. Perhaps best known in the area of risk assessment for violence (see Guy, Packer, & Warnken 2013; Pedersen, Rasmussen, & Elsass 2010), the underlying assumptions of SPJ apply equally well to the area of applied crime analysis.

Today, SPJ is seen to be the favored alternative to unstructured judgments or purely actuarial approaches that have historically driven risk assessment practices during different periods. Ogloff and Davis (2005) provide a useful summary of the different "generations" of risk assessment. First generation instruments were those that were conducted by clinicians with little guidance, relying solely on their experience with patients, and their education and training. This was characteristic of risk practices during the 1970s and 1980s. "The essential failure of…first generation…assessments led many to assert that clinicians were unable to make accurate decisions…and, as such, any attempt to make such decisions was unethical and beyond the ken of psychiatrists and psychologists" (Ogloff & Davis, 2005, p. 306).

During the second generation of risk assessment (after 1981), the pendulum swung the other way. Here, the focus was on statistical prediction with a greater focus on short-term predictions, situational variables, and specified populations. Because of the failure of clinicians in assessing risk, it was felt that taking the decision making out of their hands and relying more on statistical models would reduce the error previously characterizing violence risk. During this time, a large number of actuarial risk assessment instruments (ARAIs) were developed that relied on the statistical association between factors that predict violence and the likelihood of violence (Ogloff & Davis, 2005).

According to Ogloff & Davis, in the third generation the prediction of dangerousness is not the task; identifying the individual's level of risk is the task. This results in more of a focus on the individual's situation and what may lead them to act violently. Third generation assessment looks both at static factors that cannot be changed (age and sex are examples) and dynamic factors (drugs and alcohol and employment are examples). As much as the goal is to assess an individual's risk factors, management of the risk is also incorporated here to assist in the process.

SPJ is essentially a combination of the first generation (clinical judgment) with the second generation (statistical or actuarial judgment). Structured approaches to decision making are of benefit because they allow the analyst or clinician to use their clinical education, training, and experience with the published research on which particular factors account for the most amount of variance in violence risk assessment. The "aim is to combine the evidence base for risk factors with individual patient assessment," and this approach is useful for supporting evidence-based practices and increasing transparency in the analysis (Bouch & Marshall, 2005, p. 85).

This same method can be used in applied crime analysis where decisions about risk and threat, relationship with the victim, relationship to the scene and other geographic assessments, criminal motivation, and staging (to name but a few) can be based not only on the assessment of evidence but also on the research that is available on whatever aspect of the case is under consideration. Beyond the obvious benefits of evidence-based practice and the need for transparency, using SPJ in crime analysis will actually enhance the validity of any analytical work because it will place the onus on the analyst to clearly identify their judgments and why they have made the decisions they have about the extant case. Specifically, should their analytical findings be contradicted by research, they will be required to provide a full accounting of the reasons for this contradiction.

Consider these few examples of when structured judgments may be made by the crime analyst.

When working on stalking cases, the question is often asked by the target of the harassment whether they are at an increased risk of homicide simply because they are the target of obsessional pursuit. There may be a tendency to err on the side of caution, to say there is a risk when there is not so that the assessor is covered should there be no adverse outcome. This is known as a false positive. How then might this overestimation be mitigated? Perhaps the best way is to anchor this professional judgment in the published research on the subject to determine not only whether the risk may be elevated but also to identify whether any of the established risk factors are present. Farnham and James (2000) presented the results of a study to the Stalking: Criminal Justice System Responses Conference held by the Australian Institute of Criminology. Their results show that, from a forensic sample in the United Kingdom of 67 stalkers, only seven cases (10.5%) committed homicide against their victim. The risk factors they identified were the absence of substance abuse, being used, an absence of psychosis, and previously turning up to the victim's home.

These risk and protective factors could then be measured against the information in the current case: Is there any substance use or abuse? Is the stalker employed and therefore has a greater stake in conformity or prosocial behavior? Is there any evidence of psychosis? Have they previously visited the victim's home? If the risk factors are present, and the protective factors are absent, we may be more justified in arguing that they may come to harm and should therefore exercise caution. Operating thusly reduces the chance of error and in turn increases the validity and accuracy of analysis.

2.8 When probability is the only possibility: analysis of competing hypotheses

This section draws heavily from Heuer's (1999) work *Psychology of Intelligence Analysis*, one of the best for explaining, in relatively intricate detail, exactly how to overcome the indecision that accompanies competing hypotheses in a given set of facts. More importantly Heuer (1999, pp. 95–96) begins the discussion by highlighting the importance of the scientific method, and what can happen when one only sets about to prove their theories:

> *When working on difficult intelligence issues, analysts are, in effect, choosing among several alternative hypotheses. Which of several possible outcomes is the most likely one?*
>
> ...
>
> *Analysis of competing hypotheses (ACH) requires an analyst to explicitly identify all the reasonable alternatives and have them compete against each other for the analyst's favor, rather than evaluating their plausibility one at a time.*
>
> *The way most analysts go about their business is to pick out what they suspect intuitively is the most likely answer, then look at the available information from the point of view of whether or not it supports this answer. If the evidence seems to support the favorite hypothesis, analysts pat themselves on their back ("See, I knew it all along!") and look no further. If it does not, they either reject the evidence as misleading or develop another hypothesis and go through the same procedure again. Decision analysts call this a satisficing strategy....Satisficing means picking the first solution that seems satisfactory, rather than going through all the possibilities to identify the very best solution. There may be several seemingly satisfactory solutions, but there is only one best solution.*
>
> ...
>
> *The principal concern is that if analysts focus mainly on trying to confirm one hypothesis they think is probably true, they can easily be led astray by the fact that there is so much evidence to support their point of view. They fail to recognize that most of this evidence is also consistent with other explanations or conclusions, and that these other alternatives have not been refuted.*
>
> *Simultaneous evaluation of multiple, competing hypotheses is very difficult to do. To retain three to five or even seven hypotheses in working memory and note how each item of information fits into each hypothesis is beyond the mental capabilities of most people. It takes far greater mental agility than listing evidence supporting a single hypothesis that was pre-judged as the most likely answer.*

From here, Heuer goes on to describe eight steps that analysts can engage and determine which of the available competing hypotheses are the most likely or the only available solution to the question under consideration. These eight steps along with relevant examples provided by this author follow.

Step 1: Identify the Possible Hypotheses to be Considered. Use a Group of Analysts with Different Perspectives to Brainstorm Possibilities

The degree to which a person could perceive the vast array of hypotheses in a case will differ depending on their education, their training, their experience (both in general, and with this crime type), and even on the amount of reading they do. The latter will expose them to other theories and situations and expand their ability to think outside of their own universe of knowledge. According to Heuer (1999), if a person cannot generate sufficient hypotheses for consideration, they have little hope of arriving at a correct answer.

Research in hypothesis generation tends to indicate that people are satisficers, being less likely to generate hypotheses when they already have a hypothesis that they believe accounts for the data (Garst, Kerr, Harris, & Sheppard, 2002). In a study examining research participants' ability to generate hypotheses, these authors found that participants handed a hypothesis generated significantly fewer hypotheses than those not given a satisficing hypothesis (the results of this study are more complex than requires treatment here, so the reader should consult Garst and colleagues for detail). The results of this study tend to suggest that people "give up" generating theories when given a theory which seems to explain the data before them.

Should the panel or group approach not be a possibility, and in many cases it will not be, using the services of any number of peer reviewers may be the best proxy. This peer review could be done at any level or stage of the analysis, such as during preliminary analysis where a series of possible theories are developed for later testing against the evidence, or at the conclusion of report authoring, where it is distributed for an assessment of content and depth. It would even be possible to use both methods; both a priori and post hoc. Ideally, this would allow for a more thorough analysis of the evidence, and consideration of what it may mean insofar as the array of possibilities goes. This may also assist where errors are noted in the report allowing these to be corrected before final submission.

Although acknowledged that this approach can bear significant fruit, it should also be noted that several potential errors may be introduced into the process at this point. This could arise from "groupthink," whereby interest in cohesion or uniformity results in an incorrect decision being reached. This can also occur where the group as a whole will not dissent from the opinion of a figure in authority, as was the case with the Kennedy administration during the Bay of Pigs disaster. Indeed, the Bay of Pigs is considered the seminal case study on groupthink by Janis, spawning mountains of later research (see Janis (1972) and Janis (1982) for seminal works in the area).

Step 2: Make a List of Significant Evidence and Arguments for and Against Each Hypothesis

As a prelude to step 3, this allows for careful consideration of the theories present, and what evidence may support or refute these theories. If the evidence leads directly to a theory about the case, this should be noted, as should a theory that is generated based on other information, such as past experience. The two types of inference should be clearly identified.

Heuer (1999) offers the following for step 2:

* List the general evidence that applies to all the hypotheses, then consider each hypothesis individually listing factors that support or contradict each one. Each hypothesis leads to different questions, which leads to seeking out different kinds of evidence.
* For each hypothesis ask: What should I be seeing or not seeing? What are all the things that must have happened, may still be happening, or that one should expect to see of the evidence? If you are not seeing all of the evidence, why not? If you are not seeing all of the evidence, why not? Has the event not happened? Is it not observable? Or has the evidence not been collected?

- Note the absence of evidence as well as its presence. Attention tends to focus on what is reported rather than what is not. Conscious effort is required to determine what is missing if a specific hypothesis was true.

In short; what one has, what one does not have, and what one should have given the case.

Step 3: Prepare a Matrix with Hypotheses across the Top and Evidence Down the Side. Analyze the "Diagnosticity" of the Evidence and Arguments; That is, Identify Which Items are Most Helpful in Judging the Relative Likelihood of Alternative Hypotheses

The matrix provides a visual representation of the information collated during step 2, and according to Heuer it is the most important step. Here, hypotheses from step 1 and evidence from step 2 are taken and placed into a matrix. Each piece of information is then analyzed according to each hypothesis. Here, the analyst works across the rows, one piece of evidence at a time, to see how consistent that item is with each hypothesis.

The following is an example of a matrix using one of the author's cases involving the death of a male victim, from Wyndham Vale, Victoria, Australia, in 1998. This victim was found with a bullet hole to the crown of the forehead (with the projectile still in the skull just above the bridge of the nose), his bathrobe cord around his neck, and half submerged in the only bathtub of the house (Figure 2.1).

Was the victim shot, strangled, or drowned first?

Although the above-mentioned evidentiary possibilities are not exhaustive (such as the victim being rendered unconscious in some other way beforehand), these certainly provide a good example of the

Hypotheses:
H1: The victim was shot first.
H2: The victim was strangled first.
H3: The victim was drowned first.

	H1	H2	H3
E1. The victim could have lived for several minutes after the gunshot.	+	+	−
E2. The victim had petechial hemorrhaging in the eyes.	+	+	−
E3. The victim had no water in the lungs.	+	+	−
E4. The victim had no scratch marks on his neck indicating an attempt was made to alleviate pressure from the ligature.	+	−	+
E5. The victim had no abrasions on his neck indicative of a struggle while being strangled.	+	−	+

FIGURE 2.1

Matrix involving the death of a male victim from Wyndham Vale, Victoria, Australia, in 1998.

process by first considering the possible hypotheses, then weighting each hypothesis with the evidence for and against each one. Let us now take a look at some of the above-mentioned factors in more depth.

At autopsy, the pathologist states that the victim could have lived for several minutes after being shot, which would certainly allow for cardiac output if the strangulation occurred second. This would also allow for the development of the petechial hemorrhages in the eyes. The lack of water in the lungs precludes the victim being placed into the half-full bath while he was still alive. Interestingly, the outcome of the above-mentioned exercise (five "tick marks" in support of hypothesis (H) 1, three in favor of H2, and only two in favor of H3), would certainly support the available evidence in terms of the sequence of wounds or injuries: the victim was shot, which rendered him unconscious but certainly did not kill him; seeing he was still alive, the offender has then removed the bathrobe cord and strangled the victim while he was unconscious but alive. Finally, for reasons that can only be speculated, the victim has then been placed into the bath (perhaps to ensure his actual death, if there had been uncertainty about past attempts, or where his death was the paramount consideration).

Step 4: Refine the Matrix. Reconsider the Hypotheses and Delete Evidence and Arguments That Have No Diagnostic Value

The definitions we use for our hypotheses and the way they are worded will be crucial in determining whether the evidence supports or refutes each and every one. For those hypotheses in which there is support or only limited support, the hypothesis or the evidence will need to be refined or reviewed so that one is a better match for the other or vice versa.

Should thinking be influenced by factors not listed on the matrix, this should be added to the matrix. Hypotheses or evidence that is unimportant or lacks diagnostic credibility should be omitted from the matrix.

Step 5: Draw Tentative Conclusions about the Relative Likelihood of Each Hypothesis. Proceed by Trying to Disprove Hypotheses Rather Than Prove Them

Step 3 considered the evidence relative to the hypotheses working across the rows. In step 5, the analyst works down the rows and examines each hypothesis as a whole. Adopting a central premise of the scientific method, the analyst should favor disproving theories and not proving them. Evidence can be stacked up ad nauseum in support of a theory, but it only takes one piece of evidence (as long as that one is valid), to disprove a theory.

When working with the matrix, the hypothesis with the least (−) signs (or whatever sign the analyst has employed), are likely to be the most accurate or probable. Those with the most (−) signs will be the least accurate or probable. This part of the analysis may be more superficial than this cursory discussion provides, because some evidence will be more important than other evidence. Put another way, certain items will carry more weight than others. For example, properly interpreted physical evidence should be given more weight than simple statement evidence, as a general rule.

Step 6: Analyze How Sensitive Your Conclusion is to a Few Critical Items of Evidence. Consider the Consequences for Your Analysis If That Evidence Were Wrong, Misleading, or Subject to a Different Interpretation

In step 3, the evidence and arguments that were diagnostic were identified, and in step 5 these findings were used to make tentative judgments about the hypothesis. Go back and question the few main assumptions or items of evidence that drive the outcome of your analysis in any given direction.

When analysis turns out to be wrong, it may be because main assumptions went unchallenged and proved invalid. The problem is in determining which of the assumptions actually merit questioning, and an advantage of the ACH model is that it can be used to tell you which need to be checked. Here, additional research may be needed to check key judgments, or evidence may need to be revisited to determine whether it has been interpreted properly or to whether any instructive value may have been overlooked.

Step 7: Report Conclusions. Discuss the Relative Likelihood of all the Hypotheses, Not Just the Most Likely One

If the report is to be used to drive decision making processes, it is necessary to inform the reader as to the likelihoods involved. Decisions need to be made on the full set of alternative possibilities, not just the most likely one. "When one recognizes the importance of proceeding by eliminating rather than confirming hypotheses, it becomes apparent that any written argument for a certain judgment is incomplete unless it also discusses alternative judgments that were considered and why they were rejected" (p. 107).

Step 8: Identify Milestones for Future Observation That May Indicate Events are Taking a Different Course Than Expected

Situations may change or remain unchanged while new information comes in that may alter the appraisal. Specify in advance things to look for or be alert to that, if seen, would suggest a change in the probabilities. Highlighting in advance what would change your mind will make it difficult to rationalize these judgments should they occur.

Conclusion

Good logic and sound reasoning is a critical component of any analysis, from the initial intake of a case where it must be determined whether the presentation is false or legitimate, through to specific assessments such as an assessment of crime scene staging, case linkage, risk assessment, or threat management. The effective employ of both inductive and deductive logic as well as the use of heuristics to aid judgment can help overcome bias and cognitive distortions. However, these heuristics can also be a pitfall in judgment, impacted by our past experiences, education, and training, and the degree to which we have learnt from them. Structured professional judgment, also discussed in this chapter, may be a safeguard against these potential errors.

REVIEW QUESTIONS

1. The statement "it is likely that the offender and victim knew one another" is an example of what type of logic?
   a. Deductive
   b. Adductive
   c. Reductive
   d. Inductive
   e. None of the above

2. Heuer's ACH methodology is comprised of how many steps?
   a. 1
   b. 2
   c. 4
   d. 6
   e. 8

3. What are the three generations of risk assessment, and what does each involve?

4. The existence of many interdependent variables in a given system is known as
   _____.

5. Name two types of heuristic and briefly explain each.

6. Knowing that one is in error and why is an area of cognitive psychology known as
   _____.

References

Aickelen, U., & Clark, A. (2011). Heuristic optimisation. *Journal of the Operational Research Society, 62*, 251–252.

Alexandra, A., Matthews, S., & Miller, S. (2002). *Reasons, Values and Institutions*. Melbourne: Tertiary Press.

Aliseda, A. (2007). Abductive reasoning: challenges ahead. *Theoria, 60*, 261–270.

Alison, L., Goodwill, A., & Alison, E. (2011). Guidelines for profilers. In L. Alison (Ed.), *The Forensic Psychologist's Casebook*. Gloucester: Willan Publishing.

Alleyne, E., & Wood, J. L. (2010). Gang involvement: psychological and behavioral characteristics of gang members, peripheral youth, and nongang youth. *Aggressive Behaviour, 36*, 423–436.

Beauregard, E., Proulx, J., & Rossmo, D. K. (2005). Spatial patterns of sex offenders: theoretical, empirical and practical issues. *Aggression and Violent Behavior, 10*, 579–603.

Bevel, T., & Gardiner, R. (2008). *Bloodstain pattern analysis: with an introduction to crime scene reconstruction* (3rd ed.). Boca Raton: CRC Press.

Bhattacharyya, S. (1958). The concept of logic. *Philosophy and Phenomenological Research, 18*(3), 326–340.

Bouch, J., & Marshall, J. J. (2005). Suicide risk: structured professional judgement. *Advances in Psychiatric Treatment, 11*, 84–91.

Coroner's Court. (2003, March 5). *Transcript of evidence: K. Illingsworth*. N. S. W. Coroner's Court.

Dörner, D. (1996). *The logic of failure: recognising and avoiding error in complex situations*. New York: Basic Books.

Farber, M. (1942). Logical systems and the principles of logic. *Philosophy of Science, 9*(1), 40–54.

Farnham, F., & James, D. (2000). Stalking and serious violence. In *Criminal Justice Responses Conference*. Canberra: Australian Institute of Criminology.

Finucane, M. L., Alhakami, A., Slovic, P., & Johnson, S. M. (2000). The affect heuristic in judgements of risk and benefits. *Journal of Behavioural Decision Making, 13*, 1–17.

Garst, J., Kerr, N. L., Harris, S. E., & Sheppard, L. A. (2002). Satisficing in hypothesis generation. *The American Journal of Psychology, 115*(4), 475–500.

Gigerenzer, G. (1991). How to make cognitive illusions disappear: beyond "heuristics and biases". In W. Stroebe, & M. Hewstone (Eds.) *European Review of Social Psychology: 2*. (pp. 83–115).

Gross, H. G. (1924). *Criminal investigation* (3rd ed.). London: Sweet and Maxwell, Ltd.

Guy, L. S., Packer, I. K., & Warnken, W. (2012). Assessing risk of violence using structured professional judgement guidelines. *Journal of Forensic Psychology Practice, 12*, 270–283.

Heuer, R. J. (1999). *Psychology of intelligence analysis. Center for the Study of Intelligence, CIA*. Available from https://www.cia.gov/library/center-for-the-study-of-intelligence/csi-publications/books-and-monographs/psychology-of-intelligence-analysis/index.html. Accessed 06.03.13.

Janis, I. (1972). Groupthink. *Psychology Today, 5*(6), 43–46. 74–76.

Janis, I. (1982). *Groupthink: psychological studies on policy decisions and fiascoes*. Boston: Houghton Mifflin.

Keller, C., Siegrist, M., & Gutscher, H. (2006). The role of affect and availability heuristics in risk communication. *Risk Analysts, 26*(3), 631–639.

Kruger, J., & Dunning, D. (1999). Unskilled and unaware of it: how difficulties in recognising one's own incompetence lead to inflated self assessments. *Journal of Personality and Social Psychology, 77*(6), 1121–1134.

McEwan, T., Mullen, P. E., & Purcell, R. (2007). Identifying risk factors in stalking: a review of current research. *International Journal of Law and Psychiatry, 30*, 1–9.

Merriam-Webster. (2011). *Webster's American English Dictionary*. Connecticut: Federal Street Press.

Meloy, J. R. (1998). The psychology of stalking. In J. R. Meloy (Ed.), *The Psychology of Stalking: Clinical and Forensic Perspectives*. Boston: Academic Press.

Moore, B. N., & Parker, R. (2009). *Critical Thinking* (9th ed.). Boston: McGraw Hill.

Musgrave, A. (2009). Popper and hypothetico-deductivism. In D. M. Gabbay, S. Hartmann., & J. Woods (Eds.). *Handbook of the History of Logic. Inductive Logic:* (Vol. 10). San Diego: Elsevier.

Nisbett, R. E., Krantz, D. H., Jepson, C., & Kunda, Z. (1983). The use of statistical heuristics in everyday inductive reasoning. *Psychological Review, 90*(4), 339–363.

Ogloff, J. R. P., & Davis, M. R. (2005). Assessing risk for violence in the Australian context. In D. Chappell, & P. Wilson (Eds.), *Issues in Australian Crime and Criminal Justice*. Sydney: Butterworths.

Pachur, T., Hertwig, R., & Steinmann, F. (2012). How do people judge risks? Availability heuristic, affect heuristic, or both? *Journal of Experimental Psychology: Applied, 18*(3), 314–330.

Pedersen, L., Rasmussen, K., & Elsass, P. (2010). Risk assessment: the value of structured professional judgements. *International Journal of Forensic Mental Health, 9*, 74–81.

Petherick, W. A. (2014). Induction and deduction in criminal profiling. In W. A. Petherick (Ed.), *Profiling and Serial Crime* (3rd ed.). Boston: Elsevier.

Popper, K. (1960). *On the sources of knowledge and ignorance*. Proceedings of the British Academy LXVI. 39–71.

Prodan, M. (1995). *Criminal investigative analysis in the homicide of Roberta Ann Butler*.

Safarik, M. (2000). *Criminal investigative analysis in the homicide of Shawna Michelle Edgar*.

Tversky, A., & Kahnemann, D. (1974). Judgement under uncertainty: heuristics and bias. *Science, 185*, 1124–1131.

van den Brink-Budgen, R. (2010). *Advanced critical thinking skills*. Oxford: HowtoBooks.

Physical Evidence and the Crime Scene

Wayne Petherick[1] and Andrew Rowan[2]
[1]*Bond University, Gold Coast, QLD, Australia*
[2]*Queensland Police Service, Brisbane, QLD, Australia*

Key Terms

Science
Forensic science
Forensic generalist
Evidence technician
Laboratory technician
Physical evidence
Testimonial evidence
Preparatory crime scene
Primary crime scene
Secondary crime scene
Validity
Sufficiency
Reliability
Crime reconstruction

INTRODUCTION

Regardless of the type of crime, there will usually be evidence available for examination. The type and amount of evidence will be dictated almost entirely by the type of crime and the level of interaction between the offender and victim, offender and crime scene (and other locations such as those used to prepare for the event), and victim and crime scene. For example, when a victim is killed in a homicide, the offender may try to set fire to the scene in an effort to destroy evidence of the interaction, such as wounds to the victim, biological evidence linking them to the victim and crime, and others such as fingerprints and bloodstains. Fire can change the nature of evidence; it can destroy it entirely, or it can leave new evidence behind (such as the fire patterns resulting from the accelerants used).

This chapter discusses physical evidence: what it is, the different types of physical evidence that may be encountered, what a crime scene is, and other aspects of evidence and crime scene evaluation. This is done to inform the reader as to the types of evidence they may encounter during analysis and to provide some cursory understanding of physical evidence, its collection, and interpretation. It is not the purpose of this chapter to make the reader an expert on all aspects of physical evidence however, because this requires considerable education, knowledge, and training.

3.1 What is science?

Science is a broad and inclusive term used to describe any discipline that has four essential components: (1) an orderly body of knowledge, with (2) principles that are clearly enunciated (Thornton, 1997); (3) reality oriented (relates to this world and obeys the laws of nature); and (4) open to testing (falsification is a fundamental tenet of the scientific method). In short, science is what we use to understand and describe the physical universe, and it is characterized by the notion of hypothesis testing (Inman & Rudin, 2001).

3.1.1 Science and forensic science

The term forensic is from the Latin *forensus* meaning "of the forum" where in ancient Rome government debates were held (Thornton, 1997). Today, forensic has taken on the general meaning of the application of a discipline to matters of the law; thus, forensic science can best be described as the application of science to legal matters. When applied to a discipline, forensics covers a broad range of fields from accounting to zoology, and Wikipedia, for example, lists 35 subdivisions of forensics. Regardless of this simple point, it should be noted that not all fields of expertise that assist the court relate to scientific fields; it would therefore be misleading to assume that the practice of forensics implies that any given endeavor is by extension a science.

Inman and Rudin (2001) describe forensic science as the application of scientific principles to solve crime, and it is therefore an applied science. Just as the engineer uses principles of physics to design a bridge, and medical doctors uses principles of biology, anatomy, and biochemistry to diagnose and treat patients, forensic scientists use the principles of biology, chemistry, and physics to provide answers about criminal events. Although the forensic sciences use experimental methods developed from the imperial or pure sciences, it must be acknowledged that experimentation in the true sense is not always possible. This is because, as noted by Inman and Rudin (2001), it is not possible to control all variables within the crime scene.

In forensic analysis, the scientist starts with a known result: a bloodstain pattern in a hallway, a wound pattern on a victim, or a particular constellation of burn patterns at the scene of an arson. Generally, the behaviors or the sequence of events that resulted in the deposition of evidence will be unknown unless there is some irrefutable chronicle of these events (closed circuit TV (CCTV), for example). Although the precise events will be unknown, the scientist will be able to develop a range of hypotheses (inductive theories) that can be measured and tested against the evidence using the scientific method. When this is done, the results of these attempts to falsify theories will produce a more informed opinion about what happened, and from here the scientist can conclude, to a varying degree of certainty, what did, or may have, occurred.

It may be argued of hypothesis testing that the only true validation of the theory would be a suspect's confession to what occurred, although this requires the assumption that the confession is true and correct in every respect. For good or bad, depending on one's point of view, any version of events given by the suspect can be treated as a theory to be tested using scientific principles. This is discussed further in the reconstruction section in this chapter.

The various aspects of hypothesis testing are complex and beyond the scope of this chapter, but all analysts should have a basic appreciation of null and alternative hypothesis testing. Although there may be talk about how this theory or that has been proven, in science we can only consistently fail to disprove and not prove a theory. As discussed by Inman and Rudin (2001), science can only provide us with the best estimate of what occurred based on the available information at the time.

We can consider crime as a problem in the negative: what better way to establish what happened in a case than to consistently fail to disprove theories?

Those from a nonscientific background may find this confronting. When involved in case work they will often assert that the scientist is doing the exact opposite of what was requested, or worse, challenge the scientist as working for the "other side." This thinking reveals two underlying problems. The first is that they have revealed an ignorance of basic hypothesis testing, and what is achievable within the confines of science and scientific reasoning. The second is that they are advocating for an outcome and subsequently believe the scientist is also an advocate. The forensic scientist must at all times remain impartial to the evidence, with no investment in the outcome (Thornton, 1997), allowing them to speak objectively for the evidence on behalf of any party. Should they abandon this position and become partisan, they are no longer a scientist.

3.2 Forensic roles

There are a variety of different roles within forensic science, each requiring different levels of training, education, and experience. As discussed, not all those involved with forensics are scientists. In fact, the majority do not hold formal qualifications in science. This does not, however, mean that these roles do not fulfill important functions or contribute nothing of value to other forensic endeavors. Much of their work is critical to the overall process and forms the foundation on which much other assessment is based. Some common roles are described below.

3.2.1 Forensic generalists

These roles are typically filled by forensic scientists broadly trained in a variety of forensic specialties. They work in a crime scene to detect and collect evidence in major cases, typically involving serious crimes such as murder and arson. These specialists assist in the reconstruction of crime and can sometimes be referred to as the case scientist who oversees the forensic investigation. The forensic generalist is well aware of the limits of their individual knowledge and refers to forensic specialists as needed. This role is very skilled, in that they require an understanding and appreciation of the work performed by other forensic roles, and they are therefore well versed in questioning or critiquing the work of others during case review. Forensic generalists are mostly based in very small, highly trained, specialist police or scientific units. Most agencies require a minimum postgraduate degree in science to gain entry into these units. The role has traditionally been performed by sworn police positions, but recently some jurisdictions have moved toward civilian positions in execution of this task.

3.2.2 Evidence technicians

Evidence technicians perform an important role in the detection, collection, and preservation of physical evidence. They are trained in the detection and collection of most common forms of physical evidence. Their primary role is the examination of volume crime where they will process numerous crimes within a single shift. Although these technicians will attend serious crimes, including murders, they will usually be under the direct supervision of a forensic generalist. Training is typically limited to "on the job" practical training after attending a short theory course. A full-time evidence technician is typically not a forensic scientist and is not necessarily qualified to provide opinions on the results of examinations of forensic evidence.

This role is typically housed within police agencies and performed by both sworn police and civilian positions, depending on the jurisdiction. The range of roles performed by an evidence technician will vary as widely as the terms used for this role and will be dictated mostly by jurisdictional arrangements. A common title for this role is scenes of crime officer (SoCO) or simply crime scene officer. In many cases, these officers, due to their experience, are able to perform many, if not all the tasks of the forensic generalists as described above.

3.2.3 Laboratory technicians

These practitioners operate the laboratory equipment and perform tests based on methods directed by forensic specialists. The results of the tests they perform are provided to the forensic specialists who adopt the results to form their conclusions. The technicians may not understand the theoretical background to the tests they perform, or have a complete understanding of science or forensic science. The training and education levels required of this role vary widely depending on the field of forensics in which they are working.

3.3 What is a crime scene?

According to Lee, Palmbach, and Miller (2001), there are many ways to classify a crime scene, including by location where the crime was committed, the physical boundaries of the crime scene, and the various activities at the scene. For example, a particular site may be identified as a residential crime scene, a vehicular crime scene, or a homicide crime scene. These descriptions do not necessarily convey the meaning the scene had to the offender if it was chosen for a particular purpose, and they do not necessarily convey the order in which actions took place over multiple sites.

A crime scene is loosely defined as any location where a criminal event occurred or where evidence of it can be found. Osterburg and Ward (2010) state that the crime scene encompasses all of the areas over which actors moved during the commission of a crime. This includes the victim, criminal, and any witnesses to the event. As such, the variety of types and locations of crime scenes are limitless and may be large (an educational institution subject to a rampage school shooting), small (a convenience store subject to a robbery homicide), mobile (a car or other mobile platform), or static (a residence where a domestic homicide occurs). Crime scenes may occupy a physical space or can occur entirely within the digital universe of the Internet not isolated to one place or jurisdiction.

3.4 Crime scene types

Defining the crime scene type by the actions that took place may not be as instructive as needed to understand the events, their sequence, and the location at which they occurred. For example, stating that a location is a homicide crime scene will be self-evident once it is established that it is the location at which the victim was killed. This may provide no further benefit to the analyst. To be of use, the classification system must have some meaning; it must provide some idea as to how much interaction took place there, and where within several different locations each resides relative to the others.

With this in mind, the following crime scene types are presented based on those presented in the literature, with minor modification and addition where necessary: preparatory, point of contact, primary, secondary, and dump/disposal site.

3.4.1 Preparatory crime scenes

The authors introduce a crime scene type not previously addressed in other literature: that of the preparatory crime scene. A preparatory scene is defined as any location where the offender(s) planned, prepared for, watched, or waited, to execute the offense or part thereof. The preparatory scene is important because it will contain evidence that may later be used in prosecution of the offense, and it may help establish planning and intent.

The type of items that may be located at the preparatory scene include diagrams, sketches, schematics, ammunition, disguises, security routines and locations, CCTV locations, photographs, or other evidence of surveillance of locations or individuals, and any other type of evidence that suggests the offender planned/prepared or watched in this location. The preparatory scene terminates at the point at which the offender actually comes into physical contact with the victim or location at which the offense is executed. At this point, the scene becomes one of the following types: point of first encounter, primary or secondary scene, or dump/disposal site.

3.4.2 Point of first encounter

The point of first encounter is a term used to describe the point at which the victim and offender first met. It does not imply a time frame in which the crime event was set in motion; rather, it identifies that moment in time at which the relationship, whatever that may be, between the victim and offender started. It is the first author's opinion that this is a more useful term than others because this term provides both the time and the context under which the relationship started, rather than that fixed point in time at which the crime event began.

For example, in one case from the first author's file, a 2.5-year-old boy died as a result of an acceleration-deceleration injury resulting in a stroke. The father confessed to and was subsequently convicted of the crime. There is, however, evidence to support the assertion that the injury was caused by another party several days before hospitalization and death. The father and mother in this case met some years before the child-victim was even born. The circumstances under which the two met (i.e., where they first encountered one another) is of greater value than considerations proximal to the death. This is important given the wealth of evidence of psychopathology, including chronic attachment issues, that ultimately lead to a failure to bond with the child. The point of contact could be described as the day the child sustained the life-ending injuries, whereas the point of first encounter was some years prior when the father and mother first met and the context under which this occurred. This can be crucial in determining relational dynamics, among others.

Of course, the above-mentioned case represents ideal situations under which to determine the point of first encounter. If the offender is unknown, or if the victim or offender has sought to actively conceal the time frame and circumstances under which he or she encountered the other party, the point of first encounter may remain a mystery. This confound would apply equally well to approaches that define and use a point of contact or other location description of first contact.

3.4.3 **Primary and secondary scenes**

Geberth (2006) suggests that the location at which the body is found is the primary crime scene, with the body location characterizing the significance of this location. He further notes that the use of this word does not imply this is the location at which the original event occurred. However, this is exactly what Lee et al. (2001) suggest the distinction of primary and secondary scenes are based on. These authors note (p. 3) the following:

> By now we have learned through experience that any area, for example the suspect's hop use, victim's house, suspect's vehicle, and the road to the garage are connected to the homicide. Therefore, the crime scene can be and often is more than one location. The victim's home, the victim's place of business, the victim's car, the road they travel, and the site where the victim's body was found – can and should all be considered possible crime scenes. In an attempt to further delineate and even organise the crime scenes, the term primary crime scene is often used to refer to where the original or first criminal act occurred and any subsequent scene(s) is considered to be a secondary scene.

Chisum and Turvey (2011) propose that crime scenes be delineated by virtue of the amount of evidence or interaction they contain. As such, they propose that the primary crime scene is the location where the offender engaged in the majority of their activity, and also perhaps spent most of their time. Basing any definition on what occurred first and then subsequent to this, they suggest, misses the point of classification and prevents meaningful comparison to other events. A secondary crime scene then is "a location where some of the victim-offender interaction occurred, but not the majority of it" (Chisum & Turvey, 2011, p. 149).

These are the definitions of primary and secondary scenes adopted in this work.

3.4.4 **Dump/disposal site**

The dump or disposal site is simply where the body was left after the offense occurred. The authors suggest two further subtypes of disposal site. The first is a *passive disposal site* where the offender makes no attempt to move or relocate the body, but it is simply left where it falls or is placed. A passive dump site may be the result of an interrupted offense, the influence of drugs or alcohol, or a failure to plan, or it may be the result of an offender who does not feel the need to distance themselves or veil their connection to the crime scene. The second is the *active disposal site* where the offender moves or relocates the body to another physical location after the offense, for a particular purpose. An active dump site therefore may be the result of planning, an attempt to delay finding the victim or recovery of their remains, or an attempt to distance the offender from other crime scenes associated with it.

Of course, classifying the locations in which events occurred does not imply that events must be spread across different physical locations. In a homicide involving an opportunistic victim (so without specific victim selection or presurveillance), the point of first encounter may occur at the same time and place as the actual crime, where the victim is also left after the offense. In this example, there is no physical or temporal difference in the location, being all of the point of first encounter, the primary scene, and the dump site. However, if the offense was planned but the victim opportunistic, there may still be another location (say, the preparatory site), and if there was any type of transfer between the victim and offender, the offender's vehicle and possibly residence may also have evidence of the act. These locations are therefore secondary crime scenes.

3.5 **What is physical evidence?**

There are a great many types of physical evidence that may be encountered, differing according to a range of influences and factors. Where offenders engage in precautionary behaviors such as the wearing of gloves, items such as fingerprints may be more unlikely. When a rapist wears a condom, semen may be unlikely except in the event of breakage or slippage where the condom comes off during the assault. Accelerants may be a common finding in an arson, and blood may be common during homicides. Any or all of the above-mentioned items may be absent in stalking, although letters, emails, phone calls, and physical intrusions of a nonsexual nature may be plentiful.

Because there are different types of evidence, it is likely that many types will be left behind in any one crime, crime scene, or crime series. In a homicide, defensive injuries (an injury sustained while trying to defend oneself or ward off an attacker) may be present, along with bloody fingerprints on the murder weapon, with bloodstains on the wall from cast off created by the offender swinging the weapon during the attack. Bloodstains may be absent during stalking, where letters and other types of communication are more common. Each type may require different recognition and collection efforts, and each may tell different things about the event or events that have occurred.

The identification, documentation, examination, and interpretation of different types of evidence are undertaken by a variety of different individuals, from crime scene technicians, to laboratory technicians, to forensic scientists or criminalists (a particular type of forensic scientist who works in a crime laboratory). As a result, a crime analyst may end up with any number of different documents that they must assess and interpret. Some of these documents will be relatively straightforward requiring no further work. For example, an autopsy report stating a particular cause of death usually requires little to no specialist education on the part of the crime analyst. However, an autopsy report that later identifies "bilateral petechial hemorrhaging in the ocular sclera" may require deeper understanding. It is incumbent then on the analyst to adequately equip themselves with the requisite knowledge to be able to understand what this means when encountered. This, however, raises an important point.

Simply knowing about the interpretation or meaning of complex terminology or principles does not mean that any given analyst is inherently qualified to perform certain types of analysis. Knowing the difference between a loop and a whorl characteristic does not automatically make an analyst a fingerprint expert any more than identifying the difference between high- and medium-velocity blood splatter would make them a bloodstain pattern expert. As a result, where the crime analyst lacks the appropriate training or education in certain aspects of evidentiary examination, it is vital that they work with someone who has the requisite training and understanding. For most types of evidence this would usually mean a forensic scientist or criminalist.

3.5.1 **Evidentiary materials**

When supplied with case materials, the actual composition will differ case by case. In some cases, there will be a wealth of documentary evidence such as letters written to various government and nongovernment departments by victims of crime, communications to victims, police briefs of evidence, trial transcripts, appeals documents, statements from witnesses, records of interview with suspects, and just about anything else that can be imagined. The source of this material will be as many and varied as the types of evidence possible. Information may come from victims, victim's families, the police, defense or prosecution lawyers, and innocence projects, among others.

The source of the evidence can be problematic, with the first author having seen numerous reports stating that the source of information has been a briefing by the prosecutor or defense lawyer and their investigator. This does not imply that all lawyers and their teams are unethical or devious, or that they would deliberately withhold evidence or information. However, letting others dictate what is important to you is fraught with peril, and the possibility of bias is very real and must be a consideration when examining a case. This potential exists regardless of the source, and not just because of unethical conduct.

Different witnesses will have different perspectives about what happened and why; a victim of crime may have an emotional or financial investment in the outcome (in many places, victim compensation is dependent on a guilty verdict); parents and friends may desperately want to keep a loved one out of prison; and investigators may desperately want to get someone they perceive to be a "bad guy" off the streets. All of these things must be considered.

As a result, one should make every effort to determine the veracity of the findings of others, and to establish the validity of statements and claims made by witnesses, victims, and others (discussed further in the following section). To assume that a professional has done their job without potential error is not only dangerous but also misinformed. To question, however, may be seen to be impolite, unprofessional, or not collegial, and it is true that the degree to which you question the findings of others will be met with different receptions. Lazy or inept examiners may respond with anger and accusations of impropriety, fearing that deficient work product will be discovered. Ambivalent examiners will probably not care that their work is being questioned, and they may or may not care about your examinations or findings. Competent and professional examiners may not mind at all that you are questioning their findings, and may, in fact, engage in the process keen to show what they did, how they did it, and how they arrived at their conclusions. The given reception may even be an indication of what you can expect to find in the work product of others at the start of your involvement.

3.5.2 Validity, reliability, and sufficiency

Before moving on to individual types of evidence, it is necessary to discuss validity, reliability, and sufficiency. These are important not only from the standpoint of the forensic scientist and investigators, but also for the applied crime analyst whose work starts some time after others have engaged in and likely completed the documentation and evidence collection. As a result, determinations must be made about the nature of the evidence and the degree to which it can be relied upon for post hoc analysis to answer the client's questions. Indeed, a part of the analysis (or a separate analysis itself) may be undertaken in order to determine what is not available and therefore what would be needed to complete the requested study.

In research and statistics, validity and reliability are two key properties of good scientific inquiry that relate to the confidence we can have about our data, and they are therefore not dissimilar to the purposes for which they are used here, with some variation. In research, validity refers to whether an instrument is actually measuring what it purports to measure, while reliability refers to whether an instrument can be used consistently across situations (Field, 2009).

In applied crime analysis, validity refers to whether the evidence being examined is actually what it represents or presents as. For example, is a suicide note actually written by a victim? Is this crimson mark on the carpet in this photograph a bloodstain or red wine? Is this the victim's brand of cigarette? Validity is important to establish because it relates to the evidence that is the subject of examination and from which any opinions will be put forward. If the evidence is not what we think it is, the conclusions will not be an accurate representation of what occurred. Validity equates to accuracy.

In research, reliability refers to the ability to replicate your findings across situations. This then determines the degree to which we can trust things such as theories that are being tested. If similar results cannot be achieved with a different sample, the theory is not likely to be valid (barring methodological problems that impact the ability to detect differences). In crime analysis, reliability refers to the degree to which we can trust the evidence on which the analysis is based: literally, is the evidence reliable and can it be trusted? A statement provided by a suspect should be considered suspect in itself until the contents are validated through other means, i.e., independent verification of each and every component. In a violent assault, the claims made by each protagonist may not be as reliable as high-definition CCTV footage capturing the event. It should stand to reason then that if we cannot trust the evidence, we should not be relying on it for any assertions, opinions, or conclusions.

In this chapter, the concept of sufficiency also is introduced. This relates to the quantity of the evidence, and whether there is enough evidence on which to base an analysis. Even in a case in which the evidence has been determined to be an accurate representation of the facts (ergo, is valid), and that it can be trusted (ergo is reliable), there may not be enough of it to perform even the most cursory of examinations. Sufficiency therefore relates to quantity. It should stand to reason that, providing validity and reliability are met, more information for the analysis will be more likely to result in a greater number of more informed opinions.

These three issues are largely summed up in an axiom from intelligence analysis: garbage in, garbage out.

3.5.3 **Two types of physical evidence**

According to Fisher (2000) physical evidence can be divided into two broad types. The first is testimonial evidence, and the second is real or physical evidence. "Testimonial evidence is evidence given in the form of statements made under oath, usually in response to questioning. Physical evidence is any type of evidence having an objective existence, that is, anything with size, shape, and dimension" (Fisher, 2000, p. 1). This chapter focuses more on physical evidence and less on testimonial evidence. It is however important to understand that testimonial evidence often provides context or relevance of physical evidence located within a crime scene. Furthermore, the physical evidence, and more importantly the results of its analysis, can be used to test the validity of the testimonial evidence.

Kirk (1974, p. 33) provides the following on the importance of physical evidence:

Today, as never before, the utilisation of physical evidence is critical to the solution of most crimes. No longer may the police depend upon the confession, as they have done to a large extent in the past. The eyewitness has never been dependable, as any experienced investigator or attorney knows quite well. Only physical evidence is infallible, and then only when it is properly recognised, studied, and interpreted. The importance of overlooking nothing that will contribute to the final solution of the crime cannot be overestimated. Of the various items capable of contributing to that solution, nothing is more important than microscopic residues. They are the hardest to locate because they cannot be seen, but it is for this very reason that they are available for examination. The criminal seeking to hide his traces will dispose of many macroscopic items, but he will not deliberately destroy evidence he does not notice.

In a similar vein, Locard (1929, p. 176) in his discussion of dust traces notes the following:

Among recent researches, the analysis of dust has appeared as one of the newest and most surprising. Yet, upon reflection, one is astonished that it has been necessary to wait until this late day for so simple an idea to be applied as the collecting, in the dust of garments, of the evidence of the objects rubbed against, and the contacts which a suspected person may have undergone. For the microscopic debris that cover our clothes and bodies are the mute witnesses, sure and faithful, of all our movements and of all our encounters.

Although this specific discussion revolves around dust, it should be noted that the transfer of evidence when two items come into contact, known as *Locard's exchange principle*, applies equally to a great many types of physical evidence. This would include fingerprint marks, blood, dirt, DNA, hair, fibers, and any number of others that is subject to transfer by virtue of contact.

It has been previously noted that the evidence in a case forms the foundation for analysis. But this simple statement alone does not do the importance of evidence justice, nor does it detail the many reasons why evidence is so important. This will now be explicated from Fisher (2000, pp. 1–5):

- Physical evidence can prove a crime has been committed or establish key elements of a crime.
- Physical evidence can place the suspect in contact with the victim or with the crime scene.
- Physical evidence can establish the identity of persons associated with the crime.
- Physical evidence can exonerate the innocent.
- Physical evidence can corroborate the victim's testimony.

The authors suggest that physical evidence can provide the following facts: *who, what, when, where,* and *how* (and sometimes how many times they did it!). Collectively, these are known as the *six investigative questions*. For the forensic scientist, this would rarely ever include the *why*, because that would be an inference of intent. This is therefore outside the experience and expertise of most crime scene practitioners but may be within the purview of the applied crime analyst or other behavioral scientist.

The following types are by no means exhaustive and not all will be encountered in every crime. Some are more common than others, and some may be seen only rarely. For more exhaustive lists the reader should consult Fisher (2000), Kirk (1974), Lee et al. (2001), and Saferstein (2010).

3.5.4 Testimonial evidence

Testimonial evidence is any written or verbal statement regarding a criminal event. As stated by Fisher, this evidence will usually be given under oath but not all testimonial evidence is the result of sworn statement. Examples include statements given by witnesses (sworn or unsworn); diaries and journals of victims and other protagonists; and letters of complaint or similar communications made to police, government departments, and organizations, or sent to individuals. For obvious reasons, any testimony that has been established to be fact through other means is the most preferential, followed by sworn statements (where individuals may be legally sanctioned for bearing false witness or statement), and then any nonsworn statement. Despite the fact that the latter should not be relied upon as concrete fact, they can provide important information about the state of mind or emotional state of the author, and what they believe to have heard, seen, felt, or believed to be true.

When testimonial evidence is handwritten or its source unknown, especially when it speaks to some material fact, it may be necessary to submit it for handwriting analysis, also known as graphology, or to a more scientific study known as *questioned document examination* (or *analysis*, QDE or QDA). Fisher (2000) states that handwriting analysis can be used to answer two questions: was a signature or document a forgery, and were two writings from the same person? Where the author is not known, this may also be used to identify the source of the document from a number of reference samples.

3.5.5 Blood and other biological fluids

Apart from blood, other biological fluids include semen, saliva, synovial fluid found in joints, and cerebrospinal fluid contained in the central nervous system (brain and spinal cord). This would also include urine, serum, and perspiration (Kirk, 1974). These may be the result of interaction (seminal fluid may be found in sexual assaults) or as a result of wounds or injuries. These may be unrelated to the crime such as where a victim has consensual sex before suffering harm or loss, or the result of unrelated injury. For this reason, it is important to account for the deposition of biological evidence so that they are not incorrectly factored into the analysis.

3.5.6 Hair evidence

Hair is important in criminal investigations and must be collected whenever it is encountered (Kirk, 1974). Hair is the result of a protein called keratin, and it is found in various parts of the body such as the head, hands and arms, feet and legs, pubic region, and elsewhere. The origin of the hair, as well as the species, can be identified such that even rudimentary analysis can be used to rule out a particular strand as involved in the crime. For example, microscopic analysis of hair morphology (the structure and function of organisms and their features) can reveal the difference between dog hair, cat hair, and human hair.

3.5.7 Fiber evidence

Textile fibers such as those from the victim or offender's clothing at the crime scene can be an important element in any investigation (Deedrick, 2000). Following Locard's exchange principle that any contact between two objects results in a transfer of evidence between them, any object containing fibers that comes into contact with the victim, the offender, or the crime scene can leave fiber evidence. This can be used to associate two or more objects based on class (membership to a group) or individuating features (belonging to only one individual in the known universe).

For example, an offender prepares at their home location and carries fibers from their carpet to the crime scene. These fibers are subsequently deposited not only at the crime scene but also on the victim during the offense. If a suspect is later identified and carpet fibers taken from their house for comparison, these fibers may be matched to those found on the victim and at the crime scene. It should be noted that in this use the term matched only implies that the fibers are *similar*, and not that they share a common source or origin (see Inman & Rudin, 2002).

3.5.8 Fingerprint evidence

Fingerprints represent the most important category of physical evidence for identification and association with the scene or victim (Lee et al., 2001). They are the result of oily deposits left when parts of the skin come into contact with objects. As a result, any part of the body that has distinguishing features and can leave an impression such as ears, feet, and other parts of the hand can leave a mark that can be identified and collected. Any or all of these may be found at a variety of crime scenes.

The results of fingerprint examinations are usually of the form of "this print was made by Person X..." and is based on the theory that no two people have the same ridge characteristics. Although it cannot be proven that no two people have the same prints (to do so would require the testing and comparison of every person in the known world), the theory holds until such time as it is disproven. Much debate in the scientific community surrounds the reporting of absolute opinions in this manner and there is a push to report a probability rather than definitive certainty as to the match. From the standpoint of the scientific method (disproving rather than proving), finding a single point at which a given print does not match will represent a more definitive finding than five points of similarity.

This is similar to the reporting of probabilities in DNA results. It would be interesting that if fingerprints were only being introduced now, would our modern society accept a definitive opinion of a match based on such an assumption? That is a very simplistic statement to a very complex issue because the probability of two people having the same fingerprint is so incredibly small that it would be a pointless process. It must be remembered that in many jurisdictions suspects are not convicted on single types of evidence such as a fingerprint, but with evidence corroborated through other forms of evidence.

3.5.9 Drugs

A drug is simply any substance that can alter the chemistry of the body. Parrott, Morinan, Moss, and Scholey (2004, p. 241–242) provide a more detailed definition of drugs, encompassing not only their origin, but their effect, use, and classification:

> Any biologically active chemical that does not occur naturally in the human body. Drugs can be used to prevent and treat disease, alter mood and cognition or otherwise change behaviour. Drugs are classified into families according to their chemical structure and/or pharmacological effect. They are commonly referred to by their generic name (e.g., fluoxetine, amphetamine), pharmaceutical company brand name (e.g., Prozac) or recreational street name (e.g., speed).

The importance of drugs is covered in some detail in forensic victimology (this volume) so they will not be covered in great depth here. Suffice to say that any evidence of drugs or drug use, prescription or nonprescription, must be fully undertaken and factored into any analysis. Drug interactions should be noted in the case of polydrug use, and the effects of individual substances should be researched and reported. For example, a drug that impacts on the prefrontal cortex may affect decision making and memory formation. Those that affect the brainstem may impact on movement and the coordination and timing of movement, as well as hypervigilance. These may impact on offender or victim decision making and response style as well as other cognitive and behavioral issues.

3.5.10 Bitemarks

A bitemark is any impression left by dentition on any other item of evidence, although commonly it refers to those teeth marks left on body parts as a result of a bite. These bitemarks can occur between victim and offender, offender and victim, and offender or victim and crime scene or evidence item. It is commonly held that an individual's dentition is relatively unique, as are fingerprints; so, dental impressions are used to identify one individual as the source of the bites.

These bites can have many behavioral or emotional origins. They can be a response to victim resistance in an attempt to reduce future resistance or regain or ensure compliant behavior. They can be punishing, as may occur when an offender retaliates against a victim for fighting back or resisting. They can also be sadistic in nature, where the offender gets sexual stimulation for victim pain and suffering. As such, the location of the injury, what was happening when the injury was inflicted, and the duration and force of the bite, among others, need to be established and factored into the analysis.

3.5.11 Arson and explosives

Fire and explosion scenes are examined to determine the area of origin and the events that caused the fire and explosion. Scientific principles involved in the combustion of fire or explosives are the same with only the speed of reaction being different. The combustion process leaves complex marks or patterns that provide information regarding the direction of travel. In many instances, depending on the level of fire damage, the area of origin can be reasonably and accurately deduced through alternate hypothesis testing. Samples of fire debris will be collected in the area of origin for laboratory testing for the presence of ignitable fluids. Only after the area of origin has been determined can the process of cause determination commence.

The cause is determined by analysis of component ignition sources available in the area of origin. This can be a very difficult process given the level of destruction caused by the fire. Generally, only via elimination can a cause be inferred. Basically, what can not be eliminated is possible. The difficulty in cause determination often centers on debates over what is a plausible cause versus what is a possible cause. Sometimes, fire debris analysis can assist in this process, although it is not always possible. It may not always be possible to locate remains of petrol throughout the bedroom of a house, and petrol found in a garage may not be an unusual finding.

The challenge regarding origin and cause determination is the fact that two conditional hypotheses have to be tested. Therefore, the opinions are also conditional, that being the opinion regarding the cause of fire is conditional on the opinion regarding the origin of fire being valid. It is for this reason that importance is placed upon testing the qualifications of those providing opinions in this field. Fire and explosion examination is prone to the well-meaning actions of amateurs, not familiar with the basic scientific principles required to understand the limitations in the interpretation of this evidence (Ireland & Kelleher, 2012).

3.5.12 Toolmarks and firearms

The examination of toolmarks (literally those left by tools such as screwdrivers, hammers, pry bar, and others) and firearms examinations follow a similar philosophy: the unique use of each individual item by those who use it will impart similarly unique features on both evidence items and cartridge cases and projectiles. The examination of these items is then premised on a comparison of source (the tool or firearm) and a reference object (projectile, cartridge casing, windowsill, paint, doorjamb, or other items on which a tool is used).

In some cases, a weapon found in possession of a suspect will be matched to a reference sample found in a victim; a knife blade that has snapped off after a violent stabbing may be matched to a broken knife part found in possession of a suspect; or the striae (lines) on a windowsill or door frame may be compared with a screwdriver or other object used to pry open a window or door.

Although tools are often matched on the basis of striae they leave behind, there are a variety of different marks imparted when firearms are examined. These marks include the rifling (helical grooves cut into the bore of a barrel; see Schehl, 2000); striae from wear or defect in the barrel imparted on the projectile; or markings on the case from the breach, extractor, bolt, or magazine, among others. So not only can a suspect's firearms be compared with crime scene samples on broad characteristics such as caliber (specific bore diameter) but also on features that are unique to individual weapons.

3.5.13 Wound patterns

Wound patterns are those distinguishing features imparted as a result of contact with a weapon. They can be found on the victim or offender, and their features will differ as a function of the type of weapon used and the location on the body they are imparted upon. Wounds can be either the result of sharp force (e.g., knives, axes, machetes), or blunt force trauma (e.g., baseball bats, logs, irons), and they can be penetrating or super-ficial. In cases where the victim has tried to defend themselves, defensive injuries may be present on the hands, forearms, or other body parts of the offender. Where the attacker initially approaches and injures the victim from the front and the victim then flees, wounds may also be present on the back of the victim.

Blunt force injuries fall into four categories: abrasions, contusions, lacerations, and fractures of the bones (DiMaio & DiMaio, 2006). Abrasions are further broken into scrape or brush abrasions, impact abrasions, and patterned abrasions (DiMaio & DiMaio, 2006). An abrasion is a type of injury that will be well known to the reader, especially those with children, and is simply a trauma caused to the outer layers of skin by friction (think what happens when a child falls from a bike, or when you fall on a hard surface such as concrete or asphalt). The leakage of blood and plasma from the site eventually dries and forms a crust that is the body's way of denying germs and bacteria access to the bloodstream.

Abrasions can be partially dated by the degree of healing that has taken place where the outer edges of the wound will pucker and move toward the center of the site of injury over time. Contusions are com-monly referred to as bruises and result from a rupture of blood vessels, and these, too, can be dated by appearance and histology (DiMaio & DiMaio, 2006). Lacerations are tearing of the tissue, and they must be distinguished from incisions that are usually accompanied by visual observation of tissue bridging (Dix, 2000). This means that although the outer layers of the injury may have separated tissue, underlying struc-tures may still connect one side of the wound with the other. Fractures require no further explanation.

Sharp force injuries include stab wounds, incised wounds, chop wounds, and therapeutic or diag-nostic injuries (DiMaio & DiMaio, 2006). A stab wound is one where the depth of the injury is greater than it's width, whereas an incised wound is one where the width is greater than the depth. A chop wound is usually caused by a heavier object with a sharp edge that cleaves the epidermis, separating tissue. Therapeutic injuries are those that result from medical intervention and must be identified and documented so that they are not erroneously included in wound pattern analysis by any of those involved in the analytic process. These injuries may include intravenous injection sites, which may be confused for drug use; damage to the esophageal tract from attempts to intubate the victim; and inci-sions made for access to airways or internal organs, such as attempts at a tracheotomy.

Acceleration/deceleration injuries may also be seen in some cases, and they are usually found in cases involving blunt force trauma or other violence to the skull. These injuries are the result of inertia where a sudden cessation of movement causes the brain to rapidly impact with the inside of the skull, causing further injury (Spitz & Fisher, 2005). The resulting internal impact injury occurs on the side opposite the initial impact, and this results in what is known as a *contre coup* lesion (Geberth, 2006). Being punched in the head, falling to the ground after a blow, or smashing someone's head into a static object can cause these injuries. A rotational injury is similar, where a sudden rotation of the head stops but the brain continues on its axis, causes a shearing of tissue or blood vessels.

3.5.14 DNA

DNA, or deoxyribonucleic acid, is a genetic code that carries instructions for the development and functioning of living organisms (including viruses and bacteria). It is considered a "magic bullet" in criminal justice terms, because it can be used to provide, to a high degree of certainty, the probability that a crime scene sample originated from a particular individual. It is rare that a police investigation would not ask whether DNA could play a role (McCartney, 2010).

DNA is available in all of the body's cells and therefore in abundant quantities in body fluids. It is usually available from bloodstains, can be found in semen, and is available in saliva from buccal swabs (here, skin cells from the mouth are sought for analysis, rather than the saliva per se). Hair can yield DNA where there is a root (skin cells attached), or in mitochondrial DNA that uses the mitochondria (the cell's power) and can at best establish maternal lineage. Although DNA is considered the holy grail of crime scene evidence, caution must still be exercised when establishing association. Those with a legitimate or innocent reason to be at the crime scene may reasonably be expected to leave their DNA without criminal association; any parties who have engaged in consensual sex with someone who later becomes a victim may be inculpated in a crime in which they had no involvement. In short, DNA evidence should not be used to automatically imply involvement in criminal behavior.

The reason for this note of caution when attempting to interpret DNA is because the result cannot assist in answering the question of *how* the sample got there. This is unlike many other forms of physical evidence that can provide a level of input into the reconstruction of events that contributed to the deposition of the mark or stain. For example, a blood mark can provide very specific information as to the event that caused the mark. A shoe impression provides directionality. A fingerprint implies only a limited number of positions available to leave the mark, including which hand was used to make it. Tool marks can indicate the position the tool was held at and how it was moved.

3.6 Should the crime analyst visit the crime scene?

The short answer is no, with few exceptions. The main question to ask with regard to the crime scene is *what is the purpose of the visit?*

If the scene has not been documented (including photo and video records) and the physical evidence has not been processed by the appropriately qualified personnel, then the answer to this question will be unequivocally no. This is because every person entering the scene has the potential to contaminate evidence, by either altering that which is in place, or by introducing something new into the scene. Thus, it is critical that this possible alteration or destruction of the evidence be minimized or negated by providing access only to those with a legitimate purpose.

If the scene has been released, there may, on occasion, be benefits in visiting locations to get a "feel" for the physical environment and to understand "the lay of the land." This can help to establish lines of sight, ambient light and noise, occupancy, and distance, among others. After reviewing the documentary evidence in a multiple homicide, the first author was given the opportunity to visit the crime scene some years after the event. Not only was this instructive in terms of being able to talk to the family but also standing within the physical environment provided a much better understanding of distances and locations relative to the areas of the house wherein the crime occurred.

3.7 Processing the crime scene

The act of processing a crime scene is typically portrayed in a television show something like the following:

> *A small, clean, well kept apartment with numerous police and crime scene technicians calmly poring over the scene in the hunt for evidence. The critical evidence to the case is usually located by the bright spotlight focused on it. Upon seeing this evidence, the crime scene technician recognizes it instantly as the missing piece to the puzzle and declares this proudly to the police investigator. This results in a complete reconstruction of the events leading up to the crime and describing with confidence, a blow-by-blow account of the crime itself.*

The complete opposite is invariably the case.

In reality, crime scenes will be processed with little to no information, other than being told to "forensic it." The scene will be within an extremely untidy house that may have a history of previous crimes where it is difficult to determine with accuracy what evidence may relate to what crime. The technician will be cold/hot, tired/exhausted/hungry, and working either on their own or within a small team in an unfamiliar location for many hours without respite. If there are other technicians present, there will be constant argument over theories of the crime and methods being used to locate the evidence. In some cases or jurisdictions, this may extend into whose job is what and who is responsible to collect what and when.

Most if not all of those working in crime scene areas are highly trained, dedicated, professional, and proud of the work they do. They are experienced at searching for and detecting various types of physical evidence: the problem is that many times no one can tell them what they are looking for. That is the problem with the general term "forensic it," which is sometimes the only direction and advice provided. This is similar to being told to search a haystack, without being told you are looking for the needle. An experienced crime scene technician will find the needle very quickly, if given that information, but without it the search will be very time-consuming and involve documentation of every stalk of hay. Suffice to say that before the crime analyst even gets any information, a considerable amount of time and effort has gone into its identification, documentation, collection, and interpretation.

The crime scene is processed in a series of steps or stages. Several different methods are described in texts (e.g., Inman and Rudin, Fisher) and all involve a system of planning, searching, collecting, and processing the evidence. The following outlines the general stages in processing a scene. These stages occur regardless of the complexity of the scene, from a simple home burglary to a terrorist event involving numerous victims. The only difference would be the detail in which each step is documented. It is recommended that scenes are processed using a methodical approach because there may only be one opportunity to recover the evidence, so every effort needs to be made to prevent contamination and damage of the evidence.

The following steps provide an understanding of the processes involved.

3.7.1 Initial assessment

Before attending a crime scene, a series of decisions need to be made. Upon receiving the initial phone call, the technician will need to ascertain basic details of the crime. Obviously, this will include the location and type of crime, number of victims, and their respective condition(s). These factors will determine the type of response provided, including what resources they will need to take along to the scene and whether any specialist forensic fields will need to attend. Typically, an experienced technician will make these decisions automatically and make any further arrangements for resources while responding to the scene.

3.7.2 Assessment upon arrival

After arriving at the scene, a further assessment occurs. At this stage, greater detail relating to the alleged offense can be obtained via the investigators or witnesses. The presence of other emergency services, such as fire and ambulance, may have complicated the scene, and it will be important to learn what actions were taken by them while in attendance. At this time, an appreciation of the size of the scene can be made, and further assessment of the available resources to adequately process the scene can be made. The attending crime scene team will establish a forensic staging area where the forensic equipment and the forensic team leader operate from. This site will be outside the crime scene but close enough to enable ready access without presenting any risk of contamination. The selection of this site will occur only after a safety risk assessment has been made to ensure any hazards to resources or personnel are identified.

3.7.3 Scene security

Having established the circumstances relating to the offense, the next phase of the crime scene process is to ensure the scene is secure. This has a twofold objective. First is to ensure the crime scene boundary is sufficient to include all areas where evidence has the potential to be located, and second to exclude unauthorized persons from entering the scene. The methods used to secure the scene will vary depending on the size and complexity of the location. Typically, police will be positioned to restrict entry to the scene and maintain a crime scene log including details and times when personnel entered and exited. Given the fragile nature of trace evidence and the potential for contamination, entry to a crime scene will be limited to only those who are required to enter for a specific purpose that relates directly to resolving the scene. Sight seeing and tours of a crime scene will not be tolerated.

To aid control of access in and out of the scene, a single defined point of entry will be established. It will be at this point that entry details will be documented and any necessary contamination controls adopted. The technicians will then follow a defined path from that point while inside the crime scene. This is sometimes referred to as the "safe walk" path or zone and will usually use stepping plates so that the scene is not being impacted directly. This safe walk is chosen to provide movement within the scene in locations that will limit the destruction of any evidence. This concept also controls access into areas that may be unsafe for personnel operating in the scene. For example, in a fire-damaged scene some parts of the structure may be at risk of collapse, and hence the safe walk will not intrude upon those areas.

The scene security is predominantly aimed at managing the movement of persons in and out of the scene to ensure potential evidence is not compromised. Obviously, as part of the process, curiosity of the public and overzealous media need to be considered, and it is recommended that crime scene-staging areas be broken into three zones. The first zone is the crime scene where potential evidence is located, and this zone is restricted to persons performing crime scene examinations only. The second zone encompasses the crime scene and includes the forensic staging area and other investigative functions and is restricted to personnel supporting the crime scene teams. The third zone is the outer crime scene area and is the limit that members of the public and the media can access the scene.

3.7.4 **General survey**

The general survey is the phase where directed entries commence into the crime scene. The purpose of this initial entry, or more accurately, walk around the perimeter of the scene, will be to identify specific examinations that will need to take place inside the scene. This survey will consider the complexity of the scene and the available resources that can assist in that process. Obviously, depending on the location of the crime scene, there is a limit to the amount of resources that can operate within a scene at any one time. It is therefore important to coordinate the activities of the various technicians that may be required and also to direct the sequence of examinations to ensure one technique or method does not destroy potential evidence. An example would be the use of fingerprint powders throughout a murder scene without considering the effects this may have upon other types of physical evidence. The use of powders before collection of trace DNA will limit the recovery of this material.

Therefore, at the conclusion of the general scene survey, the crime scene team leader will formulate a plan to ensure an effective and safe resolution of the crime scene. This plan will involve the types of examinations that will be carried out, including the resources and personnel that will be tasked, and the sequence of examinations. This sequence will be used not only to limit the destruction of evidence as discussed but also to take into consideration priorities directed by investigators.

It will be during this phase that the scene is documented by photography and notes are made detailing the conditions of the scene. During this process, nothing is touched or moved, because the scene needs to be recorded in minute detail. The ultimate purpose of the documentation is to be able to replicate the scene if it was ever required.

3.7.5 **Detecting and preserving physical evidence**

Only after a detailed plan has been formulated and brief provided to all those entering, the scene has been secured, safety aspects have been considered for both the evidence and the responding personnel, and a sequence of tasks has been outlined and delegated will the actual physical process of directed crime scene entries commence. The searching for, and detection of, physical evidence is a critical first step in the process of evidence reporting. The following statement by Inman and Rudin (2001, p. 207) should be impressed permanently on the minds of anyone entering a crime scene: "If you don't recognise it, you cannot collect it; if you don't collect it, you cannot analyse it; if you cannot analyse it, you cannot interpret it."

The search for evidence is not merely a casual observation of the scene, depending on the type of evidence that is being searched for, this process may require specialist equipment and very detailed and time consuming examinations. The key being that the search must be conducted with a purpose, based on a particular theory of the events, thought to have occurred. For example, if the victim has been shot, it is obvious that the teams will not be searching for a knife or other weapon with blood on it. The search for evidence is based on the context of the scene, and hence the technicians are attempting to establish the veracity of a theory, or as Inman and Rudin note, test a working hypothesis.

Experienced crime scene technicians will test any alternate hypothesis to determine whether this theory is sound. There is debate in academia regarding the bias of the examiner by allowing them context, but without context and a working hypothesis the search for evidence will be a never-ending search, collecting endless data with no purpose. Of course, this will result in examiner bias, and it would be ridiculous to suggest that the results of any examination are not based on assumptions and bias. The important factor is to recognize what factors have influenced the examination so that controls can be implemented to ensure they are reasonable.

Many search techniques are available in various texts, with some very practical methods, but all have as a theme a system to ensures all aspects of the item or scene being examined have been covered. For the same reasons that a systems approach is recommended for crime scene examination, a systems approach needs to be adopted for the searching of evidence, to ensure that it is not overlooked, destroyed, or contaminated. Not only must the search technique be appropriate but also the personnel must be capable of recognizing the evidence. That requires appropriately trained persons conducting the search for evidence as the critical factor is not just knowing what to look for, but where to look (Inman & Rudin, 2001, p. 208). Depending on the type of evidence, enhancements via light source or chemical reaction may be required to visualize the evidence.

3.7.6 Collection of physical evidence

Once potential evidence has been located during the detection phase, a series of questions now need to be answered, including *what is it*? And *how best to collect and preserve it*? It first has to be established whether the item is what we believe it to be. This is quite obvious with fingerprints and shoe impressions where there is little confusion as to what they represent. However, in the case of trace evidence that is invisible, particularly body fluids such as blood and semen, a presumptive test will be conducted to determine whether the sample is worth collecting. These tests are simple and easily carried out at the scene, but they are not conclusive and are prone to false positives (i.e., the incorrect identification of a sample as relevant). As a result, recommended analyses are conducted in the laboratory to verify the substance in question.

For obvious reasons, it is not possible to collect the entire crime scene, although many wish they could and have attempted this end. It is, however, a futile process that results in the collection of useless items that only have the potential to hide the critical evidence and overburden the administrative processes that inevitably flow from the collection of evidence. At some point, critical decisions need to be made regarding what items will be collected and what will be left, and this decision is typically what defines a good crime scene technician. This decision will need to be made with the value of experience within the context of information provided by the witnesses and investigators.

When a decision is made that a particular piece of evidence is going to be collected, the question of how to collect it, and preserve it, is critical. The technician will be trained in the appropriate collection techniques for the various types of evidence, because most evidence is fragile and minute, so it is important to prevent damage or destruction in the collection process. The technician will also need to consider what the item is being collected for, that is, what type of examination and analysis will occur with the item? This may dictate how the item is collected and more importantly how it is packaged, stored, and preserved.

3.7.7 Distribution/analysis of physical evidence

After the items have been collected from the crime scene, they are typically submitted to the forensic laboratory for analysis. It is this final phase in the life of the physical evidence where value can be assigned to the item via analysis of the evidence. Again, it is important for the forensic scientist to have some direction as to the types of examinations that are required and ultimately the purpose of the examination because this will form the basis of the scientific question they will endeavor to answer. For example, segments of glass may be submitted to the laboratory, but without information relating to the examination required, the scientist will not be able to commence examinations. The investigator may require the scientist to determine the source/manufacturer of the glass, but this is not a valid method without a control sample from a known source.

Given the specialist nature of analysis required for most forms of physical evidence, rarely will this occur within one organization. Therefore, with many different providers of service, it is important that requests are clear and the methods used are valid for the purposes and standard of proof required.

3.7.8 Interpretation and reconstruction

Crime reconstruction is "the logical analysis of physical evidence and other facts into the formulation of a theory regarding the actions that took place during the commission of a crime" (Chisum, 2006, p. 64). Crime reconstruction is necessary to know what happened in a crime, who was involved, what they did, and where certain things happened. Chisum (2006, p. 71) then states that the information derived in the reconstruction can be used in the following ways:

- Investigators conducting interviews to test the veracity of the statement.
- Criminal profilers in making a "profile" of the perpetrator.
- District attorneys or defense attorneys to determine how to prepare and argue their case in court.
- The court in determining sentences.

Despite the use of the blanket term *crime reconstruction*, it is actually composed of many individual types of reconstruction depending on the type of evidence and what questions need to be answered. For example, the order of events relies on a different type of analysis than whether a particular object, such as a gun or speed camera, worked the way it should. These individual types of reconstruction are detailed in Chisum and Turvey (2011, pp. 187–192):

- Sequential evidence: anything that establishes or helps establish when an event occurred or the order in which two or more events occurred.
- Directional evidence: anything that shows where something is going or where it came from.
- Location/positional evidence: that which shows where something happened, or where something was, and its orientation with respect to other objects at the location.
- Action evidence: anything that defines anything that happened during the commission of a crime.
- Contact evidence: something that demonstrates whether and how two persons, objects, or locations were at one point associated with each other.
- Ownership evidence: something that helps answer the "who" question with a high degree of certainty.
- Associative evidence: usually a form of trace evidence that can be identification or ownership evidence. The finding of common materials on the suspect and victim, the suspect and the scene, or the scene and the victim is used to suggest contact.
- Limiting evidence: that which defines the nature and boundaries of the crime scene.
- Inferred evidence: anything the deconstructionist thinks may have been at the scene when the crime occurred but was not actually found.
- Temporal evidence: anything that specifically denotes or expresses the passage of time at the crime scene relative to the commission of the crime.
- Psychological evidence: any act committed by the perpetrator to satisfy a personal need or motivation.

To this list the authors would add functional evidence, defined as anything that demonstrates how something worked, or did not work, and the manner in which it was used during the crime. This may help establish how something was used during the crime, and how it may have subsequently impacted victim and offender behavior, and the choices that were made regarding use of the item.

These different types of analysis raise the general question as to who should perform any of the above-mentioned reconstructions within the general purview of crime reconstruction. The short and simple answer to this is someone who is trained within the type of science that relates to the required analysis. In some cases, this will be solely within the remit of the forensic scientist, whereas in others, the criminologist, crime analyst, police investigator, or psychologist or psychiatrist may be more suited. This would be especially true of behavioral evidence such as that in the inference of motive, where the forensic scientist may not be well suited to understand the psychological nature of human drives. Temporal reconstruction, the assembly of evidence items into a timeline of events, could be accomplished by virtually anyone.

3.8 Reporting results

3.8.1 Overview of classes of opinions and conclusions

At the conclusion of examinations, results will be reported. Depending on the purpose of the examination, this may form a simple summary of conclusions in an email; or if the results are required to be admitted in evidence, a more detailed statement of witness or technical report will follow. The contents of the report will vary depending on the field of science and the institution issuing the report, although the content and layout of any forensic report may not differ greatly to that of the applied crime analyst's report, covered in detail in Chapter 12 of this work. Regardless of this variation, the report will have the following information:

Brief Overview: This brief summary of results is similar to the abstract of a research paper, with sufficient information to convey why the examination was performed and what the results were. It will not have details relating to methods and is intended to provide a quick overview of the results only.

Training and Experience: The report writer will outline their experience and qualifications and any authorizations they may have requested in order to complete the analysis conducted. The organization that is issuing the report may take part in third party accreditation processes and this will be detailed in the first part of the report.

Background Information: This section includes information provided to the examiner by others. For example, any versions supplied by witnesses or directions given by investigators should be included that have relevance to the examination. If any information was provided to the examiner that has influence on the opinions they have formed, this may be the appropriate place to list these.

Purpose or Aim: Early in the report the examiner will outline the reasons or purpose for the examination. This may include information regarding the field of science and the limitations of any examinations made.

Items Received/Items Examined: Depending on the purpose of the examination, this section will either list the items collected as a result of a crime scene search, or list the items received at the laboratory. In cases where the examiner has both collected the item and examined it, the sections will appear in chronological order throughout the report. This provides an audit trail of the evidence, when it was received, and by who. It is important, in relation to physical evidence, to maintain a clear continuity of the evidence until the conclusion of any inquiry.

Results of Examination: As the items are examined, the results of each should appear in the report. The results of examination should reflect the purpose that was stated earlier in the report. This section will briefly outline the methods used. A common error made by inexperienced practitioners is simply to regurgitate their complete examination notes into the report believing this will prevent them being called to court to explain their methods. All this will achieve is to confuse the reader due to the large amount of useless information they are now presented with, and ultimately cause them to be called upon to explain their results.

Opinions and Conclusion: The conclusion will summarize the results and outline the opinions formed in relation to the case. It will be important to state any assumptions that have been relied upon in the forming of those opinions. This is the opportunity to neatly wrap up the examinations that were conducted or outline any limitations in the opinions and what further examinations would be required to provide more certainty in the findings.

Conclusion

It could be said that physical evidence is the basis and cornerstone of any crime analysis for any purpose. This is true whether doing an assessment of staging, linking cases, assessing a matter as a false report, or conducting a risk assessment or threat management. As such, it is vital that the different types of evidence are known and understood, an well as the conditions under which the evidence is deposited, collected and interpreted, and reported.

Although the applied crime analyst may not be responsible for the collection and interpretation of the initial evidence at the scene, they most certainly will be involved in determinations that arise from it. As a result, they need to be aware of the detection, documentation, collection, and interpretation of evidentiary items they rely on. The types and limitations of each evidence item need to be appreciated and understood if they are to make any meaningful interpretation from them. This applies whether it is DNA, fingerprints, blood spatter, hair or fiber, or ballistics evidence that forms the core of examination.

REVIEW QUESTIONS

1. According to Fisher, the two types of evidence are _____ and _____.

2. The transfer of evidence or material from one object to another is based on _____ _____ _____.

3. The species of the origin of hair can be determined from rudimentary analysis. True or false?

4. Name and briefly describe three different types of wounds.

5. The applied crime analyst should always visit the crime scene. True or false?

6. Evidence that defines the nature and boundaries of the crime scene is known as
   a. Action evidence
   b. Ownership evidence
   c. Inferred evidence
   d. Contact evidence
   e. Limiting evidence

References

Chisum, W. J. (2006). Crime reconstruction. In A. Mozayani, & C. Noziglia (Eds.), *The forensic laboratory handbook: procedures and practice*. New Jersey: Humana Press.

Chisum, W. J., & Turvey, B. E. (2011). *Crime reconstruction* (2nd ed.). San Diego: Elsevier Science.

Deedrick, D. W. (2000). Hairs, fibres, crime, and evidence. *Forensic Science Communications, 2*(3). np.

DiMaio, V. J., & DiMaio, D. (2006). *Forensic pathology* (2nd ed.). Boca Raton: CRC Press.

Dix, J. (2000). *Colour atlas of forensic pathology*. Boca Raton: CRC Press.

Field, A. (2009). *Discovering statistics using SPSS* (3rd ed.). London: Sage Publications.

Fisher, B. A. J. (2000). *Techniques of crime scene investigation* (6th ed.). Boca Raton: CRC Press.

Geberth, V. J. (2006). *Practical homicide investigation: tactics, procedures and forensic techniques* (4th ed.). Boca Raton: CRC Press.

Inman, K., & Rudin, N. (2001). *An introduction to forensic DNA analysis* (2nd ed.). Boca Raton: CRC Press.

Inman, K., & Rudin, N. (2002). The origins of evidence. *Forensic Science International, 126*, 11–16.

Ireland, K., & Kelleher, J. (2012). Fire investigation. In I. Freckelton, & H. Selby (Eds.), *Expert Evidence*. Australia: Thomson Reuters.

Kirk, P. (1974). *Crime investigation* (2nd ed.). Florida: Krieger Publishing Company.

Lee, H. C., Palmbach, T. M., & Miller, M. T. (2001). *Henry Lee's crime scene handbook*. San Diego: Academic Press.

Locard, E. (1929). The analysis of dust traces: Part I. *Revue Internationale de Criminalistique I, 4–5*, 176–249.

McCartney, C. (2010). Understanding the role of forensic DNA: a primer for criminologists. In W. A. Petherick, B. E. Turvey, & C. E. Ferguson (Eds.), *Forensic Criminology*. San Diego: Elsevier Science.

Osterburg, J. W., & Ward, R. H. (2010). *Criminal investigation: a method of reconstructing the past* (6th ed.). Boston: Andersen Publishing.

Parrott, A., Morinan, A., Moss, M., & Scholey, A. (2004). *Understanding drugs and behaviour*. West Sussex: John Wiley & Sons.

Saferstein, R. (2010). *Criminalistics: an introduction to forensic science* (10th ed.). New Jersey: Prentice Hall.

Schehl, S. A. (2000). Firearms and toolmarks in the FBI laboratory. *Forensic Science Communications, 2*(2). np.

Spitz, W., & Fisher, D. J. (2005). *Spitz and Fisher's medicolegal investigation of death: guidelines for the application of pathology to crime investigation* (4th ed.). Springfield: Charles C. Thompson Publishing.

Thornton, J. I. (1997). The general assumptions and rationale of forensic identification. In D. L. Faigman, D. H. Kaye, M. J. Saks, & J. Sanders (Eds.). *Modern scientific evidence: the law and science of expert testimony:* (Vol. 2). St Paul: West Publishing.

Forensic Victimology

Wayne Petherick[1] and Claire Ferguson[2]
[1]Bond University, Gold Coast, QLD, Australia
[2]University of New England, Armidale, NSW, Australia

Key Terms

Victimology
General victimology
Interactionist victimology
Critical victimology
Victim precipitation
Risk factors
Protective factors

INTRODUCTION

From the Latin *victima*, a victim was an animal or person who was ceremonially put to death for some power or deity (Karmen, 2007). Since these early roots, the term *victim* has adopted a more general meaning to include anyone who has suffered harm or loss, most usually at the hands of another. Forensic victimology, therefore, is the thorough study of a victim of a crime, including examination of their demographic features, lifestyle, habits, routine, personality traits, sources of risk, and various histories such as medical, psychological, and financial, among other things. The findings of this examination are usually presented in a report as a victimology. For the purposes of this work, the discussion is limited to victims of crime and does not include those who experience loss through natural disaster or other non-crime-related phenomena. Moreover, our concern here lies more exclusively with victims of violent interpersonal crime.

This chapter defines victimology from a historical perspective and examines victim risk, including those factors that contribute to harm or loss. What should be included in a victimology report is outlined with proposed contents in the suggested approach to preparing a victimology.

4.1 Victimology defined

Victimology is a subdiscipline that exists within the general field of criminology and is defined as the "scientific study of the physical, emotional, and financial harm people suffer because of illegal activities" (Karmen, 2007, p. 2). In general, victimologists belong to one of three main subgroups (Ferguson, Petherick, & Turvey, 2010): general victimology, interactionist victimology, and critical victimology.

General victimology refers to the broader study of those who have suffered harm or loss, whether as a result of a crime, traffic accident, disaster, or any other cause (Wilson, 2009). General victimologists identify and develop preventive measures and tools for victims and focus on remedies as well as causes (Ferguson et al., 2010). *Interactionist victimologists* (also known as penal or positivist victimologists) are primarily concerned with the interactions between victim and offender. Indeed, much of the early work of victimologists such as Mendelsohn, a Romanian lawyer; von Hentig, a German lawyer; and Schaefer, a Hungarian lawyer, was about the interaction between the victim and offender and the levels of culpability that could be attached to the relationship between the two. *Critical victimology* developed largely in response to the shortcomings of interactionist/positivist victimology, which suffered from too much concern about culpable victims (Wilson, 2009). Within this paradigm, the term *victim* itself becomes the subject of analysis; those falling outside the traditional mold of victim are at risk of being ignored, denied services, and marginalized (Wilson, 2009).

In one of the first seminal texts on the subject, Stephen Schafer (1977, p. 1) states of his work:

> *This book is an introduction to the study of criminal-victim relationships. This branch of learning has been called "victimology," but it is questionable whether denoting it as a specific doctrine or science is justified, or whether it should be considered as an integral part of the general crime problem.*

This conflict over the place of victimology within the grand scheme of criminology has since been resolved, with a number of journal articles, texts, and professional organizations dedicated to the subject. Therefore, it could be said that victimology is now a discrete discipline of its own merit, with theories and research that form a core body of knowledge. However, when related specifically to crime victims, it would still be logical to argue that it sits within the general field of criminology—the study of crime and criminals.

4.2 **Victim precipitation**

The term *victim precipitation* was first used by Marvin Wolfgang to describe a situation in which the victim initiated the actions that ultimately led to their harm or loss (Petherick & Sinnamon, 2014). Cases precipitated by victims are traditionally viewed as those in which the protagonist, who became the victim, initially used a deadly weapon or struck a blow (Brown, Esbensen, & Geis, 2010). More generally though, victim precipitation can include any number of situations in which the ultimate victim initiated or instigated the events that led to their demise. These do not always include actual physical violence. Despite the protestations of many, victim precipitation is not the functional equivalent of blaming the victim. Instead, it represents an attempt to fully and completely understand the criminal event and the behavior of both the victim and the offender and how this contributed to the victimization overall. Most importantly, the elements that contributed to or precipitated the crime may further contextualize the offender's behavior, such as what they were doing or reacting to at a given moment, or other important analytical or evidentiary considerations such as the offender's emotional state or their motivation.

In an attempt to determine the number of homicides that were precipitated by victims, Wolfgang studied homicides in Philadelphia, gathering data between 1948 and 1952 and finding that more than a quarter of those homicides studied were precipitated by the victim (Doerner & Lab, 2005). About two decades later, Curtis (1973; 1974) found that, in the United States, 22% of all clearances and 14% of nonclearances were precipitated by the victim. Pesta (2011) found that 18% ($n=35$) of those cases studied in Ohio were precipitated by the victim. It would seem, then, that precipitation plays a signifi-cant role, at least in homicide (for investigations into precipitation in other crimes, see Petherick & Sinnamon (2014) for further details).

Siegel (2008) discusses two variants of precipitation. The first is *passive precipitation*, where the victim has characteristics that evoke a response from their attacker. The victim may be introverted or emotionally withdrawn or may belong to a particular group with whom the offender takes issue. Alter-natively, the characteristics are imagined, and the perceived provocation is a fabrication or misrepresen-tation of reality by the offender.

The second type is *active precipitation*, where the victim takes an active role and engages in behav-ior that elicits a response. This would include those cases in which the victim is the initial aggressor but suffers the most harm or loss. Depending on the type of behavior involved, if the victim instigated the event, it may be difficult to determine what each party did, why they did it, and the lawfulness of their actions. Arguably, it is this type that most frequently comes to mind during discussions of precipitation.

Polk (1997) is critical of the concept of precipitation and how it is used within this field, suggesting that the time has come to reexamine the concept of victim-precipitated homicides. He notes that many of the problems of precipitation revolve around extracting the relevant data from case files to properly determine the presence and nature of precipitation. This problem exists not only in those cases where limited information is available and precipitation difficult to determine but also in cases where informa-tion is plentiful and detail is rich. Polk notes that these concerns do not represent a problem with the theory of precipitation itself but with difficulties in its measurement, empirical inquiry, and analysis. As such, these problems may apply more to the ex post facto domain of studying precipitation more so than the idiographic study of single cases.

4.3 Risk factors for victimization

Numerous works have detailed at length a number of proposed risk factors for becoming a victim of violent crime. These include the original works of Schafer (1977), von Hentig (1948), and Karmen (2007). Collectively, these works suggest that there are certain factors that may predispose one to become the victim of a crime, ranging from provocation, victim precipitation, and biological weakness (Schafer, 1977) to mental retardation or being an immigrant, poorly educated, and inexperienced, among others (von Hentig, 1948). While some of these explanations may have been historically cogent, others have become redundant over time in tandem with the expanding amount of research and under-standing in the area.

When it comes to victim risk, the reality is that some of our attitudes and beliefs may be misplaced given the reality of crime figures from around the world. For instance, it is commonly believed that being among friends or within the safe confines of one's own home are protective factors against vio-lence. These assertions would seem entirely plausible given perceived risk factors, such as the preva-lence of stranger violence and what is psychologically "safe." As social beings, we do not like to think that our own friends or acquaintances may pose a risk to our safety, and "a man's home is his castle" is an axiom that should afford every level of safety that we might desire or deserve. However, official crime figures suggest that we are statistically most at risk of certain crimes in the company of those we know and in a residential dwelling.

To illustrate this, data from the Australian Institute of Criminology (Australian Crime Facts and Figures, 2011) reveal that of the homicides committed in 2010, 61% occurred within a residential dwelling; 63% of sexual assaults occurred within same. For murder, 54% of victims were killed by someone classified as an intimate family member or friend. In the United States in 2010, the largest proportion of offenders were known by the victim in some capacity (coworker, intimate, family member, etc.), accounting for 6294 incidents, whereas the second largest proportion of offenders were classified as unknown (5944 incidents). Strangers perpetrated only 2211 incidents. Data from the Crime Survey for England and Wales (Office for National Statistics, 2012) show that women, compared to men, are more at risk of being killed by someone they know (78% vs 54%, respectively); most women are killed by a current or ex-partner (51%), whereas males are more likely to be killed by a friend or acquaintance (39%).

While these are but a few examples, they illustrate the misplaced nature of our fears about where exactly risks lie. Despite this statistical probability, however, it is not appropriate in any case to simply identify someone known to a victim as the potential perpetrator until after any and all possibilities have been exhaustively examined and alternate theories excluded. These figures do serve as a useful statistical anchor, though, as discussed in Chapter 2 of this work.

Further study of risk factors by main types of offenses follows here. The reader is cautioned that this is by no means the final word on the matter; local statistics should be sought to understand the base rates for offenses by type, by location, and by relationship.

4.3.1 Homicide

The terms *homicide* and *murder* often are used interchangeably, although the two acts are distinct by both law and definition. *Homicide* is a clinical term that simply refers to instances when one person kills another, without considerations of intent or lawfulness. *Murder*, on the other hand, is a legal term that describes the killing of one person by another with intent and without legal justification (unlike, for example, killing in self-defense or during war).

In countries where firearms are common and/or allowed by law, one argument for retaining the right to keep firearms is premised on the notion that firearms are both a global and specific deterrent to crime. For this to be true, would-be offenders would have to know that a firearm may be present (a global deterrent) or is present in a given residence or in the possession of a particular individual (specific deterrence). Another common argument is that possession of a firearm provides self-defense—leveling the playing field between victims and armed assailants.

Evidence for this effect is less than convincing, however, with many studies showing that the presence of a firearm in a home increases the risk of victimization. In one review of the literature, Hepburn and Hemenway (2004, p. 417) note that "households with firearms are at higher risk for homicide, and there is no net beneficial effect of firearm ownership." This increased exposure presumably would be the result of the statistical probability of victims being killed by those known to them and the greater potential for a fatality to result from a gunshot wound relative to other types of weapons.

An earlier study by Kellerman, Pivara, Rushforth, Banton, Reay, Francisco, Locci, Prodzinski, Hackman, and Somes (1993) examined the risk of homicide in 420 cases in Shelby County, Tennessee; King County, Washington; and Cuyahoga County, Ohio. In their univariate analysis, alcohol was consumed by more members of the households where homicides occurred compared to matched controls. Illicit drug use also was more commonly reported. Prior violence was reported in 31.8% of the homicide households compared to 5.7% of the controls. Prior arrests or physical fights outside the home also were related to homicide by firearms, as were multiple weapons kept in the home and having a weapon loaded or unlocked. For the multivariate analysis, six variables were retained: the home is rented, the case subject or control subject lived alone, any household member ever hit or hurt during a fight in the home, any household member ever arrested, any household member used illicit drugs, and one or more guns kept in the home. Many of these variables seem to be related to a general antisocial attitude or pattern of behavior, meaning they may simply be observable characteristics of an underlying psychological or emotional pattern among victims. Put another way, it may be that houses in which firearms are kept may be more likely to be involved in other criminal enterprises or come into contact with people who are.

Based on the work of Margo Wilson and others, Campbell (2012) sought to identify the risks of homicide to women through the development of the Danger Assessment (DA) tool. Based on previous work in the domestic violence field, Campbell (2012, p. 441) "became convinced that [she] needed to conduct a data-based exploration of the risk factors for intimate partner homicide to validate the DA, determine a weighted scoring, and identify which other risk factors for intimate partner homicide might be important." The purpose was to identify risk factors over and above any prior intimate violence, which was already considered a major factor in intimate partner murders.

The first identified factor was a stepchild of the abusive partner in the home; this variable increased the odds of homicide by 2.4. Drawing on perspectives from Dawkins' *The Selfish Gene* and from Daly and Wilson's work in Canada on evolutionary perspectives, Raine (2013, p. 23) explains this increased risk from nonbiological parents in evolutionary terms:

> *What they [Daly and Wilson] demonstrated was an inverse relationship between the degree of genetic relatedness and being a victim of homicide. So the less genetically related two individuals are, the more likely it is that a homicide will take place. For example, in Miami, 10 percent of all homicides were the killings of a spouse-a family killing- but of course, spouses are almost always genetically unrelated. In fact, Daly and Wilson found that the offender and victim are genetically related in only 1.8 percent of all homicides of all forms. So 98% of all homicides are killings of people who do not share their killer's genes.*
>
> *Selfish genes in their strivings for immortality wish to increase- not decrease- their representation in the next gene pool. Hence this inverse relationship between genetic relatedness and homicide. On the other hand, if you are living with someone not genetically related to you, you are eleven times more likely to be killed by that unrelated person than by someone genetically related to you.*

The second factor is being estranged from the abusive partner in the past year (increasing the odds of femicide by 3.5). Jealousy interacted with estrangement, increasing the odds of homicide by 5.5. Research on the existing characteristics of the DA yielded an accuracy of 0.73, although with the addition of these extra factors this accuracy increased to 0.92 (studied by Campbell, Webster, and Glass (2009)).

4.3.2 Assault

Depending on jurisdiction, the definition of *assault* includes any application of force, violence to a person, or psychological harm and may include threats. Some legislation also includes the application of heat, light, electrical force, or gas, among others.

The Australian Institute of Criminology (2012) data on assault reveal that an attack is most likely to occur within a residential dwelling (48%), followed by the community (32%), then retail (12%), recreational (5%), and other spaces and places (3%). For both males and females, the age group at highest risk is 15- to 24-year-olds, followed by 25- to 44-year-olds. Collectively, the years between 15 and 44 represent the greatest risk. Very young children and the elderly had the lowest risk of assault. Known other, stranger, and family comprise 95% of the relationship types; 6% of the relationship types are unknown. Females are most likely to be assaulted by a male family member or known other, whereas males are most likely to be assaulted by a stranger, followed by a known other.

Perhaps one of the greatest risk factors for assault is alcohol. This feature is almost universal, and the age of risk for both perpetration and victimization closely aligns with the age at which most alcohol consumption occurs. Although causation cannot be directly inferred, the link is certainly strong and the evidence is considerable. When considering the relationship between alcohol and assault, a guide written for American police dealing with assaults in and around licensed premises explains that (Scott, 2002, p. 3):

> Alcohol contributes to violence by limiting the drinker's perceived options during a conflict, increasing the drinker's willingness to take risks, and impairing the drinker's ability to talk his or her way out of trouble. Many of the alcohol related problems police deal with can be attributed to ordinary drinkers who go on binges, drink more than they usually do or drink on an empty stomach. In general, those who drink a lot are more aggressive and also get injured more seriously than those who drink moderately or not at all. Moderate drinkers do not appear to be at significantly higher risk of injury than nondrinkers.

Dearden and Payne (2009) note that many research agendas have singled out alcohol as a factor related to increased levels of violence and that many victims and offenders report alcohol use before or during instances of violence. The British Home Office (Budd, 2003) conducted an analysis of the link between alcohol and violence between 1995 and 1999, with an estimated 1.2 million instances of alcohol-related assault in England and Wales. This amounts to 40% of violent instances involving alcohol. Although this also included "snatch thefts" and robberies, it was found that one-third involved stranger assaults, one-third involved acquaintance assault, and one-quarter were domestic assaults. It should be noted that these latter categories may best be described as domestic violence (discussed next), but they remain in the broad class of assault offences.

4.3.3 **Domestic violence**

For the purpose of this discussion, domestic violence is defined as any violence occurring in a residential setting among parties among whom there is a current or previous intimate or familial relationship. Certain stereotypes emerge, such as that of the aggressive and overbearing husband and the subjugated and submissive female victim, although the reality of this crime is likely far less polarized. While much research has focused on male offenders and female victims, reports of female offenders with male victims, and same-sex domestic violence among both males and females, is available.

Victims often are subjected to a range of physical injuries (which may or may not be reported), as well as a host of psychological and emotional traumas, from depression and anxiety to posttraumatic stress disorder, becoming not only a criminal justice problem but also a health care problem. Adding to this, victims of domestic violence often feature within other types of crime such as stalking (see Douglas & Dutton, 2000) and homicide (see Lau, 2012). Interestingly, the link between domestic violence and homicide is further suggested in the figures reporting a decline in homicide around the world, associated in part with the decline in domesticity (higher divorce rates and decreasing marriage rates) (Brown et al., 2010; Dugan, Nagin & Rosenfeld, 1999). It should be noted, though, that with an increase in access to primary emergency medical care (paramedics and other first-response medical personnel) and better second-tier medical care (emergency departments and surgical personnel), the rate of death from violent interpersonal crimes will decrease, taking with it deaths that would otherwise lead to classifications of homicide, murder, manslaughter, and so on.

A review of the literature reveals that, while a number of purported risk factors have been identified, not all are the result of empirical analysis, and there is some inconsistency among factors that identify risk of offending or victimization (Riggs, Caulfield, & Street, 2000). This research has identified previously experiencing violence, victimization during childhood, substance abuse, psychopathology, and perceived danger as potential risk factors, although it should be noted that the results of many investigations are equivocal at best.

4.3.4 Sexual assault

While historical definitions of rape included cases of sexual knowledge of a female forced against her will, more recent definitions have incorporated gender-neutral language (Brown et al., 2010), including heterosexual and homosexual acts other than penile penetration. Rape, therefore, is a specific type of sexual assault involving penile penetration.

In Australia in 2001 the Australian Institute of Criminology summarized several risk factors for sexual violence and sexual assault from various sources (Cook, David, & Grant, 2001). They determined that young females were most often victims of sexual assault: females represented 79% of victims who reported to police, and 46% were females younger than the age of 20. In the same data, close to half of victims were assaulted sexually by someone known to them (usually not a family member), such as a boyfriend, husband, date, previous partner, or other known man. However, offenders were more likely to be family members than strangers in reported sexual assaults. These assaults most often took place in private residential locations such as houses, apartments, or hotels (65%). Forty percent of women surveyed reported that alcohol was a factor in their assault, meaning that either the victim or perpetrator (or both) were drinking before the assault or the victim believed that alcohol was a factor in her victimization (Cook et al., 2001).

International surveys of violence against women, including in Canada, the United States, New Zealand, and Australia, have generally found that women are more likely to be sexually victimized by someone they know; significant numbers are victims of more than one sexual assault; young women are at greater risk; physical injury and weapons often are not present; few women tell anyone about the assault; and even fewer report to police (Cook et al., 2001). There is also some indication that people with intellectual disabilities (Johnson, Andrew, & Topp, 1988), those with a mental illness (Davidson, 1997), and those in correctional facilities are at an increased risk (Heilpern, 1998). Gay, lesbian, bisexual, and transgender individuals are also at an increased risk for sexual violence (Commonwealth Department of Human Services and Health, 1994), as are sex workers who work on the streets rather than in brothels (Perkins, Prestage, Sharp, & Lovejoy, 1994).

4.3.5 Stalking

Regardless of whether they actually suffer physical assault, stalking victims often are subject to a range of psychological and physical effects (Blaauw, Winkel, Arensman, Sheridan, & Freeve, 2002), such as chronic stress responses (Pathé & Mullen, 1997) that may lead to other effects such as declined immune function or an increase in anxiety over time.

In one of the earliest works on stalking, Meloy (1998) notes that at least one-half of stalkers threaten their victims, although most stalkers (25–35%) are not physically violent. This is mirrored by Miller (2012), who notes that the general consensus is that between 30% and 60% of victims are threatened with violence and that about 25–50% of stalkers actually attack their victims. Meloy further notes that the homicide rate among victims of stalking is less than 2%. This suggests that threats occur more frequently than actual violence and that the homicide rate for this crime is low. However, this needs to be tempered by the fact that those charged with homicide will not always be charged with other offenses they have committed; prosecutors may opt for the most serious offense(s) or those that are easiest to prove in a court of law. Despite this, the frequency of physical violence noted above has been consistently established in other studies.

In Australia, Thomas, Purcell, Pathé, and Mullen (2008) randomly surveyed 3,700 men and women from the electoral role in Victoria to determine various facets of their stalking victimization. Of these, 432 respondents reported being stalked (12%), and 75 (17.4%) were subjected to physical attack by a stalker. These victims were more likely to have been stalked for longer or from a younger age and were more likely to be followed, spied on, loitered near, or subject to unwanted approaches. Of those who were attacked, 53 (71%) were female; the stalkers were predominantly male ($n=66$; 88%). In cases of violence, ex-intimates were most at risk ($n=31$; 41%), followed by acquaintances ($n=14$; 19%) then family/friends ($n=10$; 13%) and work colleagues ($n=6$; 8%). Strangers exceeded work colleagues with 14 cases (19%). Most victims of violence were threatened as well ($n=55$; 73%), and in a minority of cases the offender had a mental illness ($n=17$; 37%). Collectively, in 61 cases (81%) the victim and the offender were known to one another, suggesting that victims of stalking violence are more at risk from those that they know on some level.

4.4 **Types of victim risk**

When determining risk in this context, we are not necessarily interested in determining the aggregate risk to a victim as a member of a particular class. While the types of risk explained above are common in the areas of sexual assault, homicide, and stalking, as forensic victimologists we are interested solely in the idiographic study of a single case to determine the actual level of risk present in a given individual(s) by virtue of the unique features of the case under study. While these may incorporate possible risk levels provided by actuarial instruments or may factor in those risks present in the research and the literature, the focus should remain on the nuances of the extant case and victim and what can be gleaned from this examination.

As a rule, there are two types of risk posed to victims of crime. These relate to those factors that are *stable* and those that are *dynamic*. Stable factors tend to change little and are relatively consistent in a victim's day-to-day being. These include work and occupational activities, typical social activities (such as the people one associates with and perhaps the places in which these associations occur), and other activities such as going to the gym or shopping. This has been referred to elsewhere as *victim lifestyle exposure* (Ferguson et al., 2010). The following are examples of stable risk factors:

- Employment
- Housing or residential status

- Personality and psychopathology
- Drug and alcohol abuse

While there are certain features of a victim's lifestyle that change little, there are others that change according to whim and circumstance. A last-minute decision to change a venue, a mechanical failure in a vehicle, or a detour on a usually well-traveled roadway are included here. These are circumstances, activities, or events that the victim may not have planned for or that may change irrespective of foresight or normal habits. As a result, these things are *dynamic* in nature and are referred to elsewhere as *victim incident exposure* (Ferguson et al., 2010). The following are but a few examples of dynamic factors:

- Emotion and mood
- Drug and alcohol use
- Illness
- Injury

There can be considerable overlap between the two, and a stable factor that contributes to risk in one situation can be a dynamic factor in another. An excellent example of this would be alcohol use and abuse. One can be an alcoholic (a stable factor) and under the influence of drugs or alcohol at the time of harm or loss (a dynamic factor), so it could be said that both types of risk were present at the time of victimization. However, one can be an alcoholic but *not* drunk at the time of victimization. As such the stable factor is present but the dynamic factor is not. Should one be a chronic drug user, then this may contribute to victimization as a stable factor by putting that person in constant contact with criminal elements. Should one be under the influence of drugs at the time of a criminal event, this may contribute to victimization by undermining rational thought processes or rendering them unable to defend themselves; for instance, becoming unconscious and leaving them vulnerable at a given moment.

When examining victim risk, consideration must be given to what risks the individual faces; these will differ depending upon the crime and/or analysis required. For example, in a missing person case, the question will likely be, "Based on this individual's circumstances and potential for harm or loss, what are the factors that may explain their missing status?" In sexual homicide, questions of risk must incorporate both domains: What is this individual's risk for encountering both sexual violence and homicidal violence? When issues of motivation come into play, such as when items are taken from a victim of homicide, careful consideration must be given to determining whether the victim was killed as a secondary motivation to the crime (ergo, a homicide concealing theft, for example) or whether those items held some special nonmonetary value. Here, we must ask, What is the likelihood of this being a theft gone wrong based on the possessions taken, their type, financial value, and for whom they hold intrinsic worth? Who else would have known about the items and their value? For whom would the items have most value? Once these questions have been answered, it is possible to estimate risk, and questions of motive may be more easily addressed via the use of sound logic and reasoning and the scientific method.

Protective factors, on the other hand, are those that mitigate risk in the victim's life and environment. While it has been established elsewhere in this chapter that being in a residential dwelling increases the chance of victimization; this risk may differ depending on whether one is in the company of known others (another factor that contributes to risk in certain types of offenses).

Together, these two situational factors compound the risk. Taken individually, the risk may be reduced.

Both risk and protective factors can arise from a number of domains including on a contextual (poverty and geographic location), interpersonal (parenting styles and peer groups), and individual level (coping strategies and emotional factors) (Reingle, Jennings, Lynne-Landsman, Cottler, & Maldonado-Molina, 2012). While there are certain risk factors that may put someone in harm's way, factors that mitigate future victimization must also be considered. Whereas some victims will internalize their experiences with a negative overall effect on their psychological well-being, others will learn and adapt, using their experience as a basis to foster growth. For these individuals, *resilience* serves as a protective factor at the individual level.

Beyond resilience, having just experienced victimization may be enough to heighten sensitivity to threats and risk factors, such as identifying potentially dangerous warning signs or attending to intuitive feelings of danger. In considering psychophysiological correlates of risk recognition, Soler-Baillo, Marx, and Sloan (2005) examined the differences in respondents who self-reported as victims of sexual assault and those who did not. Among other results, victims reported that a vignette was more unpleasant than nonvictims, with victims rating the scenario as more realistic than nonvictims. This may suggest that, providing sufficient insight into the reasons for past victimization, prior victims may be more likely to avoid situations in which there is a possibility of risk. Put another way, their fight or flight mechanisms may be more well attuned to remove themselves from potentially dangerous situations.

Victim adaptations such as insight and resilience must be factored in when conducting a victimology, as this will help to better inform risk assessment, threat management, and the total context in which victimization occurred. This allows for a better and more informed assessment.

4.5 **Victimology: a suggested approach**

As with many areas, the amount of information available and the quality of that information is dictated by the case at hand. This includes the care and attention previously paid by others involved in the case, the amount of information collected and retained by the victim themselves, and the amount of information available from other sources. For example, unless the victim kept documents relating to medical and psychological history, or they have been subject to discovery from other parties, this information may at best be received third-hand, having been sourced from those who heard from someone, who heard from someone else, aspects of what is traditionally considered confidential information. Because of this, there are no hard and fast rules about what will be available, but there are good general guidelines by which we can determine what should be sought out. As with all areas of evidentiary consideration, issues of *validity*, *reliability*, and *sufficiency* (as discussed in Chapter 3) must be at the forefront of any analysis.

Should there be more than one victim, or victim precipitation, a victimology would need to be created for each person involved. What is known about each and the quality and quantity of the information that can be determined are dictated by the source of the information along with a plethora of other factors. Each of the sections below provides a guideline for a type of content that should be included in a forensic victimology and what specific information should be chronicled under each.

The source of the information should be clearly identified, and the differences between facts, opinions, and assumptions should be stated. Similarly, those things that are facts of the case should be delineated from the analyst's opinions about that fact.

4.5.1 Administrative inclusions

Before outlining the victim's characteristics and the assessment of risk posed to the victim when the crime occurred, it is important to first address a number of administrative elements affecting the findings. The victimologist should include their own name and contact details, the date of the examination, as well the details of the party who has asked them to compile the report. Specific questions asked of the victimologist should be included, such as, "Was the assault precipitated by the victim?" In addition, each piece of information the victimologist has access to should be listed, as well as any information they requested but were not given. This is a crucial step as it allows both the analyst and those utilizing the report to sequence when it was written relative to when other evidence was discovered, analyzed, and so on. Such an inclusion is also helpful for the victimologist should they eventually be asked to elaborate on or testify to their findings. This inclusion also will speak to whether a chain of custody for the evidence was maintained. A running header and page numbers should always be included in reports to avoid confusion should pages be lost or mixed with other forensic reports.

4.5.2 Demographics

Basic information about the victim should be recorded here, including, at the very least, name(s), age, sex, height, weight, hair color, eye color, and identifying marks/features/tattoos.

Any names, previous names, or aliases or other identifying particulars of the victim should be investigated and recorded. Should the victim have a number of aliases, the times at which they have been known by certain names should be detailed. Where *justice system history* may be present, information would need to be sought and chronicled about whether criminal history checks had been undertaken under these other names.

The victim's age should be recorded based on established or valid sources, such as a birth certificate or a driver's license. Caution should be exercised when relying on the word of others because they may not be privy to factually accurate information or they may simply be reporting their perception of age or other demographic material.

While seemingly common sense, the adoption by some of gender-neutral Christian names or abbreviations (such as Sam, Casey, Charlie, and Jamie, among others) and other circumstances (such as cross-dressing or transgender victims, for example) means that it is prudent to record the victim's biological sex and their identified gender in any other records.

The victim's height should also be included. If the victim is deceased this will usually be reported in the autopsy (if it is available to the analyst) or may be reported on their driver's license. Height tends to fluctuate little over time, so this figure should be relatively reliable as long as the source is accurate and not too dated.

As with height, weight also will usually be reported in the autopsy, so if the victim is deceased this would be a good source. Unlike height, however, weight can fluctuate greatly over time, so current weight should be established compared to past weights and, if reliable indicators are available, to mean or average weights. Times of weight loss and weight gain may be indicative of psychological or emotional turmoil, past victimization such as molestation or sexual assault, drug use and abuse, or use of prescription medications.

Hair color can fluctuate greatly over time as the result of natural aging processes or artificial mechanisms such as hair coloring. As with weight, it is wise to cover any array of hair colors, ending with what the victim had at their death. In serial cases, this may help establish patterns based on victim selection criteria along with other demographic information.

Other basic demographics such as eye color should be included to complete the picture of the victim's general appearance, including details about any identifying marks, tattoos, or other distinguishing features.

4.5.3 Medical history

One aspect of the victimology that may be difficult to acquire is the victim's medical history because this is considered a form of privileged information that is private and confidential. Some victims who survive their victimization may be entirely forthcoming with this information, whereas others may be reluctant to provide it. Should the victim be deceased in a jurisdiction where medical confidentiality extends beyond death, information from third parties may be the limit of knowledge about any medical issues that contributed to the victim's harm or death. As such, the analyst must exercise caution in any interpretations made as a result of this information. In one case, a female victim of a victim-precipitated homicide claimed to have autism. However, no actual medical files reflected this diagnosis, and the only source of the information was the victim's own claims, which were told to a number of others.

4.5.4 Psychological history

Like medical history, psychological history can be difficult to get because it is also subject to confidentiality. Also like medical history, some victims keep a detailed record of their contact with psychologists, psychiatrists, and other health professionals. This may be especially true when other psychopathology exists, such as when the victim believes that they are being poisoned, targeted, or harassed and that the way to chronicle this and expose the offender is by keeping extensive documentation. Detailed notes and histories may serve to further reinforce in their own mind their status as a victim or target.

Others, such as those who were in relationships with the victim in the past, can provide good general information regarding how the victim responded to certain challenges in life, drug and alcohol use, general personality characteristics, and a host of other features of importance. Once again, caution should be exercised when taking this information at face value. The most ideal information is that which is corroborated by a number of independent witnesses or for which other confirmation exists.

4.5.5 Justice system history

As discussed above in relation to the various types of crime, a history of antisocial behavior or contact with the justice system can increase the risk of victimization for certain crimes. Previous contact with the criminal justice system may represent a general antisocial attitude or indicate a more diverse array of criminal behavior, pointing to an increased risk of victimization due to associations of exposure to criminals and/or other at-risk populations.

History of offenses may be used to educate the analyst as to the type of behavior that a victim or offender engaged in during past events and the types of things that they may have been exposed to. Involvement in violent crime may be out of character for someone who had only previously committed acts of minor property crime, and this change needs to be investigated and accounted for as it may answer critical questions about time, place, location, or associates.

Evidence of a history as either a victim or an offender may also be instrumental in mitigating estimations of a particular person's risk. Should a homicide occur and the offender was involved in an alcohol-related assault some 20 years earlier, this represents volumes of difference from an offender who is on remand for an assault only 2 months ago.

This section should consider not only the first and most recent instance of any events but also the particular crime types and their frequency. An examination of this history can be highly instructive in determining criminal specialization and versatility and may provide further insight into criminal networks and associates.

4.5.6 Drug and alcohol history

Drugs and alcohol play a role in both victimization and offending; the relationship between alcohol and violence was discussed previously. As such, it is critical to determine the degree to which drugs and alcohol may have played a role as a contributing factor, such as blunting a victim's ability to make rational judgments, their ability to gauge risks in their environment, or their ability to defend themselves.

This history would include both licit and illicit substances. Not only should record be made of any illegal or illegally acquired drugs (such as prescription medications obtained outside of legitimate medical services as well as narcotics). Both legally obtained prescription medications and nonprescription medications and their source, where possible, should be itemized, including the type, dosage, and history of ingestion, particularly if nonadherence to a prescriptive/recommended dosage is indicated.

In addition, prescription medications should also be listed, including, where possible, the type of drug(s), the strength of the drug(s), how often prescriptions are filled, when the last prescription was filled, and whether prescriptions have been used relative to their date of issue. This may provide insight into use and potential abuse. Where there is evidence of multidrug use, drug interactions should be sought and listed. If the medication is legitimately prescribed, then the condition for which this was given should be identified if possible. This should also be checked against the toxicology at autopsy or during medical intervention after the offense, so any and all substances in the victim's system are identified and their effects accounted for.

4.5.7 Relationship history

Criminal and victim behaviors do not occur in a vacuum, and they do not occur in isolation. Occasionally, the event that leads to a victimologist's involvement will not be the first of its kind. As a result, it is possible—even likely—that the same behaviors that culminated in a serious or violent criminal event have been evident, seen, or experienced by others previously.

There really is no limit to the number of individuals who may have witnessed, heard of, or experienced a victim's relevant relationship history. Witnesses may include those privy to the victim's reactions to rejection and abandonment, violence history, disposition, and temperament. Past intimate partners can provide insight into coping and aggressive behaviors, risk, and protective factors; security staff may give insight into a patron's behavior while under the influence of alcohol; employers and colleagues may give vital information on work performance; friends and family can inform the analyst about past relationships and their function or dysfunction. Of course, appropriate caution and no small amount of discretion needs to be used with regard to the source and certainty of the information provided.

This information should be compiled chronologically in as much detail as possible, and it may parallel that of medical, drug and alcohol, past victimization, and other victimological information. Where possible, primary documentation should be used to supplement information provided verbally by those known to the victim or witnesses of the crime.

4.5.8 Residential history

When possible, every effort should be made to establish a residential history that includes the address, type of dwelling, and owner/rental status. The method, source, and destination of payments made should be determined through primary documentation such as bank and credit card statements. The victimologist is looking for information that may have put the victim in contact with offenders or put them in an environment in which certain crimes may be more common. Base crime rate data for the areas in which the victim has resided should be sourced and factored into any determinations of risk and threat.

The name and identity of any individuals who shared accommodation with the victim will assist with a number of determinations, and efforts should be made to establish the criminal and relational history of these individuals as well. It may also be prudent, in circumstances where it is permitted or possible, to interview neighbors as well as real estate agents or landlords who have rented properties to victims. These can often provide valuable information as to various facets of the victim's life, and possibly death.

4.5.9 Employment history

Current and former employment information must be sought out and fully scrutinized. This will help to establish such things as base levels of income, whether a victim was living above their means, potential sources of threats or harm, and vital personality and behavioral information from those who have encountered the victim in a number of different contexts. As each person typically has a number of different personality domains (work persona, family persona, social persona, etc.), talking to work colleagues and employers can provide a valuable piece of the victim puzzle.

The information collected should be compared to the victim's curriculum vitae or resumé, if available. The information obtained through employment history can often be cross-checked with other financial information or residential history to help establish an overall picture of who the victim was and how they lived their life. Threats from the environment can, in theory, come from any direction with any possible number of influences, so the more various the sources of information, the more complete the overall picture.

4.5.10 Financial history

Financial history and employment history are considered to be linked on many levels for many purposes. One's financial status is dictated largely by income from employment. However, supplementary income from other sources (legitimate and illegitimate) must be identified and accounted for. Any "off the books" or "under the table" income may help to explain victimization or to identify other avenues of inquiry that must be considered and excluded as relevant.

If the victimologist has the chance to spend time with the victim's belongings, then such things as bank and credit card statements may yield information about what was coming in, what was going out, where it was coming from, and who it was going to.

4.5.11 Technological history

In a connected world, a variety of technologies are a potentially rich source of information about victims and victimization. Our digital footprint is usually far more extensive than we believe; a variety of information is held online for retrieval by virtually anyone with access to or knowledge of computers and the Internet. With an increase in the variety of modalities by which victims connect to the digital world, more sources of information about them and their behavior become available.

The number of devices and types of connection are dictated by financial resources, employment, technical knowledge, and need, among others, so a crucial first step is to determine the number and type of devices that a victim has or has access to. In some cases, access to the device itself may be limited, such as when it is owned by an employer or when the device is taken by an offender because it has information on it about them or their relationship to the victim. Even though information about browser activity may be available on the device itself, other information, particularly that from telecommunications companies, will be hard to come by without legal cause and the necessary paperwork. As such, as detailed a picture as possible will have to be compiled from the information that is available.

Computers, tablet devices, smart phones, other cellular phones, and even game consoles can provide a picture of online activity and behavior that may be pivotal in determining some material information about the victim and the harm that has come to them. Historically, email was most commonly the richest source of this information, though any number of social networks, messenger clients, or web browsers can yield voluminous and valuable materials intelligence. Their social networking friend's lists, cookies, caches, browser favorites, emails, and text messages are instrumental in providing a holistic picture of who the victim was, information about their cyber-self, and how they conduct themselves online. It may sound logical to assume that because a person does not use the Internet, own a computer, or post anything about themselves online, this information may not be available. This assumption would be incorrect. Information about anyone can be posted by anyone else, on their behalf and without their consent, for good purposes or bad. One example of this comes from the case files of a cyberstalking case: The victim vehemently stated there would be no information about her on the Internet. She was adamant that because she had not posted anything, that there would not be anything. It took less than 10 seconds to find a picture of her online linked to a modeling profile. (This case occurred before popular contemporary social networks were used.)

4.5.12 Past victimization

In some cases, the victimization under study will not be the only instance, event, or type of victimization that an individual has been exposed to. Some victims will have suffered extensively at the hands of others, either in only one type of crime (repeat victims of domestic violence) or in multiple crimes (child abuse, domestic violence, stalking, and rape). Because one's reaction to current conditions may be influenced by past experiences, it is necessary to extensively chronicle any history of harm or loss. This should, however, be limited to those types of past behaviors that may be or are relevant to the case. For example, a charge of unlicensed driving may not be relevant to stalking unless it has been demonstrated in some way that both are part of a general pattern of antisocial behavior.

According to Tseloni and Pease (2003, p. 196) "victimisation is a good, arguably the best readily available, predictor of future victimisation." It is therefore paramount that this be documented in as much detail as possible, with supporting documentation provided when possible and necessary.

4.5.13 Wounds

The nature, location, and severity of any wounds must be noted and accounted for. An assessment of wounds can be incredibly instructive for making other decisions about how the victim reacted or what took place during the attack. This information may be taken from the victim when supporting documentation is available (such as photographs of wounds), from a medical practitioner or emergency medical personnel, or at autopsy should they be the victim of homicide. Various documentary evidence—from traumagrams to photographs and video—may also be available.

The types of wounds that can be inflicted, categorized, and documented are as many and varied as the nature of the weapons that can make them. A detailed discussion of wound types is provided by DiMaio and DiMaio (2001) and Fisher (2012); some information is provided in Chapter 3 in this volume.

When an individual attempts to fend off an attack, it is not uncommon to find *defensive injuries* on the victim. Should the victim have a weapon of their own, fight back, retaliate, or disarm the attacker, the initial aggressor may also exhibit such injuries. The nature of defensive injuries is dictated by the type of weapon, the manner in which it was wielded, and the victim's specific movements and contact with their environment, the offender, or the weapon. For example, if a knife was wielded in a downward stabbing motion, wounds on the victim's hands may be present where they attempted to grab the knife or may be seen on the interior aspect of the forearms. A blunt weapon, such as a baseball bat, may cause crushing wounds on the victim's head where the bat was swung in a downward motion onto the head, or may break the victim's nose by smashing the end of the bat into their face. An assessment of these wounds by a medical professional, and their opinion as to whether the wounds are defensive or offensive in nature, can be very instructive when reconstructing the behavior that occurred at the scene.

Regardless of the nature or the type of the injury, it is necessary to examine all injuries to any involved parties and to account for the source, including both the individual responsible and the weapon involved. Defensive injuries may speak to the manner in which the victim was approached, the degree of familiarity between the victim and the offender, or the amount of warning the victim had before the attack. Similarly, the age of injuries must be determined when possible. There is a difference between an abrasion that is new and one that has aged; the latter will be puckered around the wound edges. A fresh bruise usually appears darker and may have well-defined edges, while an older bruise will be more yellow and diffuse.

4.5.14 Risk assessment

The risk assessment should include a determination of both the level of risk presented by the (typically) victim's stable risk and protective factors and those dynamic risk and protective factors that were present at the time of the offense. Generally speaking, the two types of risks (stable and dynamic), as well as any protective or mitigating factors, should be assessed together, giving one complete assessment of overall risk, which is usually presented as high, medium, or low risk.

After presenting the above information and the victim's risk assessment, the victimologist should sign and date the report and retain copies for their own records.

Conclusion

Crime victims come in all forms. Some unknowingly contribute to their own victimization, some are the initial aggressor, and some have no idea when they are in harm's way until it is too late. Assessing victims based on their true characteristics and the series of events that led to their victimization is not synonymous with victim blaming; rather, it is an important step in understanding the nature of the events surrounding their demise and to contextualize the offender's actions.

Risk factors for violent crime are many and varied, but they often relate to being in the presence of criminals and illegal behaviors; being exposed to alcohol, drugs, and violence; being a previous victim, and/or being threatened with violence in the past. Protective factors for many types of crime also exist, including resilience, interpersonal, and contextual elements.

When assessing a victim of violent crime, forensic victimologists are charged with compiling many details from many sources to understand the exact nature of the person and their victimization. Demographic, psychological, medical, criminal, and relationship histories are important elements to include in a victimology, along with any previous victimization, an assessment of injuries, and a risk assessment incorporating both dynamic and stable factors. If compiled in a thorough and diligent fashion, these reports can assist in forming conclusions about how the victim came to be in the violent or criminal situation, where in their lives the risks were likely coming from, and thus who the potential suspects may be.

REVIEW QUESTIONS

1. True or false? *Victima* is a word originally used to describe an animal or person who was put to death for some power or deity.

2. In _____ victimology, the term *victim* itself is the subject of analysis.

3. What is the main concern expressed by some authors regarding the use of victim precipitation to understand the context of victimization?

4. Injuries that are temporary in nature are related to victim _____ exposure more so than victim _____ exposure.

5. Name three parts of the victimology report and briefly state what each should include.

6. True or false? It is not necessary to outline the nature and type of wounds in a victimology because this will have been done by someone else.

References

Australian Insititute of Criminology. (2011). *Australian crime: facts and figures 2011*. Canberra: Australian Institute of Criminology.

Australian Institute of Criminology. (2012). *Australian Crime: Facts and figures*. Available from http://www.aic.gov.au. Accessed 05.05.13.

Blaauw, E., Winkel, F. W., Arensman, E., Sheridan, L., & Freeve, A. (2002). The toll of stalking: the relationship between features of stalking and psychopathology of victims. *Journal of Interpersonal Violence, 17*, 50–63.

Brown, S. E., Esbensen, F., & Geis, G. (2010). *Criminology: explaining crime and its context* (7th ed). New Providence: Andersen Publishing.

Budd, T. (2003). Alcohol related assault: findings from the British crime survey. *Home Office Online Report, 35*. Available from http://www.homeoffice.gov.uk/rds/pdfs2/rdsolr3503.pdf. Accessed 22.12.2013.

Campbell, J. C., Webster, D. W., & Glass, N. E. (2009). The Danger assessment: validation of a lethality risk assessment instrument for intimate partner femicide. *Journal of Interpersonal Violence, 24*, 653–674.

Campbell, J. C. (2012). Risk factors for intimate partner homicide: the importance of Margo Wilson's foundational research. *Homicide Studies, 16*, 438–444.

Commonwealth Department of Human Services and Health (CDHSH).1994). *Transgender lifestyles and HIV/AIDS risk*. Canberra: Australian Government Publishing Service.

Cook, B., David, F., & Grant, A. (2001). *Sexual violence in Australia*. Canberra: Research and Public Policy Series; no. 36, Australian Institute of Criminology.

Curtis, L. A. (1973–1974). Victim precipitation and violent crime. *Psychological Bulletin, 130*(2), 392–414.

Davidson, J. (1997). *Breaking every boundary: sexual abuse of women patients in psychiatric institutions.*. Rozelle, New South Wales: Women and Mental Health Inc.

Dearden, J., & Payne, J. (2009). *Alcohol and homicide in Australia. Trends and Issues in Crime and Criminal Justice* . (372). Canberra: Australian Institute of Criminology.

DiMaio, V. J., & DiMaio, D. (2001). *Forensic Pathology* (2nd ed.). Boca Raton: CRC Press.

Douglas, K. S., & Dutton, D. G. (2000). Assessing the link between stalking and domestic violence. *Aggression and Violent Behavior, 6*, 519–546.

Dugan, L., Nagin, D. S., & Rosenfeld, R. (1999). Explaining the decline in intimate partner homicide: The effects of changing domesticity, women's status, and domestic violence resources. *Homicide Studies, 3*(3), 187–214.

Doerner, W. G., & Lab, S. P. (2005). *Victimology* (4th ed.). Newark: Anderson Publishing.

Ferguson, C. E., Petherick, W. A., & Turvey, B. E. (2010). Forensic Victimology. In W. A. Petherick, B. E. Turvey, & C. E. Ferguson (Eds.), *Forensic criminology*. San Diego: Elsevier Science.

Fisher, B. (2012). *Techniques of crime scene investigation* (8th ed.). Boca Raton: CRC Press.

Heilpern, D. M. (1998). *Fear or favour: sexual assault of young prisoners*. Lismore, New South Wales: Southern Cross University Press.

Hepburn, L. M., & Hemenway, D. (2004). Firearm availability and homicide: A review of the literature. *Aggression and Violent Behavior, 9*, 417–440.

Johnson, K., Andrew, R., & Topp, V. (1988). *Silent victims: a study of people with intellectual disabilities as victims of crime*. Victoria: Office of the Public Advocate.

Karmen, A. (2007). *Crime victims: an introduction to victimology* (6th ed.). Belmont: Thompson Wadsworth.

Kellerman, A. L., Pivara, F. P., Rushforth, N. B., Banton, J. G., Reay, D. T., Francisco, J. T., Locci, A. B., Prodzinski, J., Hackman, B. B., & Somes, G. (1993, October). Gun ownership as a risk factor for homicide in the home. *New England Journal of Medicine*, 1084–1091.

Lau, C. L. (2012). Family violence accounts for 25% of homicides in Hong Kong. *Hong Kong Medical Journal, 18*(4), 351.

Meloy, J. R. (1998). *The psychology of stalking: clinical, legal, and forensic perspectives*. London: Academic Press.

Miller, L. (2012). Stalking: patterns, motives, and intervention strategies. *Aggression and Violent Behaviour, 17*, 495–506.

Office for National Statistics. (2012). *Crime survey for England and Wales, violent crime and sexual offences, 2011/2012*. Available from http://www.ons.gov.uk/ons/dcp171778_298904.pdf. Accessed 12.05.2013.

Pathé, M., & Mullen, P. (1997). The impact of stalkers on their victims. *British Journal of Psychiatry, 170*, 12–17.

Perkins, R., Prestage, G., Sharp, R., & Lovejoy, F. (1994). *Sex work and sex workers in Australia*. Sydney: University of New South Wales Press.

Pesta, R. (2011). *Provocation and the point of no return: an analysis of victim-precipitated homicide. Thesis submitted in partial fulfilment of the degree of Master of Science in the Criminal Justice Program*. Youngstown State University. July, 2011.

Petherick, W. A., & Sinnamon, G. (2014). Motivations: victim and offender perspectives. In W. A. Petherick (Ed.), *Profiling and serial crime: theoretical and practical issues* (3rd ed.). Boston: Anderson Publishing.

Polk, K. (1997). A reexamination of the concept of victim-precipitated homicide. *Homicide Studies, 1*, 141–168.

Raine, A. (2013). *The anatomy of violence: the biological roots of crime*. London: Penguin.

Reingle, J. M., Jennings, W. G., Lynne-Landsman, S. D., Cottler, L. B., & Maldonado-Molina, M. M. (2012). Toward an understanding of risk and protective factors for violence among adolescent boys and men: A longitudinal analysis. *Journal of Adolescent Health, 52*, 493–498.

Riggs, D. S., Caulfield, M. B., & Street, A. E. (2000). Risk for domestic violence: factors associated with perpetration and victimization. *Journal of Clinical Psychology, 56*(10), 1289–1316.

Schafer, S. (1977). *Victimology: the victim and his criminal*. Virginia: Reston Publishing Company.

Scott, M. S. (2002). *Assaults in and around bars. Problem Oriented Guides for Police Series, 1*. Washington: U.S. Department of Justice.

Siegel, J. (2008). *Criminology: The Core* (8th ed.). Belmont: Wadsworth.

Soler-Baillo, J. M., Marx, B. P., & Sloan, D. M. (2005). The psychopshysiological correlates of risk recognition among victims and non-victims of sexual assault. *Behavior Research and Therapy, 43*, 169–181.

Thomas, S. D. M., Purcell, R., Pathé, M., & Mullen, P. E. (2008). Harm associated with stalking victimization. *Australian and New Zealand Journal of Psychiatry, 42*, 800–806.

Tseloni, A., & Pease, K. (2003). Repeat personal victimization. *British Journal of Criminology, 43*, 196–212.

von Hentig, H. (1948). *The criminal and his victim*. New Haven: Yale University Press.

Wilson, J. (2009). *The Praeger handbook of victimology*. Santa Barbara: ABC-CLIO.

Detecting Staged Crime Scenes: An Empirically Derived "How-to"

Claire Ferguson
University of New England, Armidale, NSW, Australia

Key Terms

Staging
Crime scene staging
Defects of the situation
Simulated crime scene
Red flags
Crime reconstruction

INTRODUCTION

This chapter provides a basis for determining how crime scene staging is accomplished by offenders in homicide cases. The purpose of the chapter is to discuss how the real nature of some criminal events is hidden through crime scene staging, as well as the personal characteristics and evidence necessary for investigators to readily detect these scenes. The literature available on crime scene staging will first be discussed, followed by the investigative red flags that identify staging. At the end of the chapter, techniques for determining whether a scene has been staged will be discussed.

5.1 Staged crimes in the literature

5.1.1 Staging and criminal investigations

Crime scene staging is a conscious attempt by the offender to thwart investigative efforts (Burgess, Burgess, Douglas, & Ressler, 1992). Staging refers to the deliberate alteration of physical evidence at the location where a crime has actually or allegedly occurred in an effort to simulate events or offenses that did not occur. Staged scenes are also called simulated scenes, and the staged elements have been referred to in many ways, including inconsistencies, contradictions, improbabilities, deceptions, and defects (Gross, 1924). Despite the fact that staged crime scenes are not uncommon (Gross, 1924; Geberth, 1996; Schlesinger, Gardenier, Jarvis, & Sheehan-Cook, 2012; Turvey, 2000), there is a dearth of published literature devoted to studying them in the forensic community (Douglas & Munn, 1992; Geberth, 1996; Gross, 1924; Soderman & O'Connell, 1936; Svensson & Wendel, 1974; Turvey, 2000). In fact, thus far only four studies have been published on staged crime scenes (Ferguson, 2014; Hazelwood & Napier, 2004;

Schlesinger et al., 2012; Turvey, 2000). Because published research efforts are few and far between, authors who have commented on simulated scenes can be discussed with some brevity. The following section outlines what has been done to date on staged scenes in homicide, including the relevant recommendations for investigators analyzing crimes.

5.1.2 Early works

Dr. Hans Gross is the most influential author when it comes to forensic criminology. It is no surprise then that he is one of the few authors who has tackled the issue of staged crime scenes, and despite his works being nearly a century old, he maintains some of the most detailed treatments of the issue. In a section dedicated to discussing the injuries present in homicides staged to appear as suicides by hanging, he states (1924, p. 430):

> It is a fair presumption that a considerable proportion of so-called suicidal deaths by hanging are really caused by another hand. Of course in such cases the murderer will not select a mode of death leaving too distinct traces. One would not hang up, under pretence of suicide, a person killed by a gunshot wound or with a fractured skull; but this is frequently done in cases of poisoning, strangling, or even killing by means of a fine and long stabbing instrument.

He noted that investigators must be constantly aware of the fact that simulated crime scenes and false reports do occur with some frequency, and therefore each case must be examined through a lens of relative skepticism. He advised investigators to consider each circumstance of the case that would signify whether the crime actually occurred the way it presents, as well as what it would signify if the crime were a false report, or something else entirely. Gross stressed that skepticism must be exercised to protect the innocent who may be accused, as well as to expose the self-made victim (nowadays, these are identified as false reporters).

To investigate these possible staged scenes, Gross first recommends that the "exterior" circumstances be examined by looking at any suicide notes left, and comparing them to known writings of the deceased. If present in the note, the reasons outlined for the suicide should be compared to the person's life to determine whether the reasons were indeed apparent (such as financial stress), and whether anyone else may corroborate any psychopathologies or emotional disturbances discussed.

His second recommendation when investigating deaths is to make detailed and exact notes on the instrument used. Investigators should write down "whence it comes, its nature and size, and the mode in which it has been used" (p. 432). This is done for two reasons: first to facilitate later investigation should further suspicion develop down the line, but also so that the weapon choice may be factored into the analysis. He notes that although some people seeking to hang themselves will choose any convenient object to use, most choose their instrument with great care. They select those ligatures that are strong and safe, and also those that will not hurt the skin. Although Dr Gross is speaking specifically of deaths involving hanging or strangulation, weapon choice should be considered when investigating all equivocal deaths.

Gross' third recommendation when investigating a possible staged scene is to document everything meticulously. He explains that doing so will allow investigators to detect any inconsistencies in the scene and in the statements of those involved.

Finally, he notes that through carrying out the above procedures the investigating officer can determine the contradictions, paradoxes, and the "defects of the situation" that lead the officer to discover the "grand blunder" which, he explains, "the most experienced and crafty criminal rarely fails to commit" (p. 433).

Although treated much more briefly, the work of O'Hara and Osterburg (1972, reprinted from 1949) should also be noted. They state that miscarriages of justice can and do occur when an offender seeks to frame another, and stages the crime to implicate that person as the offender. This is the first treatment of the issue in relation to homicides being staged as other types of homicides in the literature. Being a book on criminalistics, this work offers advice on how this simulated evidence can be detected using forensic science. They explain (1972, p. 683):

> *The laboratory in the majority of these cases will be able to detect the simulated evidence. The reasons for this are simple. The criminal is frequently suffering an emotional disturbance when committing the crime and while substituting the fraudulent clue materials. This in addition to the fact that he usually has little, if any, experience in the appearance or requisites of physical evidence, enables the deception to be uncovered.*

These authors go further, providing case examples and warning criminalists that those attempting to frame others may do so by starting rumors and innuendo against the person they wish to frame. This highlights the potential for verbal as well as physical staging.[a]

5.1.3 Contemporary works

In terms of the more recent references to staged crime scenes, there are several that bear mentioning. These are generally works dedicated to criminal investigation, or some part thereof, that have a small section or chapter devoted to staged evidence. Each will be discussed in turn.

In the FBI's Crime Classification Manual, Douglas and Munn (1992) and Douglas and Douglas (2006) discuss staged crime scenes as occurring for one of two reasons, "to redirect the investigation away from the most logical suspect or to protect the victim or the victim's family" (Douglas & Munn, 1992, p. 251). In no other literature is the second part of this definition endorsed, because these actions do not involve criminal intent on the part of the person employing the "staging." In other definitions, staging is a separate intentional act on the part of the offender to alter the interpretation of the circumstances of a crime; it is not simply a concealment of the circumstances of a prior act or event perpetrated by someone else.

In their discussions of staged scenes Douglas and Munn (1992) and Douglas and Douglas (2006) offer a list of questions to ask and matters for investigators to be cognizant of in determining when a crime scene has been staged, including red flags at both the crime scene and in the laboratory. They first note, though, that often the inconsistencies noticed by investigators are because the offender staged elements at the crime to appear as he/she *thinks* they should, not as they necessarily would appear if the scenario were legitimate. This is reminiscent of the work of O'Hara and Osterburg (1972). For instance, if staging a domestic homicide to look like a stranger burglary/homicide, the offender may have no insight into how a real stranger burglary/homicide actually presents, having never committed, investigated, or been the victim of one. They will be forced to stage the scene to resemble how they think a burglary would look, all while under the stress of having committed, or being about to commit, a homicide. Because of this stress and possible panic the pieces may not fit together in any logical way.

---

[a] Verbal staging has been discussed by Hazelwood and Napier (2004), Schlesinger et al. (2012), and Ferguson (2011). It refers to those cases where the offender makes self-initiated contact with police to report a victim missing and file a missing-persons report. Verbal staging may also involve physical manipulation of the body or crime scene, although this is not always the case.

According to these authors, the queries that need to be made include (adapted from Douglas and Munn (1992, p. 253–255) and republished in Douglas and Douglas (2006)):

Red Flags at the Crime Scene

Did the subject take inappropriate items from the crime scene if burglary appears to be the motive?

Did the point of entry make sense?

Did the perpetration of this crime pose a high risk to the offender?

Did the offender first target the person posing the greatest threat?

Forensic Red Flags

Do the injuries fit the crime?

Does the account of a witness/survivor conflict with forensic findings?

Are there signs of close offender association with the victim?

Was the victim conveniently discovered by a third party?

In staged scenes in fatal and false report cases, Hazelwood and Napier (2004) canvassed 20 consultants who testified as experts or worked cases involving staged scenes. The rationale behind this study is in agreement with other authors (for example Geberth, 1996, discussed below), who opine that staging behaviors are on the rise due to the effect of mass media and the portrayal of forensic techniques therein.

Of the 20 consultants, Hazelwood and Napier (2004) asked how many cases they had worked, how many were staged, and the types of staging that were commonly found. The authors report (p. 754–755):

The 20 law enforcement professionals that participated in the survey reported that in their experience, nonfatal false allegations of sexual assault were the most common form of staging, followed by staging homicides as burglary-related or robbery-related crimes, staging the manner of death, and finally staging the homicides as sexually-related crimes.

This survey also found that approximately 3% of violent crimes are staged (Hazelwood & Napier, 2004).

In terms of how to investigate staged scenes, Hazelwood and Napier (2004) provide many recommendations and general commentary similar to that provided by Douglas and Munn (1992). They note that an investigator has two main sources of information for any given crime, those being the scene and the victim. In order to determine whether and what inconsistencies are present that may indicate staging, they recommend a careful victimology be undertaken. They also postulate that there are three areas where inconsistencies may be discovered: victim-centered, immediate location, and distant locations. They expound (p. 757):

All behaviours and other significant facts about the crime are placed in one of these three categories. The investigator then compares what he observes in and across each category with what he would expect to observe in similar situations, basing those expectations on his education, training and experience. In other words, does what he observes make sense? If the investigator observes inconsistencies, they must be explained.

Interestingly, this work also includes a "profile" of an offender who stages a crime scene, noting that the offender is usually someone known to, or an intimate partner of, the victim and is a white male between the ages of 26 and 35. Although, based on the "observations of highly trained and experienced investigators," it is unknown exactly how this profile, or the recommendations above, can be put to use in real case work.

Geberth (1996) examines staged crime scenes in the context of equivocal death analyses, that is, determining whether a death was accidental, natural, suicide, or homicide. He discusses staging in terms of how a criminal investigation should proceed at a death scene. In his illustration of staging, several cases are used to warn investigators to be aware of the possibility that crime scenes may be staged to mislead them, or redirect an investigation. In defining this concept, he notes that staging is a "conscious criminal action on the part of an offender to thwart an investigation" (p. 22). He also notes that previous authors (Douglas & Douglas, 2006; Douglas & Munn, 1992; Hazelwood & Napier, 2004; Meloy, 2002) refer to actions of embarrassed family members as staging. Geberth disagrees with this, and explains that these actions are understandable in the bereaved. They have a completely different intent, and therefore cannot be considered staging.

Geberth further notes that in his experience he has come across three types of staged crime scenes, which are as follows (adapted from Geberth, 2006, p. 23):

1. The most common types of staging occur when the perpetrator changes elements of the scene to make the death appear to be a suicide or accident in order to cover up a murder.
2. The second most common type of staging is when the perpetrator attempts to redirect the investigation by making the crime appear to be a sex related homicide
3. Arson represents another type of staging. The offender purposely torches the crime scene to destroy evidence or make the death appear to be the result of an accidental fire.

In his book on homicide investigation, Geberth (1996, p. 23) also recommends that investigators remember "things are not always what they appear to be" and to listen to gut feelings. He adds that in his experience these events are increasing due to the public having better knowledge of death investigations through the media, television, movies, and books. This potential issue has also been addressed in other works (Hazelwood & Napier, 2004; Ferguson, 2011).

In terms of strategies for investigating these scenes, Geberth (2006, p. 26) provides a checklist to assist investigators. He states:

1. Assess the victimology of the deceased.
2. Evaluate the types of injuries and wounds of the victim in connection with the type of weapon employed.
3. Conduct the necessary forensic examinations to establish and ascertain the facts of the case.
4. Conduct an examination of the weapon(s) for latent evidence, as well as ballistics and testing of firearms.
5. Evaluate the behavior of the victim and suspects.
6. Establish a profile of the victim through interviews of friends and relatives.
7. Reconstruct and evaluate the event.
8. Compare investigative findings with the medicolegal autopsy and confer with the medical examiner.
9. Corroborate statements with evidential facts.
10. Conduct and process all death investigations as if they were homicide cases.

At a glance these guidelines seem fairly thorough; however, they have been heavily criticized elsewhere. Turvey (2000) has critiqued this checklist as being vague, redundant, and out of order, and therefore offering very little to the investigating agency. Similar to the authors mentioned previously, Geberth's advice is based on his years of experience investigating homicides, rather than any empirical examination of cases involving staging.

Citing Geberth (1996), Keppel and Weis (2004) discuss the rarity of staging as well as posing of bodies. They first differentiate between staging and posing behaviors, and then give case examples and common characteristics of each.

In this research, posing is viewed as discrete from staging. Unfortunately, the authors fail to address the fact that posing a body can be utilized as an element of staging. The authors also recognize that staging involves many more behaviors than simply repositioning the body, and in fact may not even involve such repositioning. However, in this research, cases were included or excluded from the sample of staged cases based on the question, "Did the offender intentionally place the body in an unusual position (e.g. staged or posed)?" Through this sampling procedure, many cases involving staging where the body was not positioned were surely missed.

This work goes on to address the characteristics common to cases involving either staging or posing according to their definitions. This is again problematic as the word staging is used to describe only those cases where the offender altered the body (not the scene) as part of their efforts to deceive investigators. Furthermore, the characteristics do not take into account that posed bodies may also be staged bodies, and therefore the characteristics may not be discrete but overlapping. In light of these issues, the usefulness of Keppel and Weis'(2004) research may be reduced. Arguably, the practical application of their findings to scenes where it is unknown whether staging or posing is present is not clear.

Upon identifying the obvious lack of systematic research in this area, Turvey (2000) conducted a preliminary study to identify common characteristics associated with staged crime scenes. The research examined only 25 homicide cases in the United States from 1980 to 2000, where crime scene staging was confessed to, witnessed, or proven using physical evidence. As a result of the link between staged crime scenes and domestic homicide, the study compared its findings to those found in similar studies (BJS, 1998; Mukherjee, VanWinkle, & Zimring, 1983) of domestic homicides in the United States.

Briefly, this study found that all offenders had a current or prior family or intimate relationship with their victim. This finding supported Geberth's (1996) hypothesis that crime scene staging is most commonly used to conceal an offender's close relationship with the victim (Turvey, 2000). The research also found that offenders were more likely to stage the homicide to appear as a stranger burglary than any other crime, and in many cases, although staged to appear as a burglary, no valuables were taken by the offender. Offenders used available weapons in about half the cases, and were often the person to discover the body. Together these findings illustrate the lack of sophistication that was present in the cases studied, and also highlighted the somewhat troubling notion that those involved in law enforcement may be more likely to stage scenes than non-law-enforcement offenders in order to thwart identification (Turvey, 2000).

In one of the most comprehensive empirical works to date, Schlesinger et al. (2012) studied staging behaviors in a sample of 79 staged homicide crime scenes in the United States. This analysis sought to determine how often these types of behaviors are perpetrated, the types of homicides involved, levels of effort expended by the offender(s), and the motives. Their findings indicate that the prevalence of staging is about 8% in their non-random sample of homicides taken from an FBI database. They determined that the most common types of staging methods employed by offenders were arson (25% of staged cases), verbal staging (22%), burglary/robbery/breaking and entering (18%), accidents (14%), suicides (8%), and homicide-suicide (5%). They found multiple victims were often present and staging behaviors were often basic and non-sophisticated, where only 12% of cases involved elaborate staging. Domestic, non-serial sexual, serial sexual, and general felony homicides were separated in this sample, and it was determined that domestic homicides often involve verbal staging, whereas non-domestic more often involve arson. Victim and offender relationship patterns were discussed and it was found there is often an intimate or familial relationship involved (77%). As discussed, these findings generally align with the anecdotal discussions the authors outlined, as well as with the descriptive findings of Turvey (2000) and those of Hazelwood and Napier (2004).

In her as-yet-unpublished doctoral thesis (early 2014), Dr. Laura Pettler conducted a qualitative content analysis of 18 staged homicide cases between 1987 and 2009 in North Carolina. Pettler classi- fied behaviors into taxonomies, including those related to cleaning (10 cases), fabricating (18 cases), concealing (18 cases), creating (18 cases), planning (10 cases), and inflicting (0 cases). She found that each category was not mutually exclusive with some offenses and offenders exhibiting behaviors from two or three types. As noted by Pettler, this calls into question the practicality of using such a typology to categorize either offenders or behaviors. Instead, she suggests using a method of modified triangula- tion to determine if staging is present (see Pettler (2011) for more information).

Pettler's major findings also warrant review. Before doing so, though, it is important to note that a very broad definition of staging was used in her content analysis, meaning cases indicated as staged would not necessarily be considered staged in other studies. Specifically, Pettler argues that staging is "the purposeful alteration of a crime scene in an attempt to mislead investigators and frustrate the crimi- nal justice process" as per Hazelwood and Napier (2004, p.745). Notably, this definition is missing the element of purposefully simulating events or offenses that did not occur (Ferguson, 2014; Geberth, 2006; Turvey, 2000), meaning that many more behaviors are included in Pettler's definition, especially those usually thought of as precautionary acts rather than staging.

In this analysis, Pettler accessed full homicide cases from prosecuting agencies and analyzed all the relevant files for 62 common characteristics. In so doing, she found the 18 cases involved 27 offenders and 19 victims. Twenty behaviors were commonly utilized by crime scene stagers, including (from Pettler, 2011, p.174) fleeing victim recovery locations; moving the deceased within the crime scene, sometimes multiple times; lying to both law enforcement and family, friends, etc., of the victim; con- tacting law enforcement after "discovering" the body; changing clothing before and/or after the murder(s); transporting the victim's body for disposal; cleaning up the crime scene; kidnapping vic- tims and transporting them alive to the crime scene; removing clothing from victims; driving evidence to a distant location; not reporting the victim missing; claiming the victim "ran off" with others; claim- ing to see the victim leave; wrapping the body in bedding; claiming to have not seen the victim leave; removing bedding from the crime scene; claiming the murders were suicides; and claiming the murders were accidents. Victims and offenders were known to each other only half the time, which is in contrast to much of the other literature and likely an outcrop of the broad definition of staging used in this analysis.

As can be seen from Pettler's (2011) common behaviors, some have more practical weight than others in terms of investigative red flags. Specifically, things like clothes being removed from the vic- tim, yet an absence of evidence of sexual assault, would be a good indicator to investigators to be suspi- cious. On the other hand, having an offender flee the body disposal site is likely common to most homicide types, and having many offenders report that they both did and did not see the victim leave is of little investigative utility. However, Pettler (2011) sought to identify and analyze behavioral patterns in staged crime scenes, which has arguably been achieved here. From her work, it is now clear that often offenders will fabricate evidence by lying to police. They commonly conceal evidence of what actually transpired, and create new evidence or crime scenes by moving victims, weapons, and so on. Offenders are less likely to clean up everything so the scene is stripped of evidence, or plan the murder beforehand. None in this sample inflicted wounds on themselves to have the murder appear as self-defense.

5.1.4 Staging and death investigations

Whereas criminal investigators usually determine whether a crime has taken place, who is responsible, where, when, why, and how it happened, they also often rely heavily on the opinions of forensic pathologists, coroners, and medical examiners to determine the manner of death in cases involving a fatality. However, investigators can often complement these strictly clinical findings with much circumstantial or contextual evidence in order to assist these experts in making such a determination. Therefore, a number of criminal investigative texts have also touched on how to examine equivocal deaths (differentiating accidents, suicides, homicides, and natural deaths from one another).

In Soderman and O'Connell (1936), there is some reference to what these authors refer to as *simulated* crime scenes in their discussion of distinguishing homicides from suicides. In this section, they treat each type of weapon or cause of death separately, including shooting, hanging, choking, slit wounds, chop wounds, stab wounds, death due to traffic accidents or leaps from great heights, and poisoning. This stresses the need to identify inconsistencies in injuries and wound patterns to the victim that do not correspond with the alleged facts of the case.

In the case of an equivocal shooting Soderman and O'Connell (1936) stress the importance of determining the distance from which the victim has been shot. They advise that victims almost always shoot themselves from a distance less than 20 in. (50.8 cm), usually pressing the weapon to the skin or holding it very close to the skin. The authors note (p. 261):

> If the direction of the canal in the body seems plausible and the wounded part of the body (heart, forehead, temple, mouth) is so situated that the suicide may have fired the shot from a comfortable position, a conclusion of suicide may be well founded, especially when the wounded part of the body has been uncovered…Naturally attention should always be paid to fingerprints, footprints, traces of violence etc., which may indicate murder. If several bullet wounds are found in a dead man a conclusion of homicide may be reasonably drawn, but it should be kept in mind that suicides may and sometimes do shoot themselves several times.

These authors go on to explain the importance of examining both entry and exit wounds in shooting cases and what can be expected.

In terms of deaths by hanging, Soderman and O'Connell (1936) note that this is a very simple method of suicide that may be carried out with any available material. They stress that the victim need not be completely off the ground in order to cause death, but that when they are, investigators should determine exactly how they were able to bring themselves into the hanging position (perhaps by stepping on a stool or something else), and may need to carry out experiments in order to determine whether and how this apparatus would have functioned. Evidence of the victim's shoes and feet should be identified on the item stepped upon. They also go into some detail on the types of markings to be expected on the skin depending on the material used for the strangulation, as well as what markings should not be expected. For instance they mention that "there will be an interruption in the mark at the place where the knot was tied" (p. 264). Perhaps more importantly, these authors explain *Goddefroy's Method*, which can be utilized to determine whether someone has been hung up by someone else.

Goddefroy's Method is one of the only specific techniques ever proposed that can address whether a crime has been staged based on known and fixed indicators. According to these authors, it works on the principle that the outer fibers of a ligature will lie in the opposite direction to which the person, either the victim or the offender, pulled. The theory is that a human body is extremely heavy, being dead weight. Maneuvering such a weight into a noose or rope that is already suspended is extremely difficult. Instead of doing this, an offender may choose to tie the ligature around the victim's neck while they are on the ground, loop it over another object, and pull down, thus hoisting the body into the air. Goddefroy postulates that in so doing, the fibers on the ligature will be directed upward, opposite the pulling, by the friction caused between the ligature and the substructure. This is a good indication that the limp body has been hoisted up, instead of stepping into the noose as one would expect in a suicide. However, the authors note that a lack of these directional fibers does not necessarily indicate suicide, as the perpetrators may have lifted the body up into the noose or already hanging rope. They also make mention of the fact that the knots utilized in these cases can be of great importance, and every measure should be taken to preserve and document them in as much detail as possible.

In the case of strangulation deaths, Soderman and O'Connell (1936) note that there is no interruption in the patterns of injuries, as can be expected with hangings. Similarly, there is also a more horizontal marking across the throat or neck. They draw attention to the fact that any indication of manual strangulation is a good indication of homicide, as it is physically impossible to manually strangle yourself to death (as soon as you lose consciousness your hands would fall away from your throat, precluding death).

In terms of slit or stab wounds, Soderman and O'Connell (1936) explain the importance of determining the handedness of the decedent. The wounds in suicides are most often found in places that can be reached comfortably, with enough pressure and in normal positions, by the dominant hand, including the front of the neck, upper arm, elbow, wrists, or thighs (Soderman & O'Connell , 1936). They add that several parallel cuts are more typical of suicide, whereas irregular and deep cuts accompanied by bruises and other injuries are more indicative of homicide. Despite the common notion that wounds to the hands, outside of the arms, and fingers indicate defensive wounds (and thus probable homicide), these authors maintain that wounds running across the fingertips are also found in many suicides (p. 267):

> These arise from the fact that the suicide, in stretching the skin with the finger of one hand over the area which he has selected to incise, involuntarily cuts them with the knife carried across by the other hand. These finger wounds may also be due to the grasping of the knife blade with both hands so as to exert more power during the performance of the act. These wounds are not to be mistaken for the 'defence wounds' found in the palm of the hand, which are signs of homicide.

These authors' treatment of determining suicide from homicide in cases involving chop wounds, traffic accidents, and leaps from great heights is fairly brief. Basically, they add that it is very difficult to tell from the wounds of a decedent whether they jumped from a height or in front of a car or train, or were pushed. They explain that in these cases, secondary information and evidence should be sought, including suicide notes and the like. They also note that chop wounds are usually not found in suicides, but when they are they are, generally more likely to be parallel and close together than if inflicted during a homicide. Presumably this is based on the notion that a non-compliant victim would be difficult to "chop" more than once in the same place because they would be moving and possibly attempting to escape or fight.

When it comes to drowning deaths, Soderman and O'Connell (1936) state that in their time, this was the most common type of suicide. Given the technologies of the day, it was nearly impossible to determine whether a drowning was the result of a suicide or a disguised homicide caused by something else. Nowadays, making this determination may not be impossible, though it can still present a challenge. Homicides are often concealed by placing the deceased into a body of water, although this may not be done as an attempt at staging the scene to appear as a homicide; rather, the intent may be to destroy evidence or delay/preclude discovery (i.e., as a precautionary act). Still today, being able to tell the difference is crucial.

Finally, these authors discuss poisoning as a cause of death for both suicides and homicides. In order to determine whether the case involves a homicide, an investigator must determine what the poison is, and what its general properties are. They explain that for the most part, murderers will not utilize poisons that have odors, colors, and tastes, because these would arouse suspicion and may not be successful. "On the other hand, the suicide may take an evil-tasting and evil-smelling poisonous substance" (p. 270).

In terms of determining between homicides and suicides, Soderman and O'Connell (1936) place great importance on the need for a proper wound pattern analysis. The role of the forensic pathologist in these cases can therefore not be understated.

In a similar vein, Svensson and Wendel (1974) stress the importance of systematic analyses and proper comparisons of physical evidence in their work on the various methods of crime scene examination. They also note that the frailties of any perpetrator of homicide can be used to the advantage of the investigation (p. 292):

> *Even when the murderer has carefully planned the crime and taken all imaginable precautions to avoid leaving traces, they are still found. As a rule, the murderer comes to a sudden realisation of the terrible results of his deed after the killing. He may then lose his head completely and try to obliterate the evidence of his act, but in his confused state of mind only works against himself by leaving new clues.*

In terms of differentiating between equivocal deaths, Svensson and Wendel (1974) note that investigators examining these scenes must always err on the side of caution and suspect the worst, for if there is any confusion, it is less harmful to rule a suicide as a homicide than vice versa. In terms of how to actually go about determining between accidents, suicides, and homicides, they stress the importance of a thorough and detailed investigation. Although not cited anywhere, this advice is reminiscent of Gross' (1924) discussion of the defects of the situation. These authors provide a list of inquiries that the investigator should make early on in the investigation to facilitate the accurate determination of the manner of death. They recommend the following queries be addressed (p. 293):

1. What are the causes of death?
2. Could the person himself have produced the injuries or brought about the effect which caused death?
3. Are there any signs of a struggle?
4. Where is the weapon, instrument or object which caused the injuries, or traces of the medium which caused death?

It is necessary to determine whether there are signs of a struggle, according to Svensson and Wendel (1974), because this may be the first indication that a death was a violent one at the hands of someone else. One of the most important signs of a struggle, whose presence will also help facilitate a crime reconstruction, is the amount, location, and distribution of bloodstains (p. 294):

> Generally, no bloodstains are produced during the first stage of the attack, before bleeding has commenced. If the victim does not immediately become unconscious at the first blow, stab, cut or shot, it can nearly always be reckoned that his hands will become covered with blood from touching the injured parts of his body. If the victim tries to escape or to put up a resistance, his blood-covered hands leave marks which often indicate his position in certain situations.

Along with bloodstains, torn-out hair, overturned furniture, crumpled rugs, and marks of weapons or parts thereof should also be examined in great detail as they can tell the investigator about the direction of movement, the behavior of the perpetrator after the fact, the location of the victim during the struggle, the escape routes they attempted to use, and how and where they fought back (Svensson & Wendel, 1974). These authors also explain the importance of distinguishing defensive injuries from accidental injuries or concurrent suicidal injuries. Although these innocent behaviors may cause suspicion, through a careful reconstruction the true series of events will be uncovered.

In terms of determining the weapon used, and its current location, Svensson and Wendel (1974) state that if the weapon or instrument is missing, a homicide has taken place. Presumably, this is because after someone commits suicide or dies in an accident, it would be impossible for them to remove the weapon from the scene. However, this may not necessarily be the case, especially in light of what is known about evidence dynamics. Also, some suicides involve weapons that were inadvertently removed by an innocent person who was unaware of the death. For example, if a person purposely drank poison and disposed of the container in a garbage bin, it may later be taken away by city workers without their knowledge. In this case, after the body is discovered, a homicide or accidental death may be rightfully suspected, although it will not always be the case.

In their discussion of the need for investigators to systematically reconstruct every crime, these authors also address the importance of step by step documentation. They recommend that everything in the scene be documented, especially the position of any clothing, its folds, twists, creases, tears, buttonholes, fasteners, stains, size, and so on. Blood stains, smears, spatters, froth, and droplets on the body or clothing must also be examined and photographed. The absence of blood should also be documented as an offender may have cleaned a victim after death. Svensson and Wendel (1974) go on to discuss each type of injury that may be present in suicides, accidents, and homicides, and what characteristics are typical for each type of injury such as stab and chop wounds, shootings, drownings, poisonings, stranglings, suffocation, and so on. This discussion is in much the same vein as the one mentioned by Soderman and O'Connell, (despite having been written nearly 40 years later). However, it should be noted that Svensson and Wendel (1974) additionally mention the importance of utilizing a qualified forensic pathologist to determine the cause and manner of death, and warn against equating experience (which seasoned investigators may have) with expertise (which is the province of the pathologist). They are also the first authors to mention the importance of what is now termed victimological information. They explain that determining whether someone has committed suicide is also based heavily on an investigation of the victim, or a psychological autopsy including interviews with friends, family, and physicians. The fact that most people who end up eventually killing themselves have previously attempted, or talked about attempting, suicide is also highlighted. This issue is addressed in more detail in Chapter 4 of this text, Forensic Victimology.

Aside from the previously reviewed texts, very little work has been done on the subject of staged crime scenes, and more notably, bar a few studies, no intensive systematic research has been conducted on the topic. This is problematic because elements of staging are such a consistent characteristic of criminal modus operandi (MO) (Geberth, 1996; Gross, 1924; Turvey, 2000), and because these determinations often necessitate successful collaborations between medical professionals and investigators. The authors noted above, with the exception of Turvey (2000), Hazelwood and Napier (2004), Pettler (2011), Schlesinger et al. (2012), and Ferguson (2014), offer suggestions on how to identify these characteristics; however, these suggestions are based solely on their expertise or the expertise of others, and therefore run the risk of being at best inaccurate, and at worst misleading and detrimental to serious criminal investigations. Certainly, more reliable and detailed research is necessary.

5.2 Red flags for staged homicides

As part of her doctoral dissertation (2011), the author undertook a detailed analysis of 115 staged homicide cases in the United States between 1973 and 2007. In 2013, these cases were reanalyzed and the following discussion will outline the major findings.[b]

Before addressing key red flags identified in this study, it is important to note that this author also found staging of homicide scenes to be increasing over time. The study examined cases between 1973 and 2007, and found that in the two full decades studied (1980–1989 and 1990–1999) the frequency of staging increased by 104% (Ferguson, 2011). This finding lends support to Geberth's (2006) assertion that offenders are staging evidence more often now than previously, highlighting the salient nature of this research.

The most common type of staging found in this sample were staged burglaries, followed by staged suicides, staged accidental deaths, and staged car accidents. The most common causes of death were injuries from firearms, multiple weapons, blunt force, strangulation, and sharp force. In a large number of cases (over 25%), some verbal or physical argument (unrelated to the homicide) was evidenced between the victim and offender immediately preceding the death. Put another way, in more than a quarter of cases the victim and offender were fighting just prior to the fatal violence, and the death was an outcrop of the anger and emotional arousal related to the conflict. This finding speaks to the possible spontaneity of these homicides, where at least the timing of the homicide may have been unplanned. In further support of this finding, opportunistic weapons were used to inflict the fatal injuries in nearly one-third of cases, while another 10% involved no weapon. In another third of cases, offenders brought the weapons with them to the scene, which may indicate forethought and planning in those cases.

Victims and offenders were involved in some sort of work, intimate, or familial relationship in all but two cases in this sample. Most were domestic relationships (72.2%; either intimate or family), and

---

[b]A more detailed discussion of the findings of this study and the methodology employed are available from the author and in a forthcoming publication (Ferguson, 2011).

a significant portion were friends or work associations (26.1%). Fifteen cases (13%) involved murders for hire, where someone known to the victim hired another person to kill them.

A summary of the staged characteristics commonly identified for the most prevalent types of staging are discussed below.

For staged burglary/homicides, the most common offender behaviors were staged point of entry/exit (34%); valuables removed (51%) or disrupted but not taken (25.5%); victim's non-valuable personal items removed (23.4%) or disrupted but not taken (36.2%); scene ransacked (40.4%); evidence of what happened cleaned up (66%); offender created an alibi (36.2%); and offender self-injured (19.1%). Interestingly, offenders did not often tamper with phones (10.6%) or lighting (6.4%); set the scene on fire (4.3%); mutilate the victim's body (10.6%); plant bloodstains (2.1%); arrange a murder weapon (4.3%); transport the body to a secondary location (6.4%); rearrange the victim's body or clothing (12.8%); or plant drugs or paraphernalia at the scene (6.4%).

For the staged suicides, evidence that the offender commonly used to stage the scene included arrangement of a weapon (81.3%); simulating self-injury to the victim (93.8%); cleaning up or destroying evidence related to the true events (50%); rearranging the body (68.8%); and removing valuables from the scene (31.3%). Offenders did not as often transport the body to a secondary scene (25%); mutilate the corpse after death (25%); plant suicide notes (18.8%) or drugs (6.3%); or arrange for an alibi (6.3%).

The evidence that was commonly manipulated by the offender in order to present the scene as an accident included arranging a weapon or implement that caused the death (56.3%); leaving the victim at the primary scene and arranging it to appear as an accident (75%); repositioning the body within the crime scene (62.5%); and cleaning up or destroying evidence (62.5%). Offenders did not as often try to make the victims' injuries appear self-inflicted (18.8%); mutilate the body after death (12.5%); plant drugs at the scene (6.3%); or arrange an alibi for him or herself (6.3%).

In the staged car accidents, the most common offender behaviors were arranging a car to appear like it had crashed (93.8%); transporting the body to the staged accident scene (93.8%); arranging/positioning the body in accordance with the accident scene (81.3%); mutilating the corpse after death by inflicting additional injuries or burning (56.3%); setting the scene on fire (37.5%); and cleaning up what actually transpired at the primary scene (62.5%). Most offenders did not attempt to secure an alibi (6.3%). Together these findings may indicate more commitment to the staging effort in staged car accidents than was seen in other types, such as by moving and arranging the body, arranging a car, mutilating a corpse, and cleaning up.

All of the findings outlined briefly above are discussed in more detail in Ferguson (2011).

5.3 Determining the presence of staging

Despite the relatively sparse literature on what to look for in order to determine the presence of staging, several techniques have been offered to assist investigators. These run the gamut from general investigative mantras to detailed and specific checklists to use in making such assessments. Two of these techniques will be discussed here, including the use of a crime reconstructionist, and the author's checklist of things to look for. It should be noted these techniques are not mutually exclusive and work on the same underlying principles of skepticism and critical thinking. They may be used singularly or in combination in any given case. For another example, see Pettler (2011).

5.3.1 Crime reconstruction

To date, crime reconstruction has been espoused as one of the most thorough and therefore effective ways of determining whether a crime scene has been staged (Chisum & Turvey, 2007b; Pettler, 2011). Through a reconstruction, investigators may understand more completely the strengths and limits of the physical evidence, allowing them to conclusively comment on whether or not a scene was staged. As Kirk (1974, p.1-2) states:

> However careful a criminal may be to avoid being seen or heard, he will inevitably defeat his purpose unless he can control his every act and movement so as to prevent mutual contamination with his environment, which may serve to identify him. The criminal's every act must be thoroughly reasoned in advance and every contact guarded. Such restraint demands complete mental control. The very fear of detection, which must almost always be present, will make such control next to impossible.

Ogle defines crime reconstruction as describing what happened during a specific crime (Ogle, 2004). Chisum and Turvey explain that "crime reconstruction is the determination of the actions and events surrounding the commission of a crime" (Chisum & Turvey, 2007a, p.21). A collaborative effort by medical examiners, forensic scientists, law enforcement, and criminalists is critical as, according to Walton (2006, p. ii):

> [H]omicide investigations reflect modern, professional investigation methods and techniques. Both entail positive interpersonal relationships among working professionals in the law enforcement, legal, and forensic spectrums. Hot or cold case investigators must exhibit teamwork that fosters a positive exchange of information and knowledge and disregards personal or personality differences.

Through this collaboration, a reconstruction helps gain information about what other evidence to investigate, who the victim is, and why the crime was committed (Crime Scene Reconstruction, 1991). According to Chisum (1999), knowing what happened and in what order helps gain information about who is capable of committing such a crime, what other evidence to investigate, how to interview witnesses and prosecute a suspect, and most importantly assists in preventing possible miscarriages of justice.

Saferstein's (2004) work on criminalistics explains that crime reconstruction is a collaborative effort that aims to objectively determine the circumstances surrounding a given crime based solely on the evidence present. Accurate reconstructions allow investigators to effectively gather victim information and determine its relevance. Reconstructions also enable an investigator to identify whose statements are accurate and whose are inconsistent with the available evidence, which is particularly relevant to staged scenes. These cases often involve offenders lying to the police, and without physical evidence these lies are very difficult to identify.

In terms of specifically identifying staged crime scenes, Chisum and Turvey (2007b) argue that good science and forensic science must be utilised in identifying, analyzing, and reconstructing possible staged crime scenes. Along with skepticism they offer several questions that the reconstructionist should address in order to determine if evidence has been staged. These recommendations are as follows (adapted from p. 463–476):

Point of Entry/Point of Exit
- Establish all points of entry and exit throughout the scene (doors, windows, paths, roads, etc.).
- Establish whether or not these locations were passable at the time of the crime (e.g. some windows and doors may be barricaded or permanently sealed, and some windows may be too high).

- Determine their involvement in the crime by virtue of documenting transfer evidence (blood, fingerprints, broken glass, dropped items, etc.) and negative transfer (the absence of footwear impressions in mud outside a window, the absence of any signs of forced entry, etc.).
- Determine whether or not entry and exit were possible in the manner required for the crime at hand, in terms of breaking in from the outside, removing any valuables and the existence of requisite transfer evidence—this may require some experimentation by the reconstructionist.

Weapons at or removed from the Scene

Of every weapon found at a crime scene, ask at least the following: Is the weapon found with the victim the one that caused the injury, and, if not, what was its purpose at the scene? Was there another weapon found at the scene? Does it have a known purpose?

…

Sometimes there is evidence of weapons use at a crime scene but no weapon can be found there. For each crime scene it must be asked whether there exists evidence that a weapon has been removed and, if so, what purpose could its removal have served?

Firearms

First, are the wounds to the victim consistent with the story presented? In suicides, could the victim have shot himself or herself?

Then we must ask whether the firearm is loaded correctly, in a manner consistent with the evidence and the statements of witnesses.

…

Next, is the hammer down on an empty casing? And is it the right casing?

Furthermore, is the rotation of the cylinder consistent with the way the shots were fired?

…

Another question to consider is whether the firearm found at the scene is defective or not? Is it capable of chambering and firing rounds?

Gunpowder Deposits

Most suicides are contact or near contact shots. The powder distribution must be something that can be caused by the person holding the gun. A lack of powder indicates that there is a greater distance or that there was an intervening target.

Movement of the Body

- Evidence of drag trails and drag stains on the ground and against environmental surfaces (i.e. bunched carpet, heels dragged across mud, bloodstains leading in from another room, etc.)
- Bunched or rolled up clothing on the victim's body
- Livor mortis inconsistent with the final resting position of the body (blood pooling against gravity)
- Rigor mortis inconsistent with the final resting position of the body (joints stiffened against gravity)
- Blood evidence in places there should not be any
- Trace evidence on the body from locations unassociated with the crime scene

Clothing

Consider also the following:

- Has the clothing been removed from the victim or the scene? What purpose may this have served?
- Have the pockets been searched? Are they pulled out even partway?

- Has the body been rolled, causing the clothing to be unevenly distributed?
- Are there smears of something on the clothes that indicate the body was dragged through (soil, vegetation, water, etc.)?
- Is there anything unusual about the clothing? Is anything inside out or backwards?
- Does it appear as though the victim may have been redressed after being attacked? If so, why were the clothes off in the first place, and why would the offender bother to redress the victim— what purpose would that serve?

The reconstructionist may need to conduct experiments in order to determine how the clothing got the way that it did.

Shoes
- Are the shoes on the correct feet?
- Do the shoes have any transfer evidence inconsistent with the scene?
- Was the victim wearing them during the commission of the crime? Or do the bottoms of the victim's feet indicate that the shoes may have been off during the crime (blood, injury, or scene transfer)?
- Where are the knots in the shoelaces?

Bloodstains
First, is the blood going in the direction it should, given the position of the body and gravity?
Next, are the bloodstains consistent with the purported actions of the victim and the suspect?

Hair
The position of the hair is a frequently overlooked clue. Decedent hair can show how the person came to the position in which she was found. This is particularly true with longer hair but not exclusively, because shorter hair may also show movement.

Along with presenting a number of case examples, Chisum and Turvey (2007b) finish their discussion by stressing the importance of testing theories against the known evidence. The presence or absence of any evidence at a crime scene can raise suspicions of staging. There are no set guidelines, and therefore all suspicious circumstances must be investigated. As noted by the authors (2007b, p. 476) "[s]uspicion justifies further investigation; it tells the investigator where to look for more evidence. Suspicious circumstances are not themselves evidence however."

5.4 Ferguson checklist: an empirically derived "how-to"

Along with crime reconstruction, and based on the red flags identified in the study discussed above, the author has developed a preliminary checklist to help determine whether or not evidence may have been manipulated by the offender(s) in any given case. This checklist is theoretical, and may be used in order to determine the next most logical step. It is recommended to be used in conjunction with more detailed treatment Chisum and Turvey's (2007b) where necessary. The following questions should be asked of the evidence early on in the investigation. Answers consistent with staging may not positively indicate the behavior; however, they will indicate that further explanation is necessary to determine why the evidence presents in that particular fashion, and what series of events the evidence supports and refutes.

5.4.1 **Preliminary inquiries**

1. *Confrontation*
    a. Is there any evidence of a verbal or physical confrontation between the victim and offender? This might include witness statements evidencing shouting or arguing, door slamming, overturned furniture, broken belongings, and so on.
    b. Was this confrontation consistent with how the scene presents? For example, in a suicide one might not expect to hear screaming and fighting immediately prior to a fatal self-inflicted gunshot wound.
2. *Discovery*
    a. Who discovered the victim? Were they going about legitimate business at the time of the discovery?
    b. Was there anyone who might have been expected to discover the victim who did not, including spouses, family members, and so on? Why did they not discover the victim?

5.5 **Staged burglary/homicides**

If the scene presents as a burglary, robbery, or home invasion, investigators should ask the following questions of the evidence:

1. *Entry and Exit Points*
    a. Do the supposed entry and exit points of the offender make sense given the time of day, accessibility and size of the window or door, debris/footprints inside and outside, other more easily accessible points, and so on?
    b. If the entry and exit points are unknown, it is possible that a door was left unlocked, or the offender was let into the scene voluntarily? Would this align with the known habits of the victim?
2. *Valuables*
    a. Are valuables missing from the scene? If so, are such items consistent with a stranger burglary?
    b. Would missing items be known (or assumed) to be present in the scene by strangers, such as tablet computers, laptops, mobile phones, and so on?
    c. Would the offender have had to know the victim in order to know that such items existed, such as large amounts of cash in a hidden safe?
    d. Are there any items remaining at the scene that would be valuable to a burglar, but have not been taken, such as cash in plain sight?
    e. Are non-valuable items missing from the scene, such as those that would not traditionally be considered of worth to a stranger burglar? Who might these items have intrinsic worth to?
    f. Is the house ransacked? Was anything taken? Were ransacked items looked through or simply dumped?
3. *Cleanup*
    a. Is there any evidence that the scene has been cleaned, tidied up, or that evidence has been destroyed? This might be indicated by the smell of cleaning products; by an inconsistency

between witness reports or wounds indicating a physical fight yet no evidence at the scene; bloodstains missing where they would be expected; recent renovating; missing clothing, bedding, carpeting; and so on.

4. *Injuries*
   a. Are injuries to other involved parties consistent with their account of events? Reported long periods of unconsciousness should be paid particular attention, especially when head injuries are seemingly minor or they refuse medical treatment.
   b. Does one alleged co-victim report going to get help while the other victims were killed or harmed?
   c. Does the behavior of any co-victims appear paradoxical, such as driving to the police station for help when they had a phone in their pocket, or failing to render first aid?

5. *Alibis*
   a. Has the alibi of any and all parties been investigated critically and thoroughly? Is there a possibility that it was fabricated?

5.5.1 Staged suicides

Should the scene present as a suicide, investigators should make the following inquiries:

1. *Weapon*
   a. Is the weapon or implement used in the fatal violence present at the scene? If not, is there a legitimate explanation for how it may have been removed and by whom?
   b. Could the weapon at the scene have caused the fatal injury?
   c. Is there a weapon at the scene that did not inflict the fatal violence? How did it come to be there and for what purpose?
   d. Is the location of the weapon in the crime scene consistent with where it would be expected to fall given the injuries to the deceased?
   e. Is there evidence of the victim on the weapon (such as bloodstains)? Are these consistent with its positioning?
   f. Is there evidence of the weapon on the victim (such as a lack of bloodstains where the weapon was positioned; gun shot residue; powder burns)? Are these consistent with its positioning?
   g. Additional questions related to weapons, firearms, and gun shot residue should be asked, as per Chisum and Turvey (2007b), explained above.

2. *Injuries*
   a. Could the injuries to the deceased have been inflicted by another?
   b. Could the injuries to the deceased have been inflicted by themselves? If so, would they have been holding the weapon in their dominant hand in a comfortable position?
   c. Was the area of the skin to be injured exposed?
   d. Is the time of death consistent with how the scene presents and when the death was discovered?

3. *Cleanup*
   a. Is there any evidence that the scene has been cleaned up, tidied or that evidence has been destroyed? This might be indicated by the smell of cleaning products; by an inconsistency between witness reports or wounds indicating a physical fight yet no evidence at the scene;

bloodstains missing where they would be expected; recent renovating; missing clothing, bedding, carpeting; and so on.

4. *Body Movement/Repositioning*
   a. Does the body indicate that it has been moved or repositioned at the scene? This could be indicated by inconsistencies in lividity or rigor mortis and the current positioning of the deceased; in gaps in bloodstains or the presence of blood where it would not be expected; unusual grasping of a weapon; drag marks; evidence that the body has been carried; and so on.
   b. Is there evidence of any other scene or location on the body?
   c. Is the victim's clothing positioned correctly on their body had they naturally moved into that position?
   d. Is the victim's hair positioned in accordance with them naturally moving into that position?
   e. Further inquiries regarding clothing, hair, shoes, and body movement should be made as per Chisum and Turvey (2007b) as outlined above.

5. *Valuables*
   a. Are valuables missing from the scene?
   b. Is there any evidence that the victim has recently given away valuable or important personal items?
   c. Would an offender have had to know the victim in order to know that such items existed, such as large amounts of cash in a hidden safe?
   d. Are there any items remaining at the scene which would seemingly be valuable to an offender, but have not been taken, such as cash in plain sight?
   e. Are non-valuable items missing from the scene, such as those that would not traditionally be considered of worth to a stranger burglar? Who might these items have intrinsic worth to?

5.5.2 Staged accidents

Should the scene present as an accidental death, investigators should ask the following questions of the evidence/victim:

1. *Weapon*
   a. Is the weapon or implement causing the fatality present at the scene? If not, is there a legitimate explanation for how it may have been removed and by whom?
   b. Could the weapon at the scene have caused the fatal injury?
   c. Is there a weapon at the scene that did not inflict the fatal violence? How did it come to be there and for what purpose?
   d. Is the location of the weapon in the crime scene consistent with where it would be expected given the injuries to the deceased?
   e. Is there evidence of the victim on the weapon (such as bloodstains)? Are these consistent with its positioning?
   f. Is there evidence of the weapon on the victim (such as a lack of bloodstains where the weapon was positioned; or gun shot residue)? Are these consistent with its positioning?

g. Additional questions related to weapons, firearms, and gun shot residue should be asked, as per Chisum and Turvey (2007b); explained above.

2. *Body Movement/Repositioning*
   a. Does the body indicate that is has been moved or repositioned at the scene? This could be indicated by inconsistencies in lividity or rigor mortis and the current positioning of the deceased; in gaps in bloodstains or the presence of blood where it would not be expected; unusual grasping of a weapon; drag marks; evidence that the body has been carried; and so on.
   b. Is there evidence of any other scene or location on the body?
   c. Is the victim's clothing positioned correctly on their body had they naturally moved into that position?
   d. Is the victim's hair positioned in accordance with them naturally moving into that position?
   e. Further inquiries regarding clothing, hair, shoes, and body movement should be made as per Chisum and Turvey (2007b) as outlined above.

3. *Cleanup*
   a. Is there any evidence that the scene has been cleaned, tidied up, or that evidence has been destroyed? This might be indicated by the smell of cleaning products; by an inconsistency between witness reports or wounds indicating a physical fight yet no evidence at the scene; bloodstains missing where they would be expected; recent renovating; missing clothing, bedding, carpeting; and so on.

5.6 Staged car accidents

Should the scene present as a car accident and suspicion be aroused that it may be staged, investigators should make the following inquiries:

1. *Vehicle*
   a. Is the position of the car consistent with a fatal automobile accident?
   b. Is damage to the car consistent with a fatal automobile accident?
   c. Is the position of and damage to the car consistent with an accident that could take place in that location, at that time, in those weather conditions, at or around the speed limit on that road?
   d. Is there evidence present at the scene that is consistent with an automobile accident happening in that location, such as tire skid marks, the smell of burning rubber, witnesses reporting hearing braking, or seeing/hearing a crash?

2. *Body Movement/Repositioning*
   a. Does the body indicate that is has been moved or repositioned at the scene? This could be indicated by inconsistencies in lividity or rigor mortis and the current positioning of the deceased; in gaps in bloodstains or the presence of blood where it would not be expected; unusual grasping of a weapon; drag marks; evidence that the body has been carried; and so on.
   b. Is there evidence of any other scene or location on the body?

    c. Is the victim's clothing positioned correctly on their body had they naturally moved into that position?

    d. Is the victim's hair positioned in accordance with them naturally moving into that position?

    e. Further inquiries regarding clothing, hair, shoes, and body movement should be made as per Chisum and Turvey (2007b) as outlined above.

3. *Injuries*

    a. Are the injuries to the deceased consistent with a car accident?

    b. Is the time of death consistent with how the scene presents and when the death was discovered?

    c. Are the injuries consistent with the known facts of this car accident, such as the speed the car was traveling, the trajectory of the victim's body related to the car, the heat/burn time of any subsequent fire, and so on?

    d. Could the injuries to the deceased have been inflicted by another?

    e. Has the deceased suffered any additional injuries after death, such as evidenced by a lack of bleeding into wounds; an absence of soot in the lungs after being on fire; a lack of blood in the car despite crashing at high speed; and so on?

4. *Fire*

    a. If a fire was involved in the car accident, has the point of origin been determined?

    b. Is the point of origin consistent with the known facts of the alleged accident?

    c. Is the point of origin consistent with the known facts of the car make and model?

    d. Did the fire behave in an expected fashion given the circumstances of the alleged accident?

5. *Cleanup*

    a. Have any other scenes that may be associated with the victim's death been investigated, such as their home or the home of their loved ones/family members?

    b. Is there any evidence that these scenes may have been cleaned, tidied up, or that evidence has been destroyed? This might be indicated by the smell of cleaning products; by an inconsistency between witness reports or wounds indicating a physical fight yet no evidence at the scene; bloodstains missing where they would be expected; recent renovating; missing clothing, bedding, carpeting; and so on.

Conclusion

Crime scene staging, although discussed relatively rarely in the crime analysis and forensic science literature, is not uncommon. Offenders may stage crime scenes to avoid detection, to distance themselves from an investigation where they know they will be suspected, or to conceal a relationship to the victim. By definition, the staging of evidence on the part of an offender is conscious and deliberate. It is carried out to not only prevent or obliterate evidence of what truly happened, but also to simulate something else in its place.

    As should now be clear, staging efforts can run the gamut from elaborate and well thought-out attempts to deceive to basic, fairly unsophisticated and spontaneous efforts. Regardless, crime analysts find themselves at a distinct disadvantage when investigating these scenes, as little empirical support exists for what to look for and how to tell whether a scene has been staged. It is hoped that this chapter may take the first step to providing investigators with empirically supported common characteristics, as well as a checklist of inquiries that should be made.

The fact remains that in order to recognize staged evidence, investigators must wear two hats simultaneously. They must determine what the evidence would mean should the scene present legitimately, but also what it would mean should the evidence be fabricated to deceive them. By undertaking a critical, skeptical, thorough, and meticulous examination of each scene, it is hoped that those seeking to literally get away with murder will be unsuccessful in their attempts. The responsibility of investigators to do their best work has never been as great as it is now, given evidence suggesting that offenders may be becoming more aware of investigative techniques, and thus more inclined to stage evidence in an effort to fool authorities.

References

Burgess, A., Burgess, A., Douglas, J., & Ressler, R. (1992). *Crime classification manual*. New York: Lexington Books.

BJS.1998). Bureau of Justice Statistics, Violence by Intimates: Analysis of data on crimes by current or former spouses, boyfriends, and girlfriends. In B. Turvey (Ed.), *Criminal Profiling: An Introduction to Behavioural Evidence Analysis* (2nd ed.). London: Academic Press (2002).

Chisum, W. (1999). An introduction to crime reconstruction. In B. Turvey (Ed.), *Criminal Profiling: an introduction to behavioural evidence analysis* (pp. 73–85). London: Academic Press.

Chisum, J., & Turvey, B. (2007a). A history of crime reconstruction. In J. Chisum, & B. Turvey (Eds.), *Crime reconstruction* (pp. 1–35). London: Academic Press.

Chisum, J., & Turvey, B. (2007b). Crime scene staging. In J. Chisum, & B. Turvey (Eds.), *Crime reconstruction* (pp. 441–482). London: Academic Press.

Crime scene reconstruction. (1991). *Journal of Forensic Identification, 41*, 248–253.

Douglas, J., & Douglas, L. (2006). The detection of staging, undoing and personation at the crime scene. In J. Douglas, A. Burgess, A. Burgess, & R. Ressler (Eds.), *Crime Classification Manual* (2nd ed.). San Francisco: Jossey-Bass.

Douglas, J., & Munn, C. (1992). The detection of staging and personation at the crime scene. In A. Burgess, A. Burgess, J. Douglas, & R. Ressler (Eds.), *Crime Classification Manual*. San Francisco: Jossey-Bass.

Ferguson, C. (2011). *The defects of the situation: an examination of staged crime scenes. Gold Coast*. Queensland: Bond University (Unpublished doctoral thesis).

Ferguson, C. (2014). Staged crime scenes: literature and types. In W. Petherick (Ed.), *Serial crime: theoretical and practical issues in behavioral profiling* (3rd ed.). Boston: Andersen Publishing.

Geberth, V. (1996b, February). The staged crime scene. *Law and Order Magazine*, 89–92.

Geberth, V. (2006). *Practical homicide investigation: tactics, procedures and forensic techniques* (4th ed.). Boca Raton: CRC Press.

Gross, H. (1924). *Criminal Investigation*. London: Sweet & Maxwell.

Hazelwood, R., & Napier, M. (2004). Crime scene staging and its detection. *International Journal of Offender Therapy and Comparative Criminology, 48*, 744–759.

Keppel, R., & Weis, J. (2004). The rarity of "unusual" dispositions of victim bodies: staging and posing. *Journal of Forensic Science, 49*, 1308–1312.

Kirk, P. (1974). *Crime Investigation* (2nd ed.). New York: John Wiley & Sons.

Meloy, J. (2002). Spousal homicide and the subsequent staging of a sexual homicide at a distant location. *Journal of Forensic Science, 47*, 395–398.

Mukherjee, S., VanWinkle, B., & Zimring, F. (1983). Intimate violence: a study of intersexual homicide in Chicago. In B. Turvey (Ed.), *Criminal profiling: an introduction to behavioural evidence analysis* (2nd ed.). London: Academic Press (2002).

Ogle, R. (2004). *Crime scene investigation and reconstruction.* Upper Saddle River: Pearson Education Inc.

O'Hara, C., & Osterburg, J. (1972). *An introduction to criminalistics.* Bloomington: Indiana University Press.

Pettler, L. (2011). *Crime scene behaviors of crime scene stagers.* Minneapolis, Minnesota: Capella University (Unpublished doctoral thesis).

Saferstein, R. (2004). *Criminalistics: an introduction to forensic science* (8th ed.). Upper Saddle River: Pearson Education Inc.

Schlesinger, L., Gardenier, A., Jarvis, J., & Sheehan-Cook, J. (2012). Crime scene staging in homicide. *Journal of Police and Criminal Psychology.* Published Online November 16, 2012.

Soderman, H., & O'Connell, J. (1936). *Modern criminal investigation.* New York: Funk & Wagnalls.

Svensson, A., & Wendel, O. (1974). *Techniques of crime scene investigation* (2nd ed.). New York: American Elsevier.

Turvey, B. (2000). Staged crime scenes: a preliminary study of 25 Cases. *Journal of Behavioral Profiling, 1*(3), 1–15.

Case Linkage

6

Wayne Petherick[1] and Claire Ferguson[2]
[1]*Bond University, Gold Coast, QLD, Australia*
[2]*University of New England, Armidale, NSW, Australia*

Key Terms

Behavioral case linkage
Case linkage
Linkage analysis
Modus operandi
Signature
Behavioral consistency
Homology assumption
Computer databases

INTRODUCTION

When investigators are faced with the possibility that two or more crimes are the work of one offender, or group of offenders, they are presented with a number of challenges. The first, and perhaps greatest, challenge is determining which cases are the work of this one person (or group), and which are not linked. When physical evidence exists, such as DNA or fingerprints, it takes time to analyze and interpret it; case backlogs can create a major delay in unequivocally confirming linkage between cases. Where no physical evidence exists or when there is some sense of urgency, investigators and analysts rely on less traditional techniques, such as examining victim and offender behavioral similarities.

The practice of behavioral case linkage has been the subject of some debate in practice (such as when databases should be used and which ones) to the theoretical underpinnings of case linkage (using behavioral consistency and the homology assumption, as will be discussed). Despite this, some research support has been found for combined case linkage techniques, using both statistical research and clinical judgment, if and when deployed correctly. This chapter will first determine what case linkage involves, and the theoretical assumptions of case linkage. It will then discuss the problem of too much reliance on databases, and finally propose an approach to case linkage, with recommended inclusions.

6.1 What is case linkage?

Case linkage, in general, refers to the practice of linking two or more known cases together in such a way as to infer they are the work of the same offender or offenders. As stated, while this may be done with DNA and other physical evidence such as fingerprints, in the absence of these types of evidence, the onus will fall to investigators to rely on a more specific type of linkage, that of *behavioral case linkage, case linkage,* or *linkage analysis*. It is this type of linkage that will be the subject of this chapter.

It should be noted that as used in this chapter, "case linkage" efforts do not include those attempts made by investigators or analysts to link an offender to a victim, an offender to a crime scene, or a victim to a crime scene that are a part of inquiries into a single crime. The interest is only in connecting multiple crimes that are the work of the same offender or group of offenders, and excludes those that are the work of other, unrelated parties.

According to Woodhams and Labuschagne (2012), case linkage is practiced by many police agencies internationally, is based on behavioral similarity or distinctiveness, and may also be called comparative case analysis. The goals and theories on which the practice is based are likely similar regardless of jurisdiction. According to Hazelwood and Warren (2003), linkage analysis involves the examination of three distinct components of a crime: (1) the modus operandi or the "how to"; (2) the *ritual* or fantasy behaviors for a type or series of sexual crimes; and (3) the *signature* or unique combination of behaviors that suggests a series of crimes is the work of the same offender.

Labuschagne (2006) suggests that linkage analyses can have various uses, depending on when they are utilized in the progression of a case from investigation to adjudication. In the initial phases, it can be a standalone report that is used by investigators to determine which, if any, of a series may be the work of one offender, and which new cases should be passed through to the team investigating that crime series. At trial, the report may assist with identification of the crimes that are the work of one offender for the purpose of prosecution, or which are not his/her work as it relates to his/her defense. This may be done so that similar fact evidence can be used against an accused, or refuted, or so that prosecutions may be combined for a number of alleged offenses. It is noted that linkage analysis is particularly useful when no eyewitnesses are forthcoming or where other evidence linking the crimes is absent.

6.2 Theoretical underpinnings of linkage analyses

There is no one theory or technique on which case linkage is based. Instead it relies on a number of theoretical assumptions and propositions that are interrelated. Because behavioral case linkage involves assessing the behaviors, or what was done before, during, and after the crime, much of the theoretical base revolves around the domain of behavior and its emotional meaning to the offender. As such, an assessment of linkage across a number of cases may revolve around uncovering similarities and differences in not only what the offender did, but also why they did it: the motive.

6.2.1 Modus operandi and signature

Law enforcement have traditionally analyzed crimes through assessing an offender's MO or modus operandi (Hazelwood & Warren, 2004) which literally translates to method of operation. MO behaviors are the functional aspects of the offense, and refer to those things that are necessary for the successful completion of the crime (Geberth, 2006; Hazelwood & Warren, 2004). While an offender's MO may remain consistent over time, it may also change as the offender learns, adapts, becomes more confident, and gets more experience (Geberth, 2006). Their MO may also deteriorate as they become more drug dependent, anxious about being caught, or mentally unstable.

Change to an offender's MO may occur in later offenses when they are confronted with a situation similar to a past situation in which actions or behaviors were not successful. For example, if a victim screaming led to the offender being chased off in their first offense, they may threaten or gag victims to prevent them from screaming in later offenses. MO may also change based on environmental events such as passersby or other observers like security guards, or by virtue of victim responses. A confident victim may require more verbal or physical control than a timid one, while an offense committed during the night would likely require a different constellation of behaviors than one committed during the day. For example, the offender may not conceal their identity at night, which affords the cover of darkness. Proximity to the victim and other witnesses may also dictate identity concealment efforts.

Modus operandi behaviors can occur before, during, or after an offense and are aimed at carrying out the offense successfully, as well as obscuring the offender's identity, their connection to the crime scene, their connection to the victim, or the victim's connection to the crime scene. As a result, there are almost limitless potential MO behaviors that can be evidenced during a crime or crime series. Some include:

Before the Crime:
- Carrying out surveillance
- Organizing in advance a means of transport to and from the scene
- Purchasing or securing a weapon to control potential victims
- Identifying the location of local police stations to increase response time to the incident
- Using other forms of planning and/or preparation

During the Crime:
- Using a weapon to maintain control of victims
- Using concealment devices such as a balaclava (ski mask) to hamper visual identification, wearing gloves to prevent fingerprints, or wearing bulky clothing to obscure physical build
- Disguising one's voice to prevent identification
- Using a getaway vehicle to increase the physical distance from the crime after it has been perpetrated
- Selecting a location with poor physical security
- Using a condom to prevent depositing DNA evidence at the scene
- Asking a victim to wash after an assault to destroy physical evidence

After the Crime:
- Disposing of masks, gloves, weapons, and any other items used during the offense that may link the offender to the crime
- Making attempts to secure or establish an alibi for the time of the offense
- Monitoring police radio, the media, or public discourse (such as press statements given by other offices or social media) about the crime and status of the investigation
- Reviewing successful and unsuccessful behaviors for future moderation and modification

Often confused with MO, an offender's signature include those behaviors not necessary to complete the offense, but represent the underlying psychological needs of the offender. While research has shown that many MO behaviors remain relatively constant over time (a few of these studies will be reviewed shortly), MO is perhaps more open to influence than signature, as discussed below. This is because the MO is not fulfilling any psychological need and may thus change by whim and circumstance, while the signature represents the reason for the offense, and is therefore intrinsically linked to motive. In short, while the modus operandi must be studied and accounted for, a change in MO should not unduly sway the analyst's determinations that cases are not linked. That is, of course, unless this change is a quantum shift from other offenses and the absence of an offender link is supported by other differences in the behavior under study.

So while modus operandi refers to that which was necessary to successfully complete the offense, signature refers to that which was not necessary, but is an expression of underlying needs (Geberth, 2006). These are derived from motivations and sexual fantasies (Hazelwood & Warren, 2004) and signature elements are said to be more unique to an offender; as such they may be more useful in case linkage attempts. While this is true, it should be noted that signature components occur with far less frequency than MO behaviors, where any individual crime may be replete with evidence of MO, but devoid of any signature beyond those that may establish general motives.

The reader is reminded that signature may not be present in any given offense or crime series, or may be present in some cases and not others even within the same series. Signature may not be unique to one offender, as the term tends to imply. As such, too much reliance on this feature alone may unduly influence decisions of linkage where no such connections exist. Some cases may be less likely to involve signatures than others, such as burglary where the offense is primarily functional rather than expressive, that is, satisfying material or monetary gain. Consider the following example.

During a residential burglary, a detailed search of the house was undertaken for valuables of varying nature. Not only was money taken, but also a laptop computer, some jewelry, and portable music devices. During a search of the occupant's bedroom, drawers were searched, including those containing lingerie. The lingerie drawers were searched more extensively than others in the room. Of particular interest was that the offender had defecated on the bedroom floor in this room. Two behaviors are important to consider further here.

The first relates to the seemingly more intensive search of the drawer containing underwear. It is possible that this was done based on a belief that this location was more likely to contain valuables, out of mere curiosity, or because the offender was looking for fetish items, even though none were stolen. The second behavior of interest is the defecation. It is possible that this offender is leaving a "calling card," a more traditional understanding of signature (see Douglas & Douglas, 2006), intended to demean and thereby humiliate the victim. It is also possible that this behavior was simply a result of the offender's parasympathetic nervous system increasing digestive activity following times of stress (see Kalat, 2013). On the surface, this behavior does not seem to be necessary for the successful completion of the crime. However, if the offender's need to eliminate waste is overwhelming, such that he/she could not continue his/her search of the home without first defecating, then this may indeed be a modus operandi behavior, and not signature at all. In this case, determining why the drawer was searched and why the offender defecated at the scene would require more information about the situation and the offender, making case linkages based on these behaviors rather tricky. In many cases, determining the functionality of behavior and whether it is an MO or signature can be very difficult indeed.

To further complicate things, even the supposed uniqueness of signature may be a dramatic overstatement, especially when considering its link to motivation. Certain signature behaviors may not be peculiar to individual offenders, but common among groups of offenders especially with similar motivations. For example, two different offenders who tell the victims of sexual offenses that they are beautiful and must receive a lot of attention from males (see Hazelwood (2009) regarding the reassurance-oriented offender) are exhibiting signature behaviors. Telling the victim that they are beautiful is not necessarily required to successfully complete the sexual assault, but these offenders are exhibiting a behavior that may lead the analyst to conclude the offences are linked simply because these two offenders have the same underlying motivational needs (for the victim to think they are a genuinely nice person).

The reader should recall the importance of structured professional judgment, or SPJ, as discussed in Chapter 2 of this work, and again in Chapter 10 on Risk Assessment. This practice relies on the incorporation, or at least the consideration, of base rates of offending and offending styles before offering a conclusion or professional opinion about an event or occurrence. So in the above case, the analyst should consider how often sexual assaults and reassurance-oriented sexual assaults take place in this jurisdiction before commenting on the peculiarity of this signature behavior to one or more offenders operating there. Without making too fine a point, SPJ is critical because it allows the analyst to integrate research into an assessment while also utilizing the value of their personal training and education. This approach is therefore more holistic than an examination based on only their assessment of the specifics of the cases.

The incorporation of a structured approach in case linkage is beneficial in that it prevents the analyst from relying too heavily on heuristics or other guides to judgment; protects them from being swayed by personal opinion as to whether a behavior is frequent or infrequent; and ameliorates the degree to which any single behavior would change an opinion about linkage in the absence of other information. For example, if investigators are suddenly confronted with six rapes that have behavioral similarities in a location where rapes occur infrequently, this may suggest that these offenses are more likely part of a series. Similarly, if a general population of stalking incidents involve reassurance-oriented motives (see Chapter 8) where a small number are retaliatory, this may contribute to an assessment that this sub-set of retaliatory offenses are more likely to be linked. Knowing the base rate at which certain behaviors occur within a population can thus be useful in determining how commonly, or by contrast how uncommonly, behavioral sets occur within the general population.

Case linkages should involve a combination of both statistical information (base rates) and clinical judgment. No matter how apparently unique, no single factor, including motive, signature, or MO, should be the sole determinant of linkage as these may change over time or with different victims. Nonetheless, the similarities or differences present in each of these could make up part of an evidentiary set of information that collectively supports or refutes linkage.

6.2.2 Modus operandi and signature literature

The following review of research provides the reader with some insight into the sensitivity of determinations of modus operandi, including a warning that we cannot consider any facet of MO as a single factor without depth or complexity. The features of each of the studies discussed, such as interactions and main effects, will not be discussed for the sake of the reader without a research background. Interested readers should consult the cited studies where appropriate.

Kaufman, Hilliker, Lathrop, Daleiden, and Rudy (1988) examined factors related to sexual offenders' MO in relation to their age (adolescent versus adult) and relationship to the victim (intrafamilial versus extrafamilial). Using the Modus Operandi Questionnaire (MOQ; and the Adult version, AMOQ) these researchers considered behaviors such as offenders offering bribes or using enticements to acquire victims (buy clothes, use alcohol and drugs, expose to porn, engage in porn, desensitize, give gifts); how threats were used to control victims (helpless, harm others, psychopathy, use weapon); and how victims were kept quiet after the offense (threaten others, threaten/harm victim, give/withdraw benefits).

Findings suggest that offenders report approaches to desensitize their victims to sexual content and contact (this has been found in other research, with the majority of cases using pornography to groom victims; see Langevin, Lang, Wright, Handy, Frenzel, & Black, 1988). The use of pornography to desensitize victims in case linkage, according to the results of this study, would not be helpful in differentiating between either adolescent or adult offenders, nor between interfamilial and extrafamilial offenders. This study found that when there were group differences, adolescents reported using more MO strategies than adults. It is suggested that adolescent offenders may need to try more to acquire victims, while adults, theoretically having age and wisdom on their side, have access to more effective strategies requiring less trial and error. For case linkage, this may indicate that for individual cases within a crime series, those employing a greater number of MO behaviors may indicate a more youthful offender, while those employing less strategies may indicate an older offender.

Further findings indicate that adolescents threatened victims with a variety of weapons more so than adults, and this may be another reflection of the above stated differences in age. Additionally, adults maintained a narrow focus when exposing their victims to pornography, whereas adolescents engaged in a wide range of behavior. These authors later note that "it is possible that some adolescents engage in coercive behaviors as a means of heightening their own arousal, rather than as strictly controlling their victims" (p. 357). This finding is just one example that potentially highlights a problem with the identification and differentiation of MO and signature behaviors. This may be especially true with those behaviors traditionally conceived of as being the province of MO, such as threatening a victim to establish or restore compliance. In this instance, coercion, as but one example, if done to heighten sexual arousal may, in fact, be more related to signature than it is to modus operandi. This even though coercion would superficially appear related to getting compliance, and therefore be necessary to complete the offense. This problem of equivocation of behavior has also been identified elsewhere in this chapter.

Looking specifically at MO in sexual offending, Fazel, Sjöstedt, LÅngstöm, and Grann (2004) studied 1,303 sex offenders released from prison between 1993 and 1997. They analyzed the offender's MO in relation to victim choice, offense nature, and severity, and compared that to prior offenses. Data for the study came from "court reports, prosecutor's plaint, parties account, and court findings" (p. 614).

The findings of this study show that victim choice is highly stable over time, with offending against males, children, family/relative, and stranger victims having Kappa values greater than 0.40 (Kappa values reflect the degree of inter-observer reliability on classification, with 0.40 reflecting fair, bordering on moderate agreement; see Viera & Garret, 2005). The risk of having a sexual offender with a prior offense against a male again choose a male victim in their first reoffense was 180 times higher than it was for those with no prior male victims. The risk of repeating a preference for a child victim was 17 times higher, 27 times higher for family/related victims, and 9 times higher for stranger victims. The lower frequency of reoffending against stranger victims may indicate that victim selection is driven largely by accessibility, and perhaps familiarity. The authors also note that sexual deviances, which are often reflected in features such as victim selection, are relatively stable over time (put another way, sexual preferences are less likely to change than other elements of MO such as the use of disguises and other methods to conceal the crime). As such, this study shows, among other things, that victim selection may be a relatively stable feature for case linkage.

6.2.3 **Behavioral consistency and the homology assumption**

The idea that the same offender will commit his or her crime in a similar fashion across offenses is known as behavioral consistency. It could be argued that this theory is a core principle of case linkage, because without consistency between offense behaviors, there would be no similarities upon which to base analyses. Woodhams and Toye (2007, p. 3) discuss this further:

> *A second hypothesis of offender profiling is the offender (behavioural) consistency hypothesis (Canter, 1995). This hypothesis predicts that an offender will show consistency (or similarity) in their criminal behaviour across their series of crimes. As explained by Mokros and Alison (2002), this hypothesis is necessary for offender profiling to work because "one person has to remain rather consistent in his or her actions if the correspondence of similarity associations holds between a person's characteristics and behaviour" (p. 26).*
>
> *The offender behavioural consistency hypothesis also underlies the practice of case linkage. If offenders were not consistent in their criminal behaviour, it would be impossible to assign crimes to a common offender on the basis of their behavioural similarity.*

Salfati and Bateman (2005) studied the degree to which consistency existed in serial homicide. Using the expressive/instrumentality model, the first three offenses of 23 American serial murderers were examined. In order to determine which characteristics were expressive versus instrumental, 61 crime scene behaviors were analyzed for frequency. By definition, those behaviors of higher frequency are not as discriminating between offenders as those that do not occur as often (because they are common to many offenses). These non-differentiating behaviors were being white (71%), the body being discovered outdoors (65%), sexual assault (61%), and the victim being female (52%). These occurred with such frequency that they are not useful for differentiating between offenders.

Less frequent behaviors include victim was male (48%), victim had an occupation (48%), victim was of medium build (46%), victim was less than 15 years old (35%), and the victim was a prostitute (32%). Planning and delaying detection (both related specifically to MO) were also assessed and included bringing a weapon (44%), hiding the body to prevent discovery (35%), and controlling (binding the body). Behaviors that occurred in 10% to 30% of cases were examined, and all relate to the actual killing: 26% involved using a knife or stabbing weapon, firearms or ligatures were found in 17% of cases, 15% of offenders used their own hands or feet, and a club or bludgeoning was used in 10% of cases. Overtly violent behaviors such as torturing were found in 12% of cases.

The infrequency of torturing, if taken as a potential signature, is a good example of the infrequent nature of signature behaviors even in the crime of serial murder. This relative rarity of torture means that it is a fairly low base rate, and thus, when present, could be an indicator on which to base a linkage. This is notable when compared to the high frequency with which MO behaviors were seen. While it would make intuitive sense to consider torture a signature behavior, it can also serve purely functional purposes, as outlined in ABC News (2005):

> *After two days of deliberations, a jury on Wednesday found 28-year-old Michael Alan Cowie guilty for his role in the kidnap and torture of the teenage tourist.*
>
> *The prolonged attack, involving four co-offenders, occurred inside an abandoned commercial building on Upper Roma Street in Brisbane in June 2003.*
>
> *The victim was bashed, burnt and whipped for his pin number.*

Cowie was also found guilty of receiving stolen property and supplying dangerous drugs.
He pleaded guilty before trial to a charge of stealing.
Cowie's co-accused pleaded guilty for their roles before trial and received sentences ranging from two to eight years in prison.

Since torture occurs in relatively few cases, it may be attractive to use for case linkage regardless of whether it is determined to be an MO, signature, or unknown behavior. However, in order for it to be useful for linkage, we need to also discover how stable or consistent torture is across offenses. That is, if torture is used in one crime in a series, how likely is it to also be used in others? If we determine that it has low consistency, its presence or absence in cases thought to be linked will be less meaningful.

To determine the consistency of the themes across offenses, Salfati and Bateman (2005) classified each crime as either instrumental (31.82%) or expressive (68.18%). After establishing the frequency of each type of homicide, consistency was tested by analyzing how often offenders offended according to the same theme or category across their series. Put another way, does an offender who is expressive in offense one remain expressive in offenses two and three? The results showed that although the expressive/instrumental classification could be used to help understand serial homicide, consistency levels across offenses were not high. This means the general theme of the homicides in a series may change across time, and as such these themes may not be helpful when attempting to illustrate or refute links between cases.

The lack of consistency shown in this analysis may be due to several factors. The results indicate a fairly high rate of classification into either category; however, the incorrect identification of a behavior as instrumental or expressive may skew the potential outcomes with regard to the allocation of cases into a category type. For example, as noted, torture could be expressive (i.e. the result of signature) or it could be instrumental (i.e. the result of MO). Both cases that are instrumental or expressive may therefore contain torture. If we assign the "torture" variable to one or the other, we may end up incorrectly labeling cases as one or the other type. As such, the deeper rationale for the offender engaging in said behavior would also have to be understood to ensure correct classification into one of the two types. Given the source of the data, this may not have been possible.

Further, the lack of support for the hypothesis of behavioral consistency may be due to the source of the data being problematic. The data were pulled from the HITS database, which has been criticized for, among other reasons, an over-reliance on newspapers as a source of information (Petherick & Turvey, 2012, p. 60–61):

One of the authors (Turvey) was retained as an expert in criminal profiling and linkage analysis to examine the Yates case, including the results of Keppel's modus operandi and signature analysis report. After requesting and examining a great deal of discovery material related to the HITS program, the author learned that most agencies in Washington State do not actually submit their cases to HITS—which is why the database has so few cases in it. The HITS budget includes subscriptions to major newspapers around Washington State, and HITS analysts cull these wpublications for cases, which are subsequently entered into HITS in order to fill the database. These and other concerns were expressed in the authors' report, which reads in part (Turvey, 2002):

6. *Unreliable HITS Data*

   The data in the HITS case database are unreliable (and subsequent conclusions drawn from the data are equally unreliable) for the following reasons:

A. *HITS data uncritically rely on information and opinions provided by the requesting agency as reliable (according to the memo from John Turner, Chief Criminal Investigator of HITS to Mary Kay High dated April 30, 2002, p. 2, Q11);*

B. *Many of the HITS form fields involve providing crime reconstruction opinions that may be beyond the ken of a given criminal investigator.*

C. *Many of the HITS form fields involve subjectively derived profiling-oriented, legal, and psychiatric opinions rather than objective facts (motive, psychopathy, victim risk, face covering, symbolic artifacts, offender anger, offender lifestyle). It should also be noted that the HITS Coding manual uses the term "Crazy People" to define the mental/insane category and inaccurately defines psychopathic as someone that commits psychotic offences;*

D. *The HITS database is apparently populated by case information at various levels of verification and reliability;*

E. *The HITS database is populated by an unknown number of unverified cases drawn from media/ newspaper accounts (according to HITS SOP, newspaper descriptions dated 9/5/95 as well as the memo from John Turner, Chief Criminal Investigator of HITS to Mary Kay High dated April 30, 2002, p. 2, Q6);*

7. *Unknown Case Linkage Error Rate*

   The case linkage error rate for HITS or those using HITS results is unknown.

A. *According to job description information provided in relation to HITS, Tamara Matheny (a HITS crime analyst) maintains a monthly log of all positive investigative analysis. I have not seen this log in the discovery material provided.*

B. *According to the memo from John Turner, chief criminal investigator of HITS, to Mary Kay High dated April 30, 2002 (p. 3, Q17), asking for a showing of the error rate of HITS is too vague. This answer seems evasive. It is apparently not known how often probabilistic HITS linkages are right and how often they are wrong.*

8. *Without this information, the reliability of HITS query results must remain in question.*

9. *False Negatives*

   The false negative case linkage rate for the HITS database is unknown. That is to say, it is not known how often HITS results, or the interpretation of HITS results, have unlinked a known offender and their known offense. This is stated in the memo from John Turner, chief criminal investigator of HITS, to Mary Kay High dated April 30, 2002 (p. 3, Q19). Without this information, the reliability of HITS query results must remain questionable.

10. *False Positives*

    The false positive case linkage rate for the HITS database is unknown. That is to say, it is not known how often HITS results, or the interpretation of HITS results, have linked a known offender and an offense known to have been committed by another offender. This is stated in the memo from John Turner, chief criminal investigator of HITS, to Mary Kay High dated April 30, 2002 (p. 3, Q20). Without this information, the reliability of HITS query results must remain questionable.

As a follow-up to Salfati and Bateman's (2005) study, Bateman and Salfati (2007) studied whether single behaviors rather than classifications remained consistent across crimes in a series, and thus whether these behaviors could be used for linkage. The rationale was that the findings of the earlier study may be due to some problem with the categories of instrumental/expressive and the behaviors they included, rather than a problem with consistency. The 2007 study thus sought to examine single behaviors, or potential signatures, in order to determine their consistency across murders in a series. Thirty-five crime scene behaviors from 450 serial murder cases committed by 90 offenders were examined. Frequency and consistency analyses were performed on each behavior to determine those uncommon to serial murders in general, and those consistent across offenses.

The victim's body being hidden, the body being moved, items other than clothing being stolen, along with other crime scene characteristics were consistent across time in this study. This means that offenders who did these did so in a number of crimes in the series. However, these behaviors, and many more that showed consistency, were also high frequency across all cases, meaning that many offenders were likely to carry them out. This high frequency indicates that all of these behaviors, while consistent, are not those that would be appropriate to use as indicators of a linkage since they tend to occur commonly across many serial murders.

On the other hand Bateman and Salfati (2007) found that bringing a crime kit to the scene, destroying evidence, oral sex by the victim, and using a ligature to kill the victim were found to be consistent across crimes in a series, and low in frequency. This means that their presence in a number of cases would be an indicator that these cases may be the work of one offender since they are relatively rare and also quite stable. However, since only four of 35 crime scene behaviors proved to be low frequency and highly consistent, support for behavioral consistency in this analysis was similar to that in the earlier study (Salfati & Bateman, 2005), low. Overall, the results of both analyses show that serial homicide offenders are not, to any necessary degree, carrying out the same crime scene behaviors across offenses. Bateman and Salfati (2007, p. 543) note "if this is true, then investigators need to develop new strategies for trying to link homicides together other than searching for commonalities amongst the homicides in question, using the behaviors delineated in this study."

Along with potential problems with databases, high frequency behaviors, and difficulties with classifying behaviors without knowing why they occurred, behavioral consistency is also hampered by factors both within and outside of the offender's control. Without any of the problems addressed above with measuring consistency, it simply may not exist in a series due to (Petherick, 2014):

- The influence of alcohol and other drugs, especially those that affect the prefrontal cortex, the part of the brain responsible for planning, problem solving, and complex actions.
- Crimes with a high emotional content such as stalking offenses and certain homicides.
- Crimes involving mental illness of those experiencing cyclical emotional states.
- Crimes involving personality disorders.
- Staged crime scenes.
- Different victim types targeted.
- Offender development over time.
- Crimes involving multiple offenders where the number of offenders is unknown.

Because behaviors change based on circumstance, learning, purposeful alteration to confuse investigators, and so on, relying on offender behavior being consistent across time in order to make linkages between cases can be very problematic. The fact that consistency is the major tenet that underpins case linkage speaks to the accuracy of these assessments, and how much weight should thus be given to them when no more objective information is available.

The other theoretical assumption underpinning case linkage theories and applied crime analysis more broadly has been referred to as the *homology assumption*. It argues for concordance between the behavior of two different offenders and their background characteristics (Petherick & Ferguson, 2014). Put simply, the assumption maintains that two different offenders who commit similar crimes will have similar characteristics. If the homology assumption is true, this may foil case linkage efforts as, in reverse, those who have similar backgrounds will also exhibit similar backgrounds. In crimes committed by multiple offenders, profiles of the offenders used to infer these background characteristics may inadvertently suggest that similar crimes are in fact linked, when they are not. The assumption of homology presents a problem for case linkage in that it can potentially incorrectly link two crimes that are unrelated based only on crime scene characteristics (which may or may not be consistent or unique), and their assumed relationship to background characteristics.

Studies on the homology assumption have largely failed to yield results that inspire confidence in its use, making the assumption not only dangerous to case linkage, but also inaccurate. These analyses have been reviewed extensively elsewhere (see Petherick & Ferguson, 2014) and thus will not be canvassed here in detail. Readers should note that assuming homology between crime scene characteristics and offender characteristics has the potential to lead to false positives in case linkage, and thus the assumption should be used with caution, if at all.

6.3 Nomothetic case linkage: the use of databases

Since examining case-specific MO and signature may be difficult in linkage analyses due to issues with discriminating between two high frequency behaviors and a lack of consistency, some seeking to make connections between cases have used databases as an alternative to idiographic methods. Although often helpful for inclusion in structured professional judgments, using information from databases alone to determine similarities between cases can also be problematic, although for different reasons. This, unfortunately, has not stopped many an enthusiastic analyst from continuing unabated. A review of how traditional violent crime databases function is provided to illustrate the danger of using them alone for linkage.

While there are numerous police databases in use throughout the world, far too numerous to list here, among the most common is the Violent Crime Linkage Analysis System (ViCLAS), developed by the Royal Canadian Mounted Police (RCMP) (Martineau & Corey, 2008). It was developed primarily in answer to an equivalent program in the United States called ViCAP, or the Violent Criminal Apprehension Program. The latter system was conceived in the 1980s by Pierce Brooks who saw that as computers were becoming smaller and cheaper, a national database could assist with the investigation and apprehension of unsolved murders that might be the work of serial killers.

Despite the reliance on these databases in most police jurisdictions around the world for intelligence gathering, communicating across jurisdictions, and so on, research is less than encouraging insofar as their utility in case linkage. The HITS database, explained above, is an illustrative example. A lack of utility with these databases may be less an artifact of the databases themselves but more a problem with the fact that they require human input or analysis at various stages. That is, a human is required to accurately input information into the database so that it can later be searched by investigators and other analysts. Therefore, the accuracy and utility of the database will be based on, at least:

1. The accurate identification and recording of information at the crime scene.
2. The accurate conveyance of this information for data entry.
3. Accurate coding of crime-related data.
4. Accurate data entry.
5. Relevant retrieval of information from the database using appropriate search terms.

It should be noted that point 3 relates to the coding of information according to the coding booklet provided with the particular database in use, while point 4 relates to physical human error involved in inputting information (that is, incorrect keystrokes). It would appear that studies on many of these facets of database utility are lacking, with Bennell, Snook, Macdonald, House, and Taylor (2012) noting deficiencies that revolve around three domains: the *reliability assumption* relating to data coding; the *accuracy assumption* related to the accuracy of data coding; and *consistency and distinctiveness assumptions*, where analysts place a high degree of reliance on results for cases that share behavioral similarity.

Of course, all of this is predicated on the database having enough information in it to compile meaningful results. In some jurisdictions, entry of case information into the database is not mandatory, and in others with high crime rates and a small workforce, entry into the database may be delayed, which will directly affect the ability of investigators to link offenses during ongoing serial offender investigations. The importance of this is discussed by Martineau and Corey (2008, p. 52) where they state:

> *Despite the fact that empirical research examining these issues is limited, such issues surrounding data quality are clearly important. For example, the quality of the data contained within ViCLAS will have an impact on crime analysis in terms of the calibre of resulting linkages. Indeed, inconsistent or inaccurate crime reporting can have a number of detrimental effects on violent crime linkage analysis.*

Two key pieces of research have specifically studied the reliability of police reporting while completing a ViCLAS report. In Martineau and Corey (2008), a total of 237 Canadian non-commissioned officers were employed for the study, with 116 completing a report in a homicide and 121 completing a report in a sexual assault. Participants were given a copy of the Field Investigator's Guide, which provides a detailed explanation of sections of the report, and were asked to fill out a ViCLAS crime report. The report itself is a 156-question booklet containing various sections such as Administration, Victim, Offender, and others. Percent agreement was used to assess the degree of inter-rater reliability in the study.

In the homicide scenario, a 79.30% agreement was reached with officers indicating the same response to items in the Crime Report. Percent agreement varied depending on the section of the report. Three out of the nine sections (Administration, Scene, and Offense Information) ranged from 90.53% to 95.08%, with two sections (Victim and Offender) scoring between 89.29% and 89.88%, and the Vehicle and Deceased Victim Information sections scoring between 71.20% and 78.66%. The last two sections of Weapon and Biological Sample had lower levels of agreement from 44.48% to 62.88%. Agreement statistic thresholds employed for the study were 70% for essential, 80% for acceptable, and 90% for high. Further Occurrence Percentage Agreement (OPA) and Non-Occurrence Percentage Agreement (NPA) were calculated to address the issue of inflated agreement. This score essentially reflects the degree to which officers believed an item did or did not appear in a crime description. While the percentage agreement was high, the NPA was only 54.67%, showing only a moderate degree of agreement that certain items did not exist. For the OPA, a smaller percentage (38.43%) agreed that items existed.

In the sexual assault scenario, officers had the same response to items in the Crime Report 87.70% of the time. Three of the seven sections could be coded very reliably (Victim, Offender, and Scene Information), ranging from 94.75% to 95.31%. Three sections were acceptable at 81.74–88.87% (Administration, Offense, and Biological Sample), and Weapon was on the low side of the necessary threshold (69.81%). The NPA was moderate at 68.80% and the OPA was very low at 25.38%. Results indicate an 8% higher agreement in the sexual assault scenario, which the authors attribute to having a living victim who could provide information not necessarily available in the homicide case.

What this study means for case linkage is that not only does the type of crime dictate the reliability of entry of information into the database, but that it will also dictate the amount of information that can be extrapolated from the database with regards to similarity and distinctiveness at a later date. Perhaps more important, this additional information does not necessarily convey accuracy, with studies on eye-witness recall showing mixed results (see Hosch and Cooper (1982), who found no connection between confidence in identification and identification accuracy; and Wells and Olson (2003); and Scheck, Neufeld, and Dwyer (2003) for discussions on DNA exonerations following prosecutions primarily based on eyewitness identification).

The second study, conducted by Snook, Luther, House, Bennell, and Taylor (2012), also looked at inter-rater reliability of the ViCLAS system in Canada. In a much smaller scale study, 10 police officers were given the same case file and asked to complete the information for input into the ViCLAS database. As in Martineau and Corey's (2008) work, the level of occurrence agreement between each officer participant was calculated, with alarming results. The level of agreement across all 106 variables was 30.77%, and agreement ranged from 2.36% to 62.87%, depending on the variable. Administrative variables showed the highest agreement (62.87%), while weapons variables the lowest (2.36%). Eleven of the 106 variables studied reached a level of agreement that was considered "acceptable," and the other 95 variables did not. Not surprisingly, this study highlighted major concerns with the validity of inferences drawn from the ViCLAS database and others of its kind.

Though addressing ViCLAS specifically, these two studies highlight the difficulties that arise when databases are relied on for determining whether a past case is similar to a current case, the base rates of various behaviors, the uniqueness of others, and the linkage of known cases to an offender. Clearly, nomothetic methods of case linkage, when used alone, can be highly problematic. It is recommended that any information drawn from databases be used with the appropriate level of caution. As seen, information contained on these databases, no matter how well organized, is highly susceptible to error and disagreement. This does not necessarily require that we throw the baby out with the bath water, however. Where possible it is recommended that detailed case files be requested and analyzed before making connections, or ruling out connections between cases found on databases and others being investigated. If assessments of linkage are made based on original information, and perhaps just identified based on the use of a database, the potential for errors is greatly reduced.

6.4 Considerations in determining case linkage

There are several suggested approaches to case linkage available in the literature, and the approach proposed herein is a distillation of the elements available from the research and literature as well as "what works" from past cases. As stated, there are no factorial or actuarial approaches to case linkage where the presence of a certain number of similar features, bypassing a threshold, provides a percentage probability that the cases are connected.

Depending on the number of cases and the amount of variables to consider, one of the most useful approaches to determining where connections exist is to write out the case features. If available, a large whiteboard may be suitable for the purpose, though a spreadsheet program can be useful in cases involving either a small or a large number of variables. The coding system will be up to the individual but it should be explained in detail so that others can follow the reasoning used. Depending on the computer skills of the analyst, they may choose to create a database specifically created for linkage in this case. That is, one database where all the possibly linked cases are represented (this is different from the databases previously describe because it is specific to an inquiry rather than a large database of all the violent crimes in the country).

In one serial rape case, investigators created a Microsoft Access database with all of the case particulars including the time of day, day of the week, location types, victim criteria, weapon type, and verbal behavior, among others. Database search results indicated that only seven of the nine offenses in the series were linked to one offender. In a later case in the series, a condom came off after the rape, spilling semen onto the bedsheets. After the sample was entered into the DNA database, a hit identified a serial rapist who had been released from prison shortly before the first offense. Once apprehended, he confessed to seven of the nine offenses—the same seven that the Access database identified as likely linked. Later investigation revealed that of the other two offenses, one was committed by the victim's husband, and the other was a false report. In this example, a database developed specifically for the case proved fruitful.

Hazelwood and Warren (2004, p. 313–314) suggest five interrelated activities involved in case linkage:

1. Gathering the necessary documents: Focuses on the victim's statement, police and medical reports, maps denoting all significant locations including locations and distances between relevant sites, and if a homicide, autopsy and toxicology.
2. Reviewing documentation and identifying significant crime features: Allows the analyst to compare significant aspects of each crime, and become intimately familiar with all offenses in the series.
3. Analyzing the crimes and identifying the MO and/or ritualistic behaviors: The analyst conducts a detailed study of the significant features and identifies those elements that form the MO and those that form the ritual.
4. Determining if the signature exists across the crimes: The analyst attempts to identify whether a unique combination of behaviors exists across crimes.
5. Preparing the opinion: Provided in writing, the opinion must include at least a listing of the materials reviewed, whether the analyst visited the crime site/s, a list of MO and ritual features, and if present, a list of behaviors that comprised the signature.

Woodhams, Hollin, and Bull (2007, adapted from p. 235) provide four steps for determining case linkage:

1. The analyst studies each crime in detail in order to understand the offender's physical and verbal behavior. This information usually comes from the victim, providing they are still alive, and may be in the form of statements made to the police or from an interview transcript.
2. From studying the available material, the analyst determines relevant behavior in each offense. These authors reiterate the MO versus signature debate, stating that research into MO has shown some promise for case linkage.

3. The third stage involves a comparison of behaviors in each offense, noting similarities and differences. The antecedents of each behavior must be noted so that each behavior is interpreted in context.
4. The final stage involves considerations of whether any similarities may have occurred by chance. Each individual behavior must be weighted in terms of the frequency of the similar behaviors in its population of crimes.

Along with keeping these points in mind in terms of procedure, the following should be included in any assessment of case linkage. Some cases will include more information than others, or more information in one domain. Having no physical evidence is a possibility, and in homicides, there may be no living primary witnesses. The appropriate weight should be given to statement evidence and consideration of the frequency of behaviors both in the crime series and in the offense category in the jurisdiction in which the crimes occurred.

6.4.1 Examine the available evidence

For reasons that should be apparent, physical evidence is the most favorable way to link cases. However, when this is absent or the results of testing are not forthcoming, behavioral considerations become the next best information on which to link cases together. Where possible, physical evidence should be used to direct case linkage efforts, and to bolster behavioral links between cases. Physical evidence can also be used to exclude the possibility of linkage, providing it has been collected and interpreted properly.

Where evidence is available, caution must be exercised so that too much weight is not given in offences where evidence is present over those where it is not. Evidence may have been missed, not collected, contaminated, or awaiting laboratory analysis. In short, for linkage purposes evidence should be one of many moderating factors.

Wound patterns should be examined, and these will likely be detailed in an autopsy report should the victim be deceased. If the victim is alive, medical examination will likely have been performed to document the wounds for trial once an offender is apprehended. Bruises, contusions, abrasions, stabs, incised wounds, and any others should be examined. Offenders who engage in repeated patterns of behavior may leave similar wounds. If they are committing similar offenses and using the same weapons over time, it is reasonable to conclude that the weapons will leave similar patterns, though this may not always be the case.

Verbal and written behavior will also inform linkage efforts. Things the offender says to the victim in terms of directions, compliments (both supportive and derogatory), orders, or general interactions should be chronicled. The exact wording is critical because individual nuances may be the difference between an indication of linkage and not. The timing of any verbal behavior is important because there is a difference between those things said to control the victim and increase compliance and those things said without cue or provocation.

When there are written communications, spelling, punctuation, and grammar must be examined. Handwriting differences should be noted, and the medium itself may provide some guidance. Is the paper exactly the same type or form factor? Is it of a similar general nature? The delivery method should be noted, and postmarks may give some idea about the geographic location from which they were sent. In a serial stalking case, letters and pornographic images sent to some of the victims were all enclosed in A4 size (8.5 × 11 in.), windowed, manila envelopes. These envelopes were all identical in type and brand, and formed one part of the linkage efforts. It should be noted that not all victims received materials in this fashion, so the absence of this in some cases should not be taken to suggest a different offender.

If a sexual crime, the nature and sequence of any sexual acts should be identified where possible. Sexual behavior can be linked to verbal behavior and so what happens during verbal interactions should be noted. This may inform the analyst as to motive. Behaviors that simply reflect sexual intercourse should be identified along with those that are punishing or derogatory in nature, as well as any that might be intimate or complimentary to victims.

The criminal's skill in executing certain actions or behaviors may indicate other types of crimes they are or have engaged in. This may also speak to other types of knowledge they have that is acquired through day-to-day activities or employment. Using the same materials such as ropes and other ligatures can not only speak to linkage but could also help identify other characteristics of the offender.

6.4.2 Perform a thorough victimology

The criteria for victim selection are important because this provides insight into fantasy elements. If victim selection is driven by psychological or emotional need, this will likely be more stable over time and will therefore be critical for linkage. Chapter 4 of this volume provides the basic structure for this, and Chapter 13 provides the structure and components of a written report.

Any hobbies, habits, or routines of the victim need to be established, such as sporting or other recreational activities, gymnasium or other club memberships, convenience and grocery stores frequented, schools or universities attended, work, family, and friends. Victimology may be less instructive in cases where victim selection criteria play less of a role.

6.4.3 Crime geography

In some cases, the offender's behavior will be limited to particular regions or geographic locations. When this is the case, the prevalence of the crime type and any individual behaviors must be established where possible. This may provide less guidance in common crimes, but this information is still necessary to gather. If there has been a rise in the prevalence of certain crimes, a commensurate rise in other crimes should also be investigated where criminal versatility may be at play. This will not only help with case linkage but may also be used to provide information about public safety, as may be the case where a sex offender is also committing break-and-enter offenses.

The exact location of the crimes will likely be available, and if there are different types of crime scenes, the location of each should be mapped out. Ideally this information should be placed onto an overhead or topographical map of the area with the different crime types and locations pinpointed. The exact distance between each location should be established and the distance between the crime site(s) and the victim's home and work addresses and other location should be provided.

General data about the area should also be gathered such as population statistics, low and high crime residential areas, and popular tourist locations. Details about how the offender got to and from the crime scene needs to be gathered, and the same information should be compiled on the victim. What was the victim doing in this location? What was the offender doing there? Was this a usual or typical place for the victim to be, and if not, what led to them being there? What type of activities usually occur at the location, and at what times of the day?

Information should also be established as to the degree to which the offender's behavior indicates familiarity with the area. Was the victim moved to a specific location once acquired? If so, does it appear that the offender was familiar with that location? Was there any evidence of pre-surveillance or did it seem that the offender already knew about the location, suggesting local knowledge?

6.4.4 Temporal data

The day of the week, time of the day, the month, and year of each individual offense must be noted. Any patterns or associations in these data must be recognized. This should be mapped to the movements of the victim using the same criteria. If the timing of offenses is part of the offender's plan and not simply opportunistic, this may be critical for case linkage.

Not only should this basic information be obtained, but any local or seasonal activities should also be identified. What type of things occur in this region around the time of the crime? Is it hunting, fishing, or fruit-picking season? Is it a holiday time?

If there is a considerable gap between offending times, the exact number of days, weeks, months, or years should be recorded. If there is a shorter time period between some offenses than others, attempts should be made to determine why this has happened. Do the victims in those cases have different characteristics to those in the rest of the series? Are they perhaps opportunistic versus targeted victims? Should the time period be significant (years as opposed to days or weeks), attempts need to be made to establish why that is the case and how it can still be established that the cases are linked. This will not always be possible but should other temporal or geographic patterns be established (such as the offenses occurring only during fishing season or holiday times), this may indicate that the offender is not local to the area.

6.4.5 Crime data

Local and regional crime data will be critical in determining the prevalence of this crime type in this area, and thereby how typical or unusual this crime is. Crimes that are low in frequency but start suddenly may be more likely to be the work of one offender, while crimes that occur regularly or are high frequency may be more likely to be the work of multiple offenders.

Crime data reflecting the demographics of the area are also critical as this may tell the analyst which of the component behaviors may be common or occur infrequently in that area. For example, the victims being white will mean little to nothing in a community that is 98% white, but the victims being black or Asian may have more meaning in linking the crimes together. The same would apply in a predominately young geographic area where elderly victims are targeted.

Conclusion

In order to link multiple offenses to a single offender or group of offenders, the applied crime analyst should assess a variety of evidentiary sources. Ideally, behaviors that show consistency across offenses and are infrequent should be used to argue linkage. The more such behaviors that are uncovered, the stronger the linkage will be. These behaviors are often those that indicate some underlying psychological need, such as those identified as signature, rather than those which simply facilitate the criminal act in the simplest possible fashion.

Crime analysts are charged with not only determining what is important in each crime, but also what core behaviors are utilized by many offenders in the physical area or in that type of crime. This means they need to use their structured professional judgment to integrate both idiographic and nomothetic information, taken from case files as well as crime statistics and databases. Without the other, either of these assessment styles will be lacking.

Analysts are hampered by the fact that consistency across a criminal series will vary for many reasons, including offender characteristics, victim reactions, and situational changes that may be completely unknown to the analyst. They must also be aware that many seemingly odd or unique behaviors actually happen quite commonly, although perhaps just not in their personal experience. Even more difficulty may be experienced when analysts are unsure of the validity of information taken from databases or crime reports from other jurisdictions or agencies.

Despite all of these hardships, the process of linking cases together is possible, and there is a considerable corpus of knowledge now in this area. Undoubtedly, case linkage is not always achievable; however, when done properly it can provide invaluable assistance to streamline investigations and solve cases. In the end, the degree to which a crime series can be linked is related to the amount and quality of evidence that is available, and the skills, knowledge, and tenacity of the analyst.

REVIEW QUESTIONS

1. Modus operandi behaviors can occur before, during, and after the crime. True or false?

2. The theory that offenders will behave in a relatively similar fashion across their offenses is known as:
   a. The similarity coefficient
   b. The homology assumption
   c. Behavioral linkage
   d. Behavioral consistency
   e. None of the above

3. Research on the homology assumption has largely failed to find empirical support for this theory. True or false?

4. What are three of the factors that may hamper examination of behavioral consistency?

5. What are three of the factors that impact on the accuracy and utility of computer databases for case linkage?

6. Case linkage revolves heavily around consistency and distinctiveness. True or false?

References

ABC News. (2005, February 17). *12 Years for torture of backpacker*. Available from http://www.abc.net.au/news/2005-02-17/12-years-for-torture-of-backpacker/1520256. Accessed 10.01.14.

Bateman, A., & Salfati, G. (2007). An examination of behavioral consistency using individual behaviors or groups of behaviors in serial homicide. *Behavioral Sciences and the Law, 25*, 527–544.

Bennell, C., Snook, B., Macdonald, S., House, J. C., & Taylor, P. J. (2012). Computerised crime linkage systems: a critical review and research agenda. *Criminal Justice and Behavior, 39*, 620–634.

Douglas, J. E., Burgess, A. W., Burgess, A. G., & Ressler, R. K. (2006). *Crimes classification manual: A standard system for investigating and classifying violent crimes* (2nd ed.). San Fancisco: Jossey-Bass.

Fazel, S., Sjöstedt, G. L., Ångstöm, N., & Grann, M. (2004). Risk factors for criminal recidivism in older sexual offenders. *Sex Abuse, 18*(2), 159–167.

Geberth, V. J. (2006). *Practical aspects of homicide investigation: Tactics, procedures, and forensic techniques* (4th ed.). Boca Raton: CRC Press.

Hazelwood, R. R. (2009). Analysing the rape and profiling the offender. In R. R Hazelwood, & A. W. Burgess (Eds.), *Practical aspects of rape investigation* (4th ed.). Boca Raton: CRC Press.

Hazelwood, R. R., & Warren, J. I. (2003). Linkage analysis: modus operandi, ritual, and signature in serial sexual crime. *Aggression and Violent Behaviour, 8*, 587–598.

Hazelwood, R. R., & Warren, J. I. (2004). Linkage analysis: modus operandi, ritual, and signature in serial sexual crime. *Aggression and Violent Behavior, 8*, 587–598.

Hosch, H. M., & Cooper, D. S. (1982). Victimization as a determinant of eyewitness accuracy. *Journal of Applied Psychology, 67*(5), 649–652.

Kalat, J. W. (2013). *Biological Psychology* (11th ed.). Belmont: Wadsworth Cengage.

Kaufman, K. L., Hilliker, D. R., Lathrop, P., Daleiden, E. L., & Rudy, L. (1988). Sexual offenders' modus operandi: a comparison of structured interview and questionnaire approaches. *Journal of Interpersonal Violence, 11*(1), 19–34.

Labuschagne, G. (2006). The use of linkage analysis as evidence in the conviction of the Newcastle serial murderer, South Africa. *Journal of Investigative Psychology and Offender Profiling, 3*, 183–191.

Langevin, R., Lang, R., Wright, P., Handy, L., Frenzel, R., & Black, E. (1988). Pornography and sexual offences. *Sex Abuse, 1*, 335–362.

Martineau, M. M., & Corey, S. (2008). Investigating the reliability of the violent crime linkage analysis system (ViCLAS) crime report. *Jounal of Police and Criminal Psychology, 23*, 51–60.

Petherick, W. A. (2014). Behavioural consistency, the homology assumption, and the problem of induction. In W. A. Petherick (Ed.), *Profiling and serial crime: Theoretical and practical issues* (3rd ed.). Boston: Anderson Publishing.

Petherick, W. A., & Ferguson, C. E. (2014). Behavioural consistency and the homology assumption. In W. A. Petherick (Ed.), *Profiling and serial crime: Theoretical and practical issues* (3rd ed.). Boston: Andersen Publishing.

Petherick, W. A., & Turvey, B. E. (2012). Criminal profiling: Science, logic, and cognition. In B. E. Turvey (Ed.), *Criminal profiling: An introduction to behavioural evidence analysis* (4th ed.). San Diego: Elsevier Science.

Salfati, C. G., & Bateman, A. L. (2005). Serial homicide: an investigation of behavioural consistency. *Journal of Investigative Psychology and Offender Profiling, 2*, 121–144.

Scheck, B., Neufeld, P., & Dwyer, B. (2003). *Actual innocence*. New York: Penguin Putnam.

Snook, B., Luther, K., House, C., Bennell, C., & Taylor, P. (2012). The violent crime linkage analysis system: a test of interrater reliability. *Criminal Justice and Behavior, 39*, 607–619.

Viera, A. J., & Garret, J. M. (2005). Understanding interobserver agreement: the kappa statistic. *Family Medicine, 37*(5), 360–363.

Wells, G. L., & Olson, E. A. (2003). Eye witness testimony. *Annual Review of Psychology, 54*, 277–295.

Woodhams, J., Hollin, C. R., & Bull, R. (2007). The psychology of linking crimes: a review of the evidence. *Legal and Criminological Psychology, 12*, 233–249.

Woodhams, J., & Labuschagne, G. (2012). A test of case linkage principles with solved and unsolved serial rapes. *Journal of Police and Criminal Psychology, 27*, 85–98.

Woodhams, J., & Toye, J. (2007). An empirical test of the assumptions of case linkage and offender profiling with serial commercial robberies. *Psychology Public Policy and Law, 13*, 59–85.

False Reports

7

Wayne Petherick[1] and Claire Ferguson[2]
[1]Bond University, Gold Coast, QLD, Australia
[2]University of New England, Armidale, NSW, Australia

Key Terms

Erotomania
False allegations
False confessions
False reports
Incidence of false reporting
Motivations for false reporting Pathological lying
Prevalence of false reporting Pseudologia fantastica

INTRODUCTION

There is often reluctance on the part of investigators, lawyers, doctors, and other professionals to question the word of those who present as victims of crime. Many feel that to do so is generally undermining the status of victim, and questioning any or all of the victim's claims may be seen to be impolite, politically incorrect, or detrimental to encouraging victims to come forward. The mere mention of potential falsities to advocacy groups may raise their ire and have you cast out as some form of uncaring sympathizer for the darker side of crime. From the point of view of investigators, a confession from the accused saves vast amounts of time and resources in an already stretched-too-thin justice system. A confession can save hundreds, if not thousands, of hours of tedious interviews, investigations, and (in the case of the crime analyst) sorting and sifting through documentary, photographic, and other information in order to give the case some semblance of meaning. However, not all reports made to police are legitimate, nor are all confessions.

Reports of victimization, allegations of the misconduct of others, and false confessions to crimes may go unnoticed at various stages of the justice system. This presents a problem on a number of fronts. For the investigator, countless hours and resources are invested in false reports, diverting attention from real crime victims; the taxpayer bears the cost of a falsely accused person going through the criminal justice system; and the falsely accused person may lose his or her job, family, and liberty. Far from being a small or insignificant problem, research (and indeed, anecdotes from a variety of sources) suggests that false reports are indeed a problem that needs to be addressed.

Because of the significant loss and harm to reputation that may arise from false reporting, it is incumbent on the crime analyst to make every effort to determine the veracity of a complaint from their very first involvement. As such, it is argued herein that any inquiry should begin with an attempt to falsify the stated complaints, whether it is a report of victimization, an allegation that another has committed a crime, or the confession of an accused. This has a number of benefits, including, but not limited to, saving time and resources expended on chasing false leads; preventing the falsely accused from entering the justice system as pseudo-criminals; and preventing false victims from accessing services intended for legitimate victims.

This chapter will examine false reports, which is a general term used to describe any false allegation or confession in which either parts or the whole account are fictitious. The discussion includes incidence and prevalence figures, the psychological backdrop to false reports of all varieties, and investigative considerations.

7.1 **The problem of false reports**

As stated, there are a number of relatively obvious problems that arise from false reporting, including the criminalization of innocent parties and the costs to justice that are ultimately borne by the taxpayer. There are a number of other indirect issues too, such as the stigma that attaches to the accused, especially in crimes involving children (see McMahon, 1999–2000; Sarnoff, 1997). In these cases, allegations made against a parent competing for custody may also weigh in the favor of the complainant parent, denying access or custody, and subsequently infringing on parental rights. False allegations may incur a financial burden on the falsely accused, wherein child maintenance payments will be dictated on the amount of time spent with each parent. In the case of falsely reported sex crimes, "life will be neither easy nor fun for the innocent man convicted of a sex crime and sent to prison" (Anderson, 1997, para 4). The same would be the case for those accused publicly, sometimes even after the allegations have been shown untrue; some may be of the opinion "where there is smoke, there is fire."

For legitimate victims, the identification of even one false report case can cast doubt on any number of other claims; there may be an existing belief that women lie about rape or are somehow complicit, either in their behavior or dress (Taylor, 1987). This may lead to a natural decline in reporting as women fear their involvement will be scrutinized or their demeanor or accusations will be disbelieved. This is clearly an untenable situation in which actual victims are denied of their rights and the due process afforded them under the law.

Where victim support groups are utilized by false reporters, it is imperative that these individuals be identified and excluded because they can be a drain on resources and time and detract from the experience of legitimate victims (Pathé, 2002). Here, they not only squander already limited resources, but often convince others that their situation is worse than that of any other in the group. These legitimate victims may then rally around the falsely reporting victim, providing emotional and other support, which they can scarce afford to surrender given their own actual victimization.

From the standpoint of good scientific inquiry, the false report can be addressed as a research problem: we should never attempt to prove an *alternate hypothesis* (that a victim is experiencing what they claim), but we should attempt to disprove the *null hypothesis* (that what they claim is not in fact happening). Should the null hypothesis be disproven, legitimate advance in the inquiry can be made on the basis that the potential for a false report has been excluded.

7.2 Incidence and prevalence of false reports

Beyond a few studies and some anecdotal reports, little is known about the frequency of false reports. A widely cited, but totally incorrect, figure is that about 2% of sexual assault cases are false. This figure has been the subject of much debate; its basis eludes precise identification, but it appears to have surfaced in a number of locations including Susan Brownmiller's book *Against Our Will* (see Turvey & McGrath, 2009). Regardless of the origin, this figure has been restated recently by Alison Saunders, Chief Crown Prosecutor for London. According to Major (2012), this was quoted in connection to the rape myth that women claim rape when they regret consensual sex or want revenge.

Greer (2000) provides a lengthy and critical evaluation of this 2% figure, examining legal dominance feminism as the lens through which this erroneous figure is promulgated. He notes that the core of the 2% argument rests on the premise that women do not lie about sexual abuse, shifting the legal burden from the woman claiming the abuse to the man in defense of it. Put another way, if women do not lie about being raped, the prevalence of false reports must be low. So, despite the frequency of the 2% false report figure in the literature, it would appear in light of critical examination to be a gross misrepresentation of the actual frequency of false complaints.

The rate at which false reports of all varieties occur will be influenced by the way in which the reports are measured. A finding of not guilty at trial is one potential way; however, given possible legal maneuvering and the tendency for juries to rely on evidence such as DNA and fingerprints (which are not be available in every case), this is perhaps fraught with the most potential error. Other studies have used a panel approach, where a case is judged to be false by the ratings of an expert panel. Sheridan and Blauuw (2004) adopted this approach and found that 11.5% of stalking cases were false. Pathé, Mullen, and Purcell (1999, p. 170) adopted a different threshold, also with stalking cases, where "the basis of the claims being clearly, and repeatedly, at odds with the available objective information." This latter study does not cite a proportion of the total sample, but 12 cases were identified from a sample of over 150 presenting to a forensic clinic. This would mean that 8% of the sample was false.

In the United Kingdom, Her Majesty's Crown Prosecution Service (CPS) and Her Majesty's Inspectorate of Constabulary published a report on false allegations of rape in 2002, which indicated that 11.8% of cases were false reports (CPS, 2002). In New Zealand, Jordan (2004) investigated rape and sexual assault files from police and determined that 48% of the claims were actually false. This included cases that were determined by the evidence to be untrue, as well as those where the complainant admitted their claim was false.[a] According to a study in the United Kingdom, false reporters of sexual assault who were eventually charged with perverting the course of justice or wasting police time are often young, vulnerable females, and many have mental health issues (Levitt & Crown Prosecution Service Equality and Diversity Unit, 2013). This is not to say that men and older individuals were not involved, just that these cases were less frequent. Alcohol and drugs were often a feature of the alleged assault, as well as sometimes the reporting of it to authorities or others.

Other factors involved in the determination of falsity include a delay in reporting, where the presenting victim does not conform to the accepted appearance of a victim of crime: with clothing in order; where their appearance is not otherwise disheveled; where they do not appear upset; or where there is a lack of serious injury (Maclean, 1979, cited in Rumney, 2006). Maclean studied 34 rape cases between

---

[a] Some true victims may eventually claim that their report was false if they believe that this is the only way to stop prosecution of the accused from proceeding.

1969 and 1974 as a police surgeon, finding nearly half of these cases to be false. Methodologically, Rumney (2006) states that this study was problematic because of the way that cases were classified. Specifically, cases in which there was a delay were more likely to be labeled as false.

Eugene Kanin (1994) studied 109 reported rapes over a 9-year period in a small community in the Midwestern United States. Kanin found that, among these cases, a staggering 41% ($n=45$) were deemed to be false. The threshold for falsity in this study was the complainant later recanting their statement and admitting that no rape had occurred. There were also yearly fluctuations, from 27% (3 out of 11) to 70% (7 out of 10). No trends were identified, but the reasons for reporting were listed as revenge seeking, providing an alibi, and attention and sympathy seeking.

Kennedy (2000) conducted a replication of the Kanin study in Michigan City, a pseudonym for a primarily Caucasian city of approximately 100,000 people. Police provided data on all forcible rape complaints from 1988 to 1997, the same years as the Kanin study. A case was deemed false if the complainant admitted during the investigation that they had falsified major allegations. Of the 68 forcible rape complaints, 22 (32.3%) were classified as false, which was in line with the findings of Kanin. Most of the false reports served to provide an alibi for the alleged victim in these cases (68%, $n=15$). This study differed on the second reason for falsely reporting, however, with attention and sympathy seeking in 27% ($n=6$) of the cases, with only 5% ($n=1$) occurring for the purpose of revenge.

There was considerable discourse in the United States during the 1980s with regards to cases of ritual satanic abuse in child care centers. Dubbed the "satanic panic" (see Glassner, 2000), these cases involved allegations of systemic and widespread abuse of children by caregivers at day care centers and preschools. However, these claims are not new and have historically also included witch hunts in various countries. Sjöberg (2002, p.132) reports on one such case from the Swedish parish of Ráttvik, involving a witch panic led by the Reverend Gustaf Elvius, involved "some 550 child witnesses and some 80 women accused of kidnapping children and bringing them to nocturnal satanic feasts". The Royal Commission of Inquiry sentenced three women to death in January 1671 as a result of these inquiries.

Petherick and Jenkins (2014) examined false reports in stalking were examined using cases drawn from the first author's case files. In this study, a total of 10 cases were compared to the Australian literature on stalking to determine whether there were differences between the features of false reports and legitimate stalking incidents. After removing two outliers (both delusional cases spanning many years), it was found that the alleged duration can discriminate between false and legitimate cases, with false complainants reporting a shorter duration of intrusion, as well as reporting intrusion more quickly than legitimate victims (reporting in the latter can lag by 18 months to 2 years). False reporters are more likely to be female (but so too are real victims of stalking) and they are likely to be older than typical stalking victims. Interestingly but unsurprisingly, false reporters tend to be employed at a significantly reduced rate than their real crime counterparts. This suggests that the amount of disposable time one has may be a factor in false reports of stalking.

7.3 False reports

A false report is a general term for any untruthful statement or claim of victimization made to another. For this work, a false report occurs when an individual files a false complaint but makes no specific reference to, or accusation of, a particular person. In fact, they may state outright that they have no idea who the offending party is. This differs from a false allegation, which is a particular type of false report that implicates a particular person, and a false confession, where the reporter accepts responsibility for a (usually) criminal act they have not done. These latter two types will be discussed subsequently.

In false reports, any aspect of the case can be misrepresented, such as the alleged victim's involvement, the location of the event, the type of crime itself, or the circumstances or context of the allegations. In some cases, certain aspects of the report may be accurate or true, but other elements are fabricated—often for reasons unknown to anyone but the claimant. Some illustrative case examples are helpful.

7.3.1 False report of stalking

Mr. M contacted the first author with a report that he was being stalked by an unknown person. His speech was florid, and he frequently changed his tone and the pace of his speech. When asked about the particulars of the case, he went into lurid detail about the alleged stalker's morbid infatuation with him. He claimed that it had been a while since he was employed, and that he did not have many friends. He was also wheelchair-bound.

Mr. M stated that, because it was difficult to get around, he spent most of his time at home. As such, it was stated that this was the location at which most of the intrusion occurred. He claimed that as a result of the stalking he had exceptional security. There was only one door into which he had fitted two deadbolts, and there were two windows: one in the bedroom that did not open, and another in the kitchen that slid open but was fitted not only with a security grille, but also metal bars.

When asked how the stalker got access to the property, he replied "I don't know. He must have a key or something." When asked who else has a key, he states no one does and that he has the only set. He later claimed that the stalker "must be coming in through the kitchen window," moving past this uncertainty when it was later stated "he[b] comes in through the kitchen window at night and watches me sleep." When asked how he could know about the nighttime intrusions if he was asleep while being watched, he responded with a grunt followed by "It's just a feeling I have when I wake up."

A drive-by inspection of Mr. M's apartment revealed that his third-floor unit was on an outermost corner of the building, with only two exposed walls. Both of these walls were, barring the windows above, featureless and composed of standard-sized house bricks. It appeared difficult to scale as there were no features to hold onto or to use to leverage the next hand-hold. That is to say nothing of the bolted-on metal bars and fastened security grill covering the window where the stalker was supposedly entering.

During subsequent phone calls, it became clear that large parts of the story were fabrications, and many of the later claims contradicted earlier statements. When asked about the difficulty anyone would face in climbing the outside of building, he said, "I don't know how he does it, but he does." At this point in the conversation, without any further inquiry or request, Mr M volunteered that the stalker was invisible and could only be seen by him. There was a sudden sharp intake of breath during this telephone call and Mr M hissed into the phone, "He's here now!" When asked how he knew the stalker was present if he was invisible, he replied, "Because he just touched my foot under the table!"

---

[b] Interestingly, it is the first author's experience that when males falsely report being stalked, they report that it is a female doing it.

7.3.2 **False report of sexual assault**

In 2004, the first author was asked by an alleged victim's daughter to consult on a case involving an alleged sexual assault. The daughter was understandably concerned about what had occurred to her mother. A meeting was arranged and Mrs H (the alleged victim) arrived with her husband, who refused to allow her to be interviewed alone, stating concerns over her welfare and her memory of the event.

Both Mr H and Mrs H worked in the service industry at a large tourist resort, and the sexual assault had occurred while Mrs H was cleaning a toilet facility. She allegedly heard someone enter despite the closed sign at the door and had a chemical sprayed in her eyes. The offender then grabbed her forcibly on both breasts from the front and pushed her back into a shower stall, where she states the rape occurred.

Her injuries were not consistent with the claims of how she was handled. A number of bruises in photographs of the injuries to her breasts were consistent with fingerprints. However, these bruises each formed crescent shapes with the opening facing up, meaning the rapist would have had to rotate both hands in an unusual and unnatural position. Additionally, while she claims that all bruises were inflicted at the time of the alleged rape, they were of varying shades, indicating the degeneration of hemoglobin (Dimaio & DiMaio, 2001). The bruises to the breasts and chest were straw-colored, indicating an older contusion, whereas those on the medial wrists and legs were deeper in color, indicating more recent trauma. In short, it was clear these injuries occurred at different times.

At interview, Mr H continually interjected into Mrs H's account and cut short the retelling at various points. Despite the fact that he was not present the day of the alleged attack, he continued to cut Mrs H off. When allowed to explicate, Mrs H would often stop herself and relate the same information using different words in a manner, which suggested that she had coached herself (or had been coached) in the specific details.

Some time later, security footage was examined by investigators, revealing crystal-clear images of the entry and exit points of the location at which the attack was claimed to have occurred. The closed sign is clearly visible, and no one entered or exited the facility during this time. Despite being given the opportunity to recant her allegation, Mrs H stood by her original claim despite a total lack of supporting information (in fact, the evidence that was available directly contradicted her claim).

At last tally, the cost of the investigation was around $30,000AUD, with several hundred thousand dollars in worker's compensation expenditure for time off work, psychological treatment, and medical treatment. Perhaps more disturbingly, the two security guards on duty the day of the alleged offense had their employment terminated on the spot as a result of Mrs H's claims.

7.3.3 **False report of Satanic child abuse**

Ms D had presented to a psychiatrist for evaluation following claims of satanic abuse of her daughter at the hands of her husband and his family. Numerous claims had been made to various government organizations, the state police, and the Australian Federal Police Commissioner. These complaints and others like them, all in writing and kept by Ms D, chronicle a long and extensive history of alleged abuse.

At interview with the psychiatrist, her daughter, KB, disclosed stories of abuse. It is noted in the psychiatric report by Dr W that it was obvious the child had been coached in what to say, and that the words used to describe events were remarkably similar to those used by the mother in separate interviews. The abuse first came to the mother's attention when, according to her account, she was told "That's where Daddy takes me to the woods" while driving on the road near the child's preschool. KB then allegedly pointed to a track where they could drive in, and at the end of the trail they found a number of cola bottles.

In this location was a cave in which there was wax from red and black candles, and a flat rock with some symbols on it just outside the entrance. KB frequently made reference to being chased through the woods, most often by bears. When asked about the bears, KB stated that her father often dressed up like a bear, and that his parents and other members of his family were also there dressed as other animals. Lions, sharks, and crocodiles were also prominent in reports.

Other reports supposedly from KB involve people being caged in wooden or wicker baskets, hung from trees, or suspended over water. The people who did this were also dressed up as sheep, goats, and other farm animals. KB also claims being hung upside down in a cave and a number of events are claimed to have occurred on dates relevant to Satanist beliefs. It is claimed by Ms D during interviews with Dr W that Satan worshipping and ritual abuse were well entrenched in the region where she lived.

After an extensive investigation, not a single claim was established to be true. It was the opinion of Dr W, a psychiatrist with over 30 years of experience in child protection matters, that KB had been extensively coached by her mother, who was subsequently diagnosed with at least three psychological disorders by two different mental health professionals.

7.3.4 False report of theft and arson

In 2008, Mr R claimed that, at approximately 8:30 am, his residence was broken into and personal items, a vehicle, and other valuables were taken before the property was set on fire. After an initial investigation, the insurer denied the claim and Mr R was suing them for a breach of contract over failure to compensate him and his partner for lost property and fire damage. The first author was commissioned to prepare a crime analysis, and the following sections from the report are included in almost their entirety.

7.3.4.1 Conclusions

1. There is some evidence of staging—that is, the deliberate alteration of physical evidence to mislead authorities. This resides in the following:
   a. In bedroom 2, drawers were pulled out but do not seem to be in disarray as though they were gone through. This suggests this was done to add to the visual presentation of the scene as that of a burglary.
   b. Each subsequent drawer was pulled out a little further than the last, not really allowing for the contents of any one drawer to be searched effectively. Again, this is more suggestive of visual presentation.
   c. Bedroom 1 shows a similar stepped arrangement to those of bedroom 2.
2. There is evidence of low criminal skill regarding the setting of the fire. This resides in the following considerations:
   a. There were multiple seats of fire, which were collectively largely ineffective.
   b. The mix of clothing and paper suggests a lack of knowledge about what material will be more combustible or a desire to try everything on hand. It would be necessary to determine the exact nature of the paper type in the basket (i.e. photocopy/printing paper versus glossy magazines) to determine combustibility. It would also be necessary to determine the fiber composition of the clothing used in an attempt to determine its combustibility.
   c. Although not clear, the photos reveal a great deal of soot deposits, suggesting that the fire never got hot enough for flashover (spontaneous combustion). All rooms seemed to suffer

minor damage, except for the master bedroom, where photographs reveal evidence of drop-down from the ceiling. This could perhaps be accounted for by the fuel load, such as composition of bedroom furniture, carpet or flooring type, use of accelerants, etc.

d. One photograph supplied depicts a lounge suite and notes that it is a possible "fourth seat of fire." On the lounge is a book of word games and some other materials, possibly including other papers. It needs to be established if these were glossy magazines or unwaxed paper, as this would speak to their combustibility. In general, these materials are not well laid out in a manner conducive to starting a fire (little to no air flow is afforded). The couch itself is noted as being "leather style" in the case summary, although it would be useful to know the exact nature of the covering materials. Interior moldings and padding are usually extremely flammable, but this would rely on the fire reaching sufficient heat to burn through the outer covering to reach the interior materials or in the fire producing enough heat to create flashover.

3. There is evidence that the alleged offender(s) had knowledge of the movement of the occupants of the house. This resides in the following:

a. There was a considerable amount of time taken at the property, a maximum of 30–35 min, by account. This suggests the alleged offender(s) knew that they would not be interrupted during the act. This is especially pertinent because Ms B was unemployed and usually home.

b. Removal of the plasma television and other cumbersome items would have taken time, including that the plasma television would have to have been dismounted from the wall bracket.

c. In general, the amount of time spent at the scene will increase the risk of discovery, not only of the occupants but of other people, such as neighbors, family, friends, or police, should the entry be reported. This risk would be further elevated by staying behind to light fires.

4. The alleged burglary presents some contradictions in terms of the goods stolen and the general features of the act. This conclusion resides in the following:

a. Although some burglaries occur while the occupants are home (approximately 30%), most occur while the premises are empty. Additionally, most occur by forcing entry of a window or door (39% and 31% respectively), while only about 4% occur through an open window (National Crime Prevention, undated).

b. A survey of burglars in the Australian Capital Territory (ACT) suggests that they are reluctant to target properties where a dog is present (75%), stating that the presence of a dog would be enough to put them off (National Crime Prevention, undated). Additionally, of interest is that a dog was present on the premises, suggesting that the dog was familiar with the entrants. However, it would need to be established if the dog usually barked at the presence of strangers to see if this is relevant to this case.

c. The neighbors were home but did not report seeing or hearing anything unusual.

d. The property taken is, in my opinion, unusual. There was a mixture of large and small items, from a 102-cm plasma television, computer monitor, Nintendo DS, and various children's toys. The exact nature and value of the children's toys needs to be established to determine if they are goods that would usually be targeted or whether they held more intrinsic value.

e. Neighbors report that when the fire was first noticed, the gate at the rear of the property was locked and the dog was left in the backyard. It was later noted that prior to the arrival of the fire brigade, the side gate was open and the dog was roaming the streets. If the side gate can be accessed through the backyard, this suggests whoever took the vehicle left through the side gate, which was closed prior to their arrival, keeping the dog inside. This would have elevated

the risk of the alleged offender(s) being identified and/or apprehended because they would have had to open the gate to facilitate removal of the vehicle. However, there are conflicting accounts of the state of the gates between the fact summary and the police report. It is also mentioned in the police report that the vehicle was taken from the backyard. It needs to be determined which gates were open/closed and which gate the vehicle was driven out of. Despite this, the above point still applies: having to alight from the vehicle to open or close any gate would significantly increase the alleged offender's risk.

f. It is, in my opinion, unusual that the house was set fire to and not the vehicle. Beyond mere vandalism, the only two reasons remaining for setting fire to the house were either to destroy physical evidence that may be used to identify the offenders (DNA, fingerprints, etc.) or for profit. In the former situation, similar physical evidence could be recovered from the vehicle as from the house. If the fire was a precautionary act, it is contradictory that a similar attempt was not made to destroy the vehicle.

In addition to the above considerations, it should be noted that the female occupant of the house and her oldest son were seen loading property into the family utility vehicle some days prior to the alleged burglary and fire. On the day of the fire, a white utility vehicle normally parked in the driveway and a vehicle owned by a family friend were seen by a number of witnesses leaving the address at speed.

7.3.5 **False report of hit-and-run**

The following information is taken from *R v. Corr* (2010). On May 12, 2008, 22-year-old Mr C joined his friend of 3 years and former girlfriend at a shopping center. Mr C's friend (hereto referred to as the mother) was accompanied by her 2-year-old son (hereto referred to as the victim) and her female friend. The victim was being pushed in a stroller as the adults lunched, went to a doctor's appointment, and shopped. While the mother and her friend went into a store, Mr C watched the victim outside. When the mother emerged from the store some 15 minutes later, she could find neither her son nor Mr C. Eventually, she observed Mr C running away from the vicinity, and she followed. When she caught up, Mr C advised that there had been a car accident, and that the 2-year-old victim had been struck by a car which subsequently left the scene. Mr C ran back toward the stores and behind them, with the mother following, where she found the victim laying in a garden, unconscious, with severe injuries to the face and head. The half-folded stroller lay nearby. The mother called paramedics who arrived shortly thereafter, along with police.

Mr C explained to the female friend that he was "doing wheelies" with the stroller when he lost control of it, sending it into the roadway where the child fell out and was hit by a passing car. The roadway was busy at that time, and despite it being broad daylight in a busy shopping center in the afternoon, no witnesses reported seeing a hit and run. Mr C reported a similar story to attending police, noting he ran away from the mother because he could not face her. Noticing that the stroller was undamaged, the officer was skeptical. Mr C also reported that he did not remember the type of car that hit the victim, nor the color.

Mr C was asked to take part in a formal police interview, where he first maintained his original account, then said the stroller had in fact collided with the corner of a building. Two days later, while still in the hospital with severe head injuries, the victim developed a high fever. It was subsequently discovered that

he was suffering from a perforated rectal wall and a linear laceration to his anus. Doctors gave the toddler a colostomy and noted the injuries were caused by the insertion of a sharp object into his anus. Eventually, Mr C admitted to throwing the child on the ground twice and punching him in the face and head twice because he was angry that the child was crying. He said the victim fell onto a sharp piece of wood as he was dropped to the ground, causing the injuries to his anus and rectal wall. Mr C made false reports to the police in the hopes that he would not be discovered as being the source of the child's injuries.

7.4 False allegations

A false allegation is a claim made by one person that they have suffered harm or loss at the hands of another; these differ from false reports in that the accused are often named or specified. Victims of false allegations (i.e. the accused) are often men who are older than the complainant and are known to them, although this is not always the case (Levitt & Crown Prosecution Service Equality and Diversity Unit, 2013). For example, in one case studied by the CPS Equality and Diversity Unit (2002) in the United Kingdom, a young woman made a false allegation of sexual assault to police and randomly selected the alleged offender on Facebook. Levitt and Crown Prosecution Service Equality and Diversity Unit (2013) found that, more often than not (56%), false reporters of sexual abuse who are prosecuted report the offender is known to them. Of these, 21% reported the rapist was an intimate partner, 9% reported it was a family member, and 26% reported it was an acquaintance. This relationship between accusers and accused can have the potential to hamper investigations into the report, especially in alleged sexual assault cases, where issues surrounding consent commonly come into play as opposed to whether or not the sex act took place. The details of these cases are subject to the perceptions of all parties involved, making it very difficult for police to determine what actually transpired and whether a crime was committed.

7.4.1 False allegation of rape

An interesting example involves Leanne Black, who was jailed after making a number of false rape allegations over a number of years. This is reproduced from the UK Daily Mail Online (2013) in its entirety:

> *A woman who made a string of false rape allegations against five men in eight years was behind bars last night.*
>
> *Leanne Black, 32, repeatedly cried rape with bogus sex assault reports to police after rowing or breaking up with her former partners.*
>
> *In one case, Black claimed she had been drugged and raped. In another she told police a boyfriend kidnapped and molested her.*
>
> *A court heard that her innocent partners would have faced up to five years in jail if they had been found guilty of such serious sexual allegations.*
>
> *However, Black was herself jailed for two years, with a judge condemning her actions, telling her that genuine rape victims would be undermined by her lies.*
>
> *The court heard that, in the most recent case in March, her boyfriend Kevin Crowley was held on suspicion of rape after he had called police to report she had thrown plates at him in their flat.*
>
> *David Wooler, prosecuting, said officers arrived at the scene of the domestic argument at the home shared by Black and her boyfriend—and she turned the tables on him.*

Mr Wooler said: 'When she was questioned by police she told them her boyfriend had raped her while she slept at his flat.'

'It was the most recent in a number of repeated false rape allegations against men since 2005.'

Newport Crown Court heard that, in June 2005, Black had made a rape allegation but the case did not proceed.

In July 2006, she accused her then partner of raping her twice and also claimed she had been kidnapped and raped. In 2009, she claimed she had been the victim of a serious sexual assault.

And in 2010, she fabricated a story about being drugged and raped. Then, earlier this year, she made the accusations against Mr Crowley.

But she finally owned up, admitting one count of perverting the course of justice against Mr Crowley.

Judge William Gaskell told Black, of Cwmbran, South Wales, she had made it more difficult for genuine rape victims to be believed. He said: 'Police have to take all allegations of rape very seriously.'

'Women who make false allegations like you undermine the whole system and police investigations.'

'It undermines the public's belief in the truth when allegations are truthfully made.'

Gareth Driscoll, defending, said Black had entered an early guilty plea and made a full admission. She will serve half her sentence before being released on license.

Inspector Rory Waring, of Gwent Police, said the sentence should act as a warning to anyone thinking about making false allegations of rape.

He said: 'As well as causing distress to innocent people accused of this terrible crime, cases like this distract officers from supporting real victims and prosecuting real offenders.'

'Those who have suffered from genuine offences are also undermined.'

Siobhan Blake, Deputy Chief Crown Prosecutor in Wales, said: 'False allegations of rape are extremely uncommon, but where they do occur they are serious offences.'

Such cases will be dealt with robustly and those falsely accused should feel confident that we will prosecute these cases wherever there is sufficient evidence and it is in the public interest to do so.

Earlier this year, the CPS published a report highlighting how rare false allegations of rape and domestic violence are.

We must not allow these cases to undermine our work to support victims of rape and domestic violence.

We want victims to feel able to report the abuse they have suffered and we are working hard to dispel the myths and stereotypes that can be associated with these cases.

One such misplaced belief is that false allegations of rape and domestic violence are widespread. We know that is not the case.

7.4.2 **False allegation of stalking**

Ms W presented to the first author with an allegation of stalking. At the first meeting, she produced a number of handwritten letters, some with ominous warnings of being watched. There were a number of photographs of her new rental property with the note of "Guess who is paying for this" written on one of the images. On another, a pornographic image downloaded from the Internet, was written "Guess who?" Also included was a photograph of the property manager of her new townhouse by the front gate. In one of the letters was information about bypassing the security system by gaining entry off a side street.

Ms W presented as a well-dressed, soft-spoken woman in her mid to late thirties. When asked whether she had any idea who may be sending these letters and images, she was somewhat vague and noncommittal, stating it could possibly be one of four people, who she then proceeded to identify. The first, Mr C, was a businessman who she had an affair with at a Christmas party some 4 years prior. She claims that it was an alcohol-fueled indiscretion that had occurred only once, and both she and Mr C were embarrassed and ashamed of the incident given that he was married. Mrs C, it was claimed, knew nothing of the affair and would therefore have no interest in the pursuit and harassment. This did not discount the possibility that Mrs C did in fact know of the affair but had not told anyone she knew. Mr D was an 80-year-old male and mutual friend of both Ms W and Mr and Mrs C. He was a possible candidate and had also written a confrontational letter to Mr C, claiming that "This came from your backyard." Another possible connection was Mrs T, a realtor who knew of the affair. Mrs T and Ms W had not spoken for many years, ostensibly because of her objection to Ms W's affair with a married man. Regardless of the fact that they had not seen each other in some time, Ms W believed her to be a likely candidate because she would have access to a tenancy database providing Ms W's latest address, apparently unknown to anyone else.

Although her initial appearance was believable, a number of flaws appeared after initial scrutiny. Ms W was asked to write her phone number and other contact details on a piece of paper and claimed she could not as she had carpel tunnel syndrome and could not grasp a pen. This handwriting sample would have been useful for comparison to those photographs that had been written on. Further, despite claiming significant anxiety over the harassment, she refused to contact the local police, despite being given specific contact details of individual officers. Ms W also stopped returning telephone calls, despite leaving many voicemail messages requesting contact.

After some time and some initial concern over the validity of the claims, the first author met with a private investigator who worked on the same case for another party, Mr D. As it turned out, Ms W had been having on ongoing affair with Mr C, which had been discovered by her current boyfriend, Mr D. Mr D, a wealthy retiree, had withdrawn financial support from Ms W after discovering the affair and had demanded that she move out of his multimillion dollar house. Without accommodation, a car, or money, Ms W panicked. Should she reveal the infidelity to Mrs C, Mr C may lose significant holdings in the family business and be left with little in the way of means to support Ms W. Should the affair be revealed by a third party, Ms W believed that this would blunt the effect, that Mrs C would leave her husband, and that he would subsequently have more chance of retaining an interest in the family businesses. This, in her mind, afforded the most opportunity for continuing the lifestyle to which she had become accustomed. Incidentally, Ms W was wanted in two other states for three counts of fraud and two of prostitution.

7.5 False confessions

A false confession is any confession given freely or under coercion where an individual or individuals accept responsibility for a crime they did not commit. Kassin, Drizin, Grisso, Gudjonsson, Leo, and Redlich (2010) note that there are two types of false confessions accepted within the psychological community: voluntary and police induced. Voluntary confessions are those that result from attention-seeking behavior, seeking notoriety, protecting the real offender, or psychopathology (resulting from an inability to distinguish between fact and fiction). Kassin and Wrightsman (1985) identify two types of police-induced confessions: compliant and internalized. Compliant confessions involve an attempt to escape

from custodial pressures, whereas internalized confessions involve manipulations about the suspect's belief about their guilt, such as overwhelming evidence of their involvement. In the latter case, this may be more common in jurisdictions where police are permitted to use coercive interview strategies, such as lying about evidence or facts of the case, as in parts of the United States (Gudjonsson & Pearse, 2011).

7.5.1 False confession to murder

A well-known false confession case revolves around a massacre at a Buddhist temple in Arizona in the United States (unless otherwise stated, all information on the case is taken from Hermann (2011)). On the 9th of August, 1992, nine victims were found laying facedown with their heads together as if spokes on a wheel. The victims suffered head wounds from a 0.22 caliber weapon, as well as shotgun blasts to the torso, arms, and legs, leaving the carpet bloodied. With no leads forthcoming, a break in the case came 1 month later when a mental patient at a local hospital, Mike McGraw, called claiming he knew who the killers were.

Leo Bruce, Mark Nunez, Dante Parker, and Victor Zarate, were named, subsequently arrested, and interrogated for 3 days. Detectives working on the case fed the suspects information about the murders, then took that information back in the form of verbal statements. At one point when being questioned about the vehicle, a suspect got the type wrong and was promptly corrected by the interviewing detective. From that point on, he correctly identified the vehicle allegedly used by the offenders.

Zarate maintained his innocence throughout and was released from custody, while the other suspects buckled under enduring and coercive police interrogation. The remaining three were charged with nine counts of murder, until approximately 1 month later when they started to protest their innocence.

On October 23, detectives on the case received a call that the murder weapon did not belong to any of the men. In an adventitious search on August 21, the weapon had been discovered when Rolando Caratachea and Johnathon Doody were stopped by police. Investigators had learned of the rifle on September 10 but had ceased any further lines of inquiry because of the call from McGraw identifying Bruce and friends. The weapon had sat in a detective's office for weeks.

Under questioning, the two boys admitted to firing the rifle with another friend, Allesandro Garcia. Garcia then admitted to the killings to escape the death penalty. The original "Tucson Four," as they had come to be known, were released on November 22.

7.6 Motivations for false reporting

The motivations for false report are many and varied. Like motivations in other domains, there may be no bright yellow lines that exist between them. A revenge-oriented false reporter can come to appreciate the financial incentives that may accompany victimization, for example. There may also be specificity in the motives, in that a serial false reporter will remain faithful to one reason, or they may generalize, employing different motives for different situations or for different purposes.

The following motives are distilled from the general literature of false reports as well as deriving from working on cases involving false reports (i.e., identifying specific false report motivations in actual claims). Although there may be some exceptions, the following will generally canvas all of the reasons an individual may have in providing a false report.

7.6.1 Delusions and factitious disorder

False reporters may be delusional, which are disorders of thought content involving strong beliefs that are misrepresentations of reality (Barlow & Durand, 2009). False allegations may be founded on delusions of persecution, where the belief is that another has an intent to harm. This and other delusions are discussed at length in the *Diagnostic and Statistical Manual of Mental Disorders, Fifth Edition* (DSM-5, American Psychiatric Association, 2013). Persecutory delusions are the belief that you will be harassed or threatened; grandiose delusions are the belief that one has exceptional abilities, wealth, or fame; erotomanic delusions are those involving the false belief that one is loved by another; nihilistic delusions involve the conviction that a major catastrophe will occur; somatic delusions focus on health and organ functioning.

Several disorders are characterized by delusions, including schizophrenia, paranoid personality disorder, borderline personality disorder, and schizophreniform disorder. Capgras syndrome is a more unusual delusional disorder in which the person believes that someone they know has been replaced by a double (Barlow & Durand, 2009). Erotomania may be especially prevalent in false allegations of stalking, where the complainant believes they are being followed by the alleged stalker, when the opposite is more likely to be true. In false reports involving delusions, the reporter believes the complaint to be true.

7.6.2 Pathological lying

In some cases, the motives for reporting will become apparent almost immediately, such as where the reporter shows excessive interest in victim's compensation or other financial incentives. This may be especially so in cases where outcomes involving money seem more important than the stated trauma and harassment that has led to their reporting in the first instance. In other cases, the motive will only reveal itself after extensive examination and review. In others still, the reason behind the reporting will not become apparent at first review, nor at any time after. Here, pathological lying or *pseudologia fantastica* may be at play, where limited factual material is mixed with colorful fantasy, with the stories not being used for profit and no distinction made between fantasy and reality (Newmark & Kay, 1999). According to Dike (2008, p. 67), pathological lying is characterized by "a long history (maybe lifelong) of frequent and repeated lying for which no apparent psychological motive or external benefit can be discerned."

Although pathological lying can be the result of delusional processes, not all results from psychopathology. We all lie at some point in our lives, but such lies usually serve a purpose (to protect the emotions of ourselves or others, for example); not so with the pathological liar, who will lie even when they do not have to. Dike (2008, p. 67) provides similar comment: "While ordinary lies are goal-directed and are told to obtain external benefit or to avoid punishment, pathological lies appear purposeless."

Pathological lying is not a motive in and of itself, rather a context in which a false report will arise. As stated, it may be difficult to discern between actual motivational elements and the pathological lie, where lying occurs without discernible benefit, or simply for the sake of it. However, it may also be part of a broader pattern of behavior and therefore every attempt to understand the context and the origin of the report must be undertaken. In the case of the serial fraudster, pathological lies may be a learned behavior resulting from past successes, or they may be the result of conditions like psychopathy. This finding too will be instructive.

7.6.3 **Revenge**

In some cases, the individual making the false allegation is "attempting revenge and retaliation, particularly in response to perceived rejection" (Mohandie, Hatcher, & Raymond, 1998, p. 234). The hope is that a pseudo-criminal is created in the accused and that they may be arrested, detained, and even charged for the offense. The range of crimes for which revenge motives come into play are vast, but domestic violence, stalking, and sexual assault most easily come to mind.

Although these accusations can stem from the emotional impact of a broken relationship, they can also be preemptive. Here, the intent is to undermine any claims made in the future about malfeasance. For example, IH and MC were engaged to be married. When IH became despondent over MC's refusal to commit to a wedding date, their relationship broke down. Despite giving the engagement ring back, they continued to see each other for a number of months, going away for weekends, and a number of other relationship-like activities. After numerous unsuccessful attempts were made to secure the return of personal belongings, IH became increasingly frustrated and threatened that she would tell MC's employer that she had completed a number of written works for promotion and university qualifications (there were numerous witnesses to this effect, and all provided sworn statements). By claiming he was a victim of harassment, MC hoped to undermine any such claims by IH in the future. In motivational parlance, these claims would be *anger retaliatory*, as discussed in Chapter 8.

7.6.4 **Need for attention and sympathy**

Included in this category are those who are crying out for help or who suffer some degree of personality or emotional dysfunction. The status of victim can be attractive to some, where others in their environment rally around them, provide support, take on work and other tasks, supervise or mind children, and excuse otherwise unacceptable behavior ("She hasn't been the same since the incident—poor thing"). Additionally, the victimization claim can arise from pending abandonment, where one party takes on the role of victim so as to prevent their partner from leaving a relationship. Here, they prey on any residual feelings their partner may have to help them cope with, respond to, or prevent the alleged rejection.

Munchausen's syndrome and its by-proxy variant are two disorders that may be seen in false reports involving the need for attention or sympathy. Although they belong to the group of disorders known as factitious disorders discussed above, the motive is often to be seen to be a loving parent of a sick child, to have friends and family rally around them during times of illness, or to provide support when they or their children are sick. Munchausen's syndrome was first named in 1951 after German Baron von Munchausen (Goodwin, 1988; Newmark & Kay, 1999), who was a pathological liar well known for his overinflated tales of adventure. Munchausen was regularly seen in military uniforms despite there being no record of any actual service.

When the focus of the disorder is not the self but another, usually a child, the disorder is called Munchausen syndrome by-proxy (MSBP) which was first described by Meadow in 1977 and is considered a rare disorder (Fisher & Mitchell, 1995). MSBP is also known as factitious disorder imposed on another (Kring, Johnson, Davison, & Neale, 2012). In both cases, whether MS or MSBP, treating medical specialists often report bizarre or unusual symptoms in the patient that may defy medical diagnosis or not be representative of any known symptomatology.

7.6.5 **Profit and personal gain**

Some people will concoct lengthy and involved allegations for personal gain or benefit. This may include those who are after something tangible, such as money provided through victim's compensation schemes, or through other benefits such as an upgrade to existing services, an increase in living conditions, an upgrade to private or domestic security, or other similar benefits. The principle aim is to convince another of victimization so as to bring about some gain.

NM was not happy with her life living in a rural area and decided she wanted to move closer to the beach. However, her welfare payment was not sufficient enough to fund her relocation, her two children, and all their possessions the 200 km or so to her preferred destination. She approached the federal government housing authority and requested that they pay for her relocation. She was informed that they were "not a travel service" and further inquired under what conditions they would cover the cost of her removal. One condition would be a direct threat to her safety, and she thus claimed she was being stalked.

7.6.6 **Mistaken belief**

In some instances, the false reporter may be operating under the false belief that they are a victim owing to a misunderstanding or misinterpretation of otherwise innocent cues. This occurs as a result of an honest mistake rather than malice. Most typically, the mistaken belief will arise as the result of heightened sensitivity to intrusion or targeting, with previous victimization presenting a significant risk for this to occur. For example, prior stalking victims may be sensitive to being followed, watched, or approached.

LB reported being stalked by no fewer than three individuals, with one allegedly crawling around on her roof. She claimed during other instances that individuals she did not know, but who were obviously working in concert with her stalkers, were following her as she went about her business. She claimed extensive surveillance at home, and noted that she had heard noises in her roof cavity and that of surrounding properties. She believed that someone was setting up cameras or listening devices. As it turned out, the property she rented was having their television aerial wiring replaced, and further inquiries about being followed revealed that she was overly sensitive to being followed up and down the aisles while food shopping. Delving into her history, LB had been the victim of severe physical and emotional violence at the hands of former partners, all of whom had stalked her at the dissolution of the relationships. She was not delusional and her beliefs, while initially resistant, were challenged and resolved over time. LB voluntary entered into counseling.

7.6.7 **Coercion or compliance**

In a coerced confession, the motive is usually to alleviate the anxiety or pressure from being interviewed about, or accused of, a criminal act. The confessor may accept liability for an act or event that they had no part in, genuinely believing that the truth will come out after an investigation. Should the existence of physical evidence be misrepresented to the falsely accused, they may even sign confessions knowing there is really no physical evidence (because they are innocent) and that their confession will subsequently be invalidated. This cognitive process follows along the lines of "they say they have physical evidence, but I did not do this; therefore, there can be no physical evidence and it does not matter if I sign this confession." Unfortunately, if the physical evidence is a fabrication, there is now a signed confession of criminal liability.

The previous example of the Buddhist temple massacre is one such example of coercion. In many jurisdictions, however, police interview and interrogation rules prohibit the use of threats or inducements in an effort to reduce this type of false admission. However, not all coerced confessions result from outright threats. They could be as simple as "Life will be a lot easier if you just confess" or "The judge will always look favorably on those that make the court's job easier." Thankfully, courts often reject confessions attained through threat of harsher punishment or lighter sentences (Kassin, 1997). Nonetheless, a simple answer to why some people confess guilt is to stop the interrogation or get out of jail (Redlich, Summer, & Hoover, 2009).

Other factors that contribute to false confessions include age and race. Gudkonsson (2003) provides an excellent overview of a large number of studies found a considerable correlation between lower age and an increased frequency of false confessions; however, not all studies showed this trend. Because of their lack of awareness of legal rights, being subordinate to adults, and not having sufficient mental maturity to identify trickery, juveniles are at an increased risk of false confessions.

Another factor that plays a significant role is a cultural issue known as gratuitous concurrence. This refers to the tendency to say "yes" to a question, regardless of the truth or accuracy of the proposition questioned (Eades, 2002). There is also a tendency to establish information through indirect means, therefore avoiding direct questions, and valuing silence in Aboriginal (or other indigenous cultures) conversations (Eades, 2004). In some interviewing contexts, however, a refusal to answer a question may be seen as a tacit admission of culpability. Whatever the reason, an increase in incarceration rates of indigenous populations may mean they are at an elevated risk of coercive tactics leading to false confessions.

7.6.8 Alibi

According to Hazelwood and Burgess (2009, p. 191), when the alleged victim is trying to establish an alibi:

> *In such situations, the pseudo victim alleges a sexual assault to avoid unwanted consequences…or to cover up for inappropriate behaviour. As an example, a married woman involved in a sexual affair might allege that she was raped after returning home extremely dishevelled.*

Alibi-related false reports may also exist when someone violates a curfew from a group home or where they violate conditions of probation or parole. Here, it is hoped that the "terrible thing" that has befallen them outweighs any wrongdoing on their part.

7.7 Managing false reporters

The logical question to arise when presented with a false report is, *"How should this be handled?"* Depending on the circumstances, the amount of effort and resources contributed, and the outcomes of any false report (whether someone has been identified, inconvenienced, arrested, or charged, for example), there may be a tendency to deal harshly with false reporters. However, depending on the circumstances, this may lead to further harm or loss, psychological trauma, and in extreme cases suicide. In revenge cases, this may result in harm to the reporter or accused as a result of further retaliation of conflict. It should also be noted that the resolution of a false report case may be dictated by the receiving agency or individual: if the complaint is made to police, they may have no choice but to prosecute depending on the laws for such an offense; if the complaint is made to a mental health professional, they may apply therapeutic interventions to address the basis for the claim; and if the report is made with a private consultant, their responses may be limited to withdrawing from the case and dealing with any subsequent aftermath.

It is critical to identify any motivations and underlying psychological issues that led to the report in the first instance because this may help dictate, in part, the appropriate response. Where there is evidence of psychopathology and the belief in victimization is rigid and inflexible, confrontation may increase paranoid or delusional thoughts (such as the analyst's involvement in a grand conspiracy) and may lead to an increased risk of suicide. Similarly, when delusion is absent but the reporter is mistakenly identifying the innocent behavior of others as intrusive and harassing, feelings of isolation and helplessness may also increase the chance for self-harm. Directly accusing a retaliatory reporter may increase the chance of anger or aggression toward the party or parties they have accused, and subsequently increase the risk of harm to others.

There is no one size fits all response because the variety of reasons for reporting and the circumstances in which reports are made will vary greatly. As such, it is incumbent on the analyst to have a comprehensive understanding of the situation so that the appropriate response can be designed to match the situation. The following are by no means template responses, and the analyst is cautioned to seek the assistance of appropriately qualified mental health and risk assessment experts as are dictated by the case. However, the following may be a useful guide in some of the situations outlined in this chapter.

> *Delusion and factitious disorders*: Provided that enough information exists to support the basis of a mental disturbance, the individual should be referred for the appropriate treatment or counseling support. Although it might not be prudent to directly challenge the belief, one must be careful not to simply feed the delusion as a reality; this will help neither the reporter nor anyone they may have accused. It may be stated that, "It is obvious this situation is causing you some distress. Might I recommend someone you can talk to for some advice?" This relatively neutral response (acknowledging the feelings without supporting or refuting their basis) may provide a positive impetus to seek help. Where possible and where this does not violate any codes of ethics of confidentiality, a phone call in advance to the referred professional may provide some necessary background, especially where they may not be given access to the material on which the original claims were made. This response may also be suitable for pathological liars.
>
> *Revenge*: Because a revenge-oriented false report is usually based on anger toward others, it is vital that this situation not be inflamed further as there is a possibility of additional retaliatory behaviors. Informing the complainant that, based on the information given by the other party, their report is false and further remediation will be sought would be most unwise. Where risk to the accused may still be a reality, they should be warned and the appropriate action taken. In extreme cases, the police, legal professionals, mental health interventions, and court orders, such as apprehended violence orders or restraining orders, should be sought.
>
> *Need for attention or sympathy*: Because the need for attention or sympathy also has psychological or emotional underpinnings, the analyst must tread carefully because of possible repercussions. Simply informing a former partner that the report is a false attempt to reestablish a failed relationship may bring rebuke and increase the possibility of retaliation. Referral for mental health interventions may again be most prudent because this will help with feelings of isolation and abandonment.

Profit and personal gain: Although the applied analyst may wear many hats, that of financial planner or advisor is not one of them. It is not the responsibility or duty of the analysts to address whatever financial concern or issue the reporting party has in profit or personal gain cases (as above where the attempt was simply to gain more favorable accommodation). However, where a party is falsely accused of a crime and may be subject to criminal or civil penalties, there may be a duty to inform others as to the factitious nature of the complaint. For example, when another is accused of a crime and victim's compensation sought for financial gain, the outcome of the compensation claim may be based on the finding of a guilty verdict in a criminal matter. If a guilty verdict is handed down where the accused has not committed the acts they are accused of, not only will the compensation be ill gotten, but an innocent party may spend time in a custodial environment as well as incur costly legal expenses to fight the claim. In such cases, it may be appropriate to inform legal counsel or investigators of the nature of the complaint. Where the matter is minor and there is no duty to report, these cases may best be left to simply fade away.

Mistaken belief: Those who suffer from mistaken beliefs may be relatively easy to deal with, but again caution must be exercised. If beliefs are rigid, the complainant may feel that they have exhausted all avenues and lose hope that their situation can be rectified. This could lead to depression and suicidal thoughts or actions. In other situations, presenting the evidence in a simple and nonjudgmental way may in fact be reassuring: if the complainant comes to accept that this is not happening, feelings of anxiety and fear may be alleviated. Referral should still be made as dictated by the circumstances to ensure that the underlying basis of the belief is addressed, which will also mitigate the possibility of relapse.

Coercion: Coerced cases may be among the most difficult, especially where a confession has been offered and accepted. These cases may best be referred to legal professionals for challenge in the appropriate arenas (e.g., the court in criminal and civil matters). The analyst's involvement may not end here, though. Their reports may be requested, and it is possible that these would be tendered into evidence to assist with freeing the false confessor from sanction.

Alibi: Unless the analyst has any involvement in the matter that leads to the need to establish an alibi, further involvement or intervention may not be warranted. If there is a duty to report or to warn others, this would override any desire or indications that the analyst should withdraw from the case. Ethical considerations may also play a role here: where there is a clear indication that the report is false and others may be directly implicated and/or adversely affected by the alibi false report, there may be a duty to inform named others.

7.8 Considerations in determining reports

Unlike risk assessment, there are no empirically derived actuarial instruments that may signal to the analyst that any given report is false. Although studies have certainly been conducted, these do not constitute a checklist per se, but rather present a number of risk factors that may tend to indicate that any given report is fallacious. To compound the problem, cultural artifacts that underpin communication may mean that a report appearing false in one location may be entirely legitimate. Add to this individual differences, such as the skill and ability of the pathological liar or psychopath in misrepresenting the truth, and the difficulty in determining falsity becomes apparent.

Although there is no measure of a report that can provide a statistical threshold for truth or lie, we can apply sound logic and reasoning and the principles of social science inquiry to any claim that is made in an effort to establish, within accepted boundaries, whether or not any report is true. Above all, the reader is reminded to treat this as a research problem: make every attempt to *disprove the null hypothesis*.

As with other areas of the analytic endeavor, assessing the case as a possible false report should follow a number of steps. Should the case prove to be legitimate, these stages also mean that much of the groundwork for the actual questions of analysis will be done. This reduces redundancy and allows for more informed analyses. The following should be included as the basis for any assessment, and there will undoubtedly be cases where more information is available in one domain than in others. In some cases, there will be little to no evidence of a certain type, whereas in others the evidence will be rich and detailed. It stands to reason that more is better; therefore, every attempt should be made to source as much material as possible and to confirm that which is offered verbally with other corroborating sources.

7.8.1 Examine the available evidence

The evidence available will be dictated by the type of false report and the reporter's motivation. As such, the type of materials available will run the evidentiary gamut. Because the analyst will possibly be dealing with evidence given to them by the reporter, they are reminded to keep issues of validity, reliability, and sufficiency in mind. For each item, determine whether the evidence represents what it is claimed to, whether the source of the evidence can be trusted, and whether there is enough of it to make a determination.

A red flag is a common term used to describe alarm or attention that is raised when two or more pieces of information are in conflict. These are raised when there are discrepancies between witness or survivor accounts and forensic results (Douglas & Douglas, 2006). If the reconstruction is at odds with the victim's statement, an investigator needs to ask why (McGrath, 2000). One red flag may be when the reporter claims to have received extensive communications, gifts, or other physical materials but cannot produce any of it. It should be noted that this is not proof of a false report, as victims who are traumatized by intrusion and harassment will often discard or dispose of items because of the negative emotions they represent. Other victims are simply careless in documenting or collecting information, or they may not fully understand the importance of evidence in establishing their claims at a later date. Others still may not understand that the evidence marks the start of what is going to be a protracted intrusion or harassment.

As with many false report indicators, context is all important. Clarify, where possible, the context under which evidence was left, discarded, or destroyed. Some digital answering machines will delete messages following a power failure, whereas some older machines may overwrite older messages when the recording medium is full. In short, not all missing evidence establishes falsity.

A detailed comparison of the evidence to all aspects of the claim should be undertaken. Which parts of the claim are substantiated and which are not? For those parts not supported by the evidence, can a legitimate reason for this be established? Compare elements for which there is no direct support (a verbal claim only) to those where there is support but that contradicts the claim (such as when bruises are claimed to be fresh but they are aged), to those where there is direct supporting evidence. Determine whether any themes exist within or between these items. Do they relate to the same things? Different things? Do they indicate something else entirely?

7.8.2 **Perform a thorough victimology**

Where possible, information about the complainant and any other parties should be chronicled as thoroughly as possible. Chapter 4 provides a good outline of those elements that should be included.

Particular attention should be given to the criminal, psychological, and medical history of the complainant (Hazelwood & Burgess, 2009; Mohandie et al., 1998). Investigators are looking for evidence of previous false allegations, deceptive or attention-seeking behavior, or evidence of psychopathology that may indicate delusional or factitious conditions. Being aware of the common indicators of delusional thinking and of the features and diagnostic criteria of many psychological disorders may help identify these as possibilities for further assessment by appropriately qualified individuals.

Criminal history may not be available to those outside of law enforcement; in the absence of other avenues, this may not be available. Occasionally the reporter will be forthcoming; however, should they have reason to conceal a criminal history, this may constitute a serious deficiency in the information. Medical history similarly may not be available for reasons of medical confidentiality. Regardless of the difficulty in obtaining information from these sources, if available, it can provide much needed information about aspects of the reporter's past.

7.8.3 **Involvement of others**

Some victims will remain quiet throughout their victimization, believing it is a personal or private matter that does not involve others or that others have no interest in. Others remain quiet out of fear that this may aggravate the situation and make matters worse. Some may have no such reservations and tell everyone in their immediate environment what is happening to them. Whether a reporter involves others is not as important as how and why they involve them. Third parties can be unwittingly dragged into a situation (in stalking, this may involve "stalking by proxy", see Mullen, Pathé & Purcell, 2008).

False victims can recruit a large network of friends, medical, and mental health professionals for the purpose of establishing their victim status. Although many actual victims may not wish to tell others what is happening or even acknowledge their victim status, some false reporters may be quick to adopt the mantle of victim (e.g., false victims of stalking, see Pathé, et al., 1999; Petherick & Jenkins, 2014).

The reverse may also be true. Some false reporters will go to great lengths to avoid involving others, especially people who may be able to refute any claims that have been made. Otherwise, the false complainant may wish to bolster their claims by presenting any possible number of witnesses who, for one reason or another, will be unavailable or unable to provide further information and therefore cannot be contacted. Examples from the first author's files in cases that later proved false include:

- A claim that various intrusions had been witnessed by the claimant's neighbors, who were apparently unavailable for comment as they were on an around-the-world cruise and would not return for 6 months.
- Despite presenting numerous photographs depicting her new residence and the property manager going about his daily maintenance, none of the photographs depicted the claimant-victim. When asked for the manager's details, she refused, noting that she was new to the property and did not want anyone thinking she was a trouble maker.
- Claiming great anxiety and fear over the intrusion, the victim was put into contact with a police officer known to the author in order to make a formal complaint. It was felt that this prior contact

would alleviate the stress of reporting the intrusion to someone who would otherwise be a stranger. Despite numerous follow-up phone calls, the claimant failed to make contact with the police.

Most typically, victimization involves a limited number of individuals, and the relationship may be as simple as victim-offender. In many interpersonal crimes, involvement is limited to a few individuals and the vast minority involve large networks involving criminal conspiracies (obvious exceptions are organized crime groups). As such, any claim in which dozens or hundreds of people are involved should be viewed with skepticism.

7.8.4 Behavioral considerations

A great deal can be learned from an interview with any claimants when performing an initial triage of the case. General appearance and disposition can be instructive, and there is some value in being able to observe body language, particularly in response to certain lines of inquiry. Because interviewing requires a certain set of skills, where bad questioning can result in bad information, it is suggested that any applied crime analyst seek out specialized training in this area. This should include basic interview skills, how to elicit information without tainting the recall, and preferably some training in detecting deception.

This factor concerns itself not only with the behavior of the alleged victim but also with the behavior of any alleged offenders. Victims may exhibit behaviors that would tend to indicate their claims are false, and it may be claimed that offenders have engaged in various courses of action that may tend to, at the least, undermine claims of action. As many false reporters may not have actual experience with the crime they are reporting, they will report a perception of how this or that should appear. Sources may include media coverage of crime events, a personal understanding of how a particular victim should act, or the behavior of victims portrayed in any number of crime programs and movies. In any or all instances, the behavior may not be reflective of actual victim behavior. Because the applied analyst will have a healthy understanding of their own experience backed up by a theoretical background, many of these inconsistencies may be easy to identify.

Supposed offender behavior that is paradoxical also warrants further attention. An attack that occurs in a busy shopping precinct during peak shopping hours should be viewed with skepticism. Other reported behaviors are also rare in reality. For example, in stalking, electronic surveillance occurs infrequently, but there is evidence it is reported more often in false reports in stalking (60% in Petherick and Jenkins (2014)) compared to 13% of legitimate cases in Sheridan, Davies, and Boon (2001)).

Obstructive behavior should also be noted, such as offering "evidence" that distracts the inquiry, failing to provide evidence that apparently exists, or continual changing of the story or statement. This must be distinguished from a change in story or events following the recall of additional or more accurate information. When asked about details of the event, they may redirect the conversation away from the particulars onto matters involving victim's support services, victim's compensation, or other topics that engender sympathy or are considered safe (Hazelwood & Burgess, 2009).

The match or mismatch between verbal and nonverbal behavior may also be useful in determining whether something is a false report. If the reporter is sitting before you recounting a tragic or threatening situation but is smiling or laughing while doing so, this outward behavior does not match the underlying emotional state. Care should be taken though to avoid confusing actual humor with awkward or uncomfortable laughter, where the aim of the latter is tension reduction as a result of anxiety.

Conclusion

False reports can take on many shapes and forms and exist in different crime types. Some are simple for the crime analyst to identify from the outset, whereas others are incredibly difficult to identify, let alone verify. Although it is known that cases such as those examined here do exist with some commonality, the prevalence and risk factors related to them is often unknown. This is an outcrop of the methodological and political minefield associated with researching them comprehensively.

Several examples of false reports, false allegations, and false confessions were described here to illustrate that motives for reporting falsely can run the gamut. As such, each and every case should be treated as a research problem where the null hypothesis must be tested before proceeding. The crime analyst's first role is to determine whether a crime has indeed occurred and, if so, what law was broken (if any), who is responsible, and why.

As a final note, we challenge you to bear in mind that it is indeed illegal to file a false report with law enforcement, and in many places it is also illegal to cause a false report to be made by another. Therefore, the crime you determine to have been committed in your analysis may be something quite different to what you were originally asked to analyze. The challenge lies in determining when and where these cases occur, then doing your best to assist with rectifying the pervasive problems they cause when overlooked.

REVIEW QUESTIONS

1. It is suggested that the first stage of determining whether a report is legitimate is to treat it like a research problem. True or false?

2. Virtually any crime can be falsely reported. True or false?

3. Which of the following is/are motives in false reports?
   a. Revenge
   b. Attention seeking
   c. Mental disorder
   d. Profit
   e. All of the above

4. *Pseudologia fantastica* is also known as:
   a. Pathological lying
   b. Revenge motivation
   c. Erotomania
   d. Mistaken belief
   e. None of the above

5. Coerced confessions do not occur. True or false?

6. When you identify a false report it is critical that the reporter is taken to task and accused of their falsity. True or false?

References

American Psychiatric Association. (2013). *Diagnostic and statistical manual of mental disorders* (5th ed.). Washington: American Psychiatric Association.

Anderson, J. D. (1997). *How to survive in prison as an innocent man convicted of a sex crime. Institute for Psychological Therapies. 3/4.* Available from http://www.ipt-forensics.com/journal/volume9/j9_3_6.htm. Accessed 04.06.13.

Barlow, D. H., & Durand, M. V. (2009). *Abnormal psychology: an integrated approach* (5th ed.). Belmont: Wadsworth Cengage.

Dike, C. C. (June 2008). Pathological lying: symptom or disease? *Psychiatric Times*, 67–73.

Diomaio, D. J., & DiMaio, V. J. (2001). *Forensic pathology* (2nd ed.). Boca Raton: CRC Press.

Douglas, J. E., & Douglas, L. K. (2006). Modus operandi and the signature aspects of violent crime. In J. E. Douglas, A. W. Burgess, A. G. Burgess, & R. K. Ressler (Eds.), *Crimes classification manual* (2nd ed.). San Francisco: Jossey-Bass.

Eades, D. (2002). Evidence given in unequivocal terms: gaining consent of Aboriginal kids in court. In J. Cotterill (Ed.), *Language in the legal process* (pp. 162–179). Hampshire: Palgrave MacMillan.

Eades, D. (2004). Understanding aboriginal English in the legal system: a critical sociolinguistics approach. *Applied Linguistics*, 25(4), 491–512.

Fisher, G. C., & Mitchell, I. (1995). Is Munchausen's syndrome by proxy really a syndrome? *Archives of Diseases in Childhood*, 72(6), 530–534.

Glassner, B. (2000). *The culture of fear: why Americans are afraid of the wrong things*. New York: Basic Books.

Goodwin, J. (1988). Munchausen's syndrome as a dissociative disorder. *Dissociation*, 1(1), 54–60.

Greer, E. (2000). The truth behind legal dominance feminism's "two percent false rape claim" figure. *Loyola of Los Angeles Law Review*, 33, 947–972.

Gudkonsson, G. (2003). *The psychology of interrogations and confessions: a handbook*. West Sussex: John Wiley and Sons.

Gudjonsson, G., & Pearse, J. (2011). Suspect interviews and false confessions. *Current Directions in Psychological Science*, 20, 33–37.

Hazelwood, R. R., & Burgess, A. W. (2009). False rape allegations. In R. R. Hawelwood, & A. W. Burgess (Eds.), *Practical aspects of rape investigation: a multidisciplinary approach* (4th ed.). Boca Raton: CRC Press.

Her Majesty's Crown Prosecution Service Inspectorate/Her Majesty's Inspectorate of Constabulary. (2002). *A report on the joint inspection into the investigation and prosecution of cases involving allegations of rape*. London: Home Office.

Hermann, W. (August 14, 2011). Valley Buddhist temple massacre has lasting impact. *The Arizona Republic*. Available from http://www.azcentral.com/news/articles/20110814buddhist-temple-murders-west-valley-impact.html. Accessed 03.06.13.

Jordan, J. (2004). Beyond belief? Police, rape and women's credibility. *Criminal Justice: International Journal of Policy and Practice*, 4(1), 29–59.

Kanin, E. J. (1994). False rape allegation. *Archives of Sexual Behavior*, 23(1), 81–92.

Kassin, S. M. (1997). The psychology of confession evidence. *American Psychologist*, 52(1), 221–233.

Kassin, S. M., Drizin, S. A., Grisso, T., Gudjonsson, G. H., Leo, R. A., & Redlich, A. P. (2010). Police induced confessions: risk factors and recommendations. *Law and Human Behavior*, 34, 3–38.

Kassin, S. M., & Wrightsman, L. S. (1985). Confession evidence. In S. M. Kassin, & L. S. Wrightsman (Eds.), *The psychology of evidence in trial procedures* (pp. 67–94). London: Sage Publishing.

Kennedy, D. B. (2000). False allegations of rape revisited: a replication of the Kanin study. *Journal of Security Administration*, 23(1), 41–46.

Kring, A. M., Johnson, S. L., Davison, G. C., & Neale, J. M. (2012). *Abnormal psychology* (12th ed.). New Jersey: John Wiley and Sons.

Levitt, A., & Crown Prosecution Service Equality and Diversity Unit, (2013). Charging perverting the course of justice and wasting police time in cases involving allegedly false rape and domestic violence allegations. *Joint report to the Director of Public Prosecutions.*

Maclean, N. M. (1979). Rape and false accusations of rape. *Police Surgeon, 29.*

Major, A. (March 2, 2012). Rape and sexual assault myths: examining their prevalence in the criminal justice system and greater society. *Slaw: Canada's Online Legal Magazine.* Available from http://www.slaw.ca/2012/03/02/rape-and-sexual-assault-myths/. Accessed 04.06.13.

McGrath, M. (2000). False allegations rape and the criminal profiler. *Journal of Behavioural Profiling, 1*(3).

McMahon, C. (1999–2000). Due process: constitutional rights and the stigma and sexual abuse allegation in child custody proceedings. *Catholic Lawyer, 39*, 153.

Mohandie, C., Hatcher, C., & Raymond, D. (1998). False victimisation syndromes in stalking. In J. R. Meloy (Ed.), *The psychology of stalking: clinical and forensic perspectives*. London: Academic Press.

Mullen, P. E., Pathé, M., & Purcell, R. (2008). *Stalkers and their victims* (2nd ed.). Cambridge: Cambridge University Press.

Newmark, A. N., & Kay, J. (1999). Pseudologia fantastica and factitious disorder: review of the literature and a case report. *Comprehensive Psychiatry, 40*(2), 89–95.

Pathé, M. (2002). *Surviving stalking*. Oxford: Oxford University Press.

Pathé, M., Mullen, P. E., & Purcell, R. (1999). Stalking: false claims of victimization. *British Journal of Psychiatry, 174*, 170–172.

Petherick, W. A., & Jenkins, A. (2014). False reports of stalking: motivations and investigative considerations. *Journal of Threat Assessment and Management, 1*(1).

Redlich, A. D., Summer, A., & Hoover, S. (2010). Self-reported confessions and false guilty pleas among offenders with mental illness. *Law and Human Behavior, 34*(1), 79–90.

Sarnoff, S. K. (1997). Assessing the cost of false allegations of child abuse: A perspective. *Institute for Psychological Therapies, 3/4*. Available from http://www.ipt-forensics.com/journal/volume9/j9_3_2.htm. Accessed 04.06.13.

Sheridan, L. P., & Blauuw, E. (2004). Characteristics of false stalking reports. *Criminal Justice and Behavior, 31*(1), 55–72.

Sheridan, L., Davies, G., & Boon, J. (2001). The course and nature of stalking victimisation: a victim perspective. *The Howard Journal of Criminal Justice, 40*, 215–234.

Sjöberg, R. L. (2002). False claims of victimization: a historical illustration of a contemporary problem. *Nordic Journal of Psychiatry, 136*, 132–136.

R v. Corr; ex parte A-G (Qld) [2010] QCA 40.

Rumney, P. N. S. (2006). False allegations of rape. *The Cambridge Law Journal, 65*(1), 128–158.

Taylor, J. (1987). Rape and women's credibility: problems of recantations and false accusations echoed in the case of Cathleen Crowell Webb and Gary Dotson. *Harvard Women's Journal, 10*, 59–116.

Turvey, B. E., & McGrath, M. (2009). False allegations of crime. In B. E. Turvey, & W. A. Petherick (Eds.), *Forensic victimology: Examining violent crime victims in investigative and legal contexts*. San Diego: Elsevier Science.

UK Daily Mail Online. (2013). *Woman is finally jailed after FIVE false rape allegations against her ex-boyfriends in eight years*. Available from http://www.dailymail.co.uk/news/article-2358759/Leanne-Black-finally-jailed-FIVE-false-rape-allegations-ex-boyfriends-years.html. Accessed 22.12.13.

CHAPTER

Motivations

8

Wayne Petherick

Bond University, Gold Coast, QLD, Australia

Key Terms

Groth typology
Behavior motivation typology
Stalking typology
Theories
Explanations
Planning
Intent
Pathways model
Reassurance oriented
Assertive
Retaliatory
Profit
Gang
Pervasive anger
Sadism

INTRODUCTION

During any inquiry, there are a number of questions that must be answered, revolving around what are commonly known as the six investigative questions: who, what, when, where, how, and why. The purpose of any crime analysis is to be able to respond to these general questions in a specific way, such as what happened and in what order, who was involved, where this happened, and when exactly the offense occurred. Questions will also arise as to why the offense occurred—this being the province of motive.

This chapter will provide a detailed examination of motive and will examine a number of motivational typologies for the crime of rape, in addition to a number developed in other areas that do not revolve exclusively around motive but include it as one axis among many. The chapter will conclude with some considerations necessary in the determination of motive.

Although important for the crime analyst, this chapter will not delve into the role of psychopathology and the ways in which this impacts motivations, such as the delusional content of the paranoid schizophrenic or the delusions of the borderline personality. This issue is sizable enough to warrant a work in its own right, although this chapter in addition to the chapter on Psychopathology and Criminal Behavior (see Chapter 11) will provide more than adequate coverage of the issue.

8.1 Motivations

Although clear definitions exist, there is still a great deal of confusion regarding the nature and types of motive and what exactly constitutes motive in an offense. Some will boldly identify a behavior and claim it was the motive for the offense, as was the case in *The Queen v. Hillier*. Hillier was accused of and tried for the murder of Ana Louise Hardwick on October 2, 2002. In 2004, he was found guilty of the murder and sentenced to 18 years in prison, with a nonparole period of 13 years (see R v. Hillier, 2007; Steven Wayne Hillier v. The Queen, 2005; Steven Wayne Hillier v. The Queen, 2008; Steven Wayne Hillier v. The Queen, 2010). Mr. Steve Longford, a "behavioral analyst" and former Queensland Police Officer, was called by the prosecutor to comment on Hillier's dangerousness at the bail hearing.

Longford based his appraisal of risk on a video of an adult male and a child, which appeared to be an interaction between Hillier and his son; Longford had not actually met or interviewed Hillier. Longford stated during the hearing that "the best indicator of future behavior is past behavior." (Because Hillier had murdered his wife, it could be assumed he posed a future risk. Note that, at this point, Hillier had only been accused of murder, not tried or found guilty). A number of other similar assumptions are present throughout (that Hillier was in fact guilty before any such legal determination had been made). However, the most notable error is the confusion of motive and behavior:

> *HIS HONOUR: Just one thing, Mr Longford. You said that you thought the threat to the children was based upon the egocentric nature of the applicant. What was the basis for your assumption— your opinion that he was an egocentric person?—Okay. If I can go back to the process I used initially to derive the characteristics from the crime scene, that gave me a look at the type of person who would not only engage in the murder but then take the extreme steps in order to try and cover up that murder. It gave me insight into the type of focus and direction. **When I understood the motive behind it, the motive for this particular offense was murder. There can be no mistake in that.** When I see that and I understand the extremes in behavior that are manifest by someone who murders, that is a very, very selfish act. On top of that then I saw a lot of indicative inputs in terms of, for example, if I can use one, a security video tape where I think it's Daniel who's controlled from outside the house after I think it's Ms Harmer leaves the establishment, and that control is physical and the way that he controls his neck and brings his head to his face is an overt characteristic of someone who wishes to have control continually, needs that control and that comes out of selfishness.*

Murder is a legal classification for the behavior in a homicide (the killing of one person by another) where there is intent. It is not a motive at all. Should the murder be carried out in a fit or revenge fueled by rage, then the motive would be retaliatory; should the death be the result of the need to cover up another crime, the motive would be crime concealment. The point being that, with regards to motive, even those who should know better often do not.

8.1.1 Definitions and importance

The term motive is used throughout the criminological literature in a number of areas, but it is rarely defined. In *Motives for Rape*, Felson and Krohn (1990) present a sociosexual model and a punishment model of rape. Although the term "motive" is used regularly throughout, no definition is offered for the term or the context in which it is used. Interestingly, at several places throughout the article, the authors use the term goal as a substitute for motive. The goal of an offense is perhaps more related to intent or the end result or plan.

Purcell, Powell, and Mullen (2005) conducted a study was conducted on clients who stalk psychologists, including prevalence, methods, and motivations, but the authors do not provide a definition for motive; rather, they explain the reasons for the stalking. This includes resentment (42%) resulting from an unfavorable outcome or termination of the therapeutic relationship; infatuation (19%) as a result of attraction to the therapist, often from clients with boundary issues; a variety of motives were noted in 17% of cases, including boredom, loneliness, testing tolerance, and the interference with relatives undergoing counseling. Interestingly, 22% of psychologists gave no motive or noted that no motive could be discerned. The motives were not linked to a particular gender of the psychologist or stalker or the client's mental status. All of the forensic psychologists stalked were pursued by resentful clients (perhaps because of the often high-stakes nature of forensic assessments), and 42% of clinical psychologists were stalked by the infatuated.

In *Sex Offenders of the Elderly*, Burgess, Commons, Safarik, Looper, and Ross (2007) studied 77 convicted sex offenders for the purpose of classifying their motives and the severity of their offenses. Using the Massachusetts Treatment Center, Revision 3 typology (MTC:R3, discussed in detail later in this chapter), these authors also fail to define motive, but classify the offenses on the basis of the four primary sexual motivations: Opportunistic, Pervasive Anger, Sexual, and Vindictive (and the subsequent subtypes). Knight (1999, p. 310) conducted a validation of the same typology, noting that "motivational themes emerged from this juxtaposition of types", but motive is not defined anywhere in this work. Nor did Knight, Warren, Reboussin and Soley (1998) in *Predicting Rapist Types from Crime-Scene Variables*, or McDevitt, Levin and Bennett (2002, p. 305) despite noting of an earlier publication that "offenders could be grouped into three major categories according to the motivation of the offenders involved".

Motive is often used, but rarely defined.

The definition of motive provided by Rockelein will be adopted for this chapter. This definition and origin of the term are as follows (Rockelein, 2006, p. 406):

> The term motivation *comes from the same Latin stem "mot-" (meaning "move") as does the term* emotion. *The term* motive *applies to any internal force that activates and gives direction to behaviour. Other related terms emphasise different aspects of motivation. For example,* need *stresses the aspect of lack or want;* drive *emphasises the impelling and energising aspect; and* incentive *focuses on the goals of motivation…In general,* motivation theories *deal with the reasons that behaviours occur and refer to the internal states of the organism as well as the external goals.*

On the importance of motive and its relevance to court proceedings, Leonard (2001, pp. 439–440) stated:

> *Motive affects behaviour. Thus, although "motive" is not an essential element of any charge, claim, or defence, evidence that a person has a particular motive can be relevant to an ultimate fact in both civil and criminal cases. The variety of circumstances in which motive might be relevant is endless, thus any effort to catalog the possibilities would fail. The principle, however, is basic and simple: When motive is relevant, evidence tending to show its existence is usually admissible, subject to exclusion if the risk of unfair prejudice is too great.*

Leonard touched upon some critical issues with regards to motive, and these will be expanded upon briefly.

The first is that motive affects behavior. This is a simple yet critical consideration in crime analysis, as there may be no direct window into motivational inference, such as interviewing an offender or through direct observation of behavior as it occurs. As the analyst often deals exclusively with documentation after an event (statements, photographs, videos, autopsy reports, and others), establishing motive may be a matter of studying this information for indications as to what emotional state led to the offense. This is to say nothing of the potential for interviewed criminals to provide self-serving responses to garner favor or to excuse the offense. This is providing they have any insight at all.

The second is that motive is not an essential element of any charge, claim, or defense. One possible exception to this would be hate (or bias) crimes, wherein it is an offense to commit a crime against another on the basis of race, religion, ethnicity, disability, sexual orientation, and others (the Anti-defamation League (2011) has an excellent online resource outlining the various hate crimes legislation across the United States).

The third, related to the second, is that if relevant to the matter, motive may be admissible. This being the case, it is essential that motive is established during the inquiry to ensure the full circumstances of the offense are established, but also because the analyst's findings may be called upon as part of the evidentiary considerations. This may be in the form of a written report or as testimonial evidence regarding opinion.

Osterburg and Ward (2010) provided a brief commentary on the difference between general and specific motives, as well as other important considerations. They noted that (p. 120):

> *From the standpoint of motive, crime may be divided into two classes. In the first class, crimes such as robbery, rape, and burglary may have a universal motive which is- in and of itself- of little value in furthering the investigation. Those in the second class, such as homicide, arson, and assault, are more likely to have a particularised motive; when one is discovered, the connection between victim and criminal may be deduced (the high clearance rate for homicide is based in part on this logic).*
>
> *Investigative experience is helpful in ferreting out the motive for a crime. In some cases, motive may be learned through adroit interviewing. In others, however, it is implied- when it can be determined who might benefit from committing the crime. Occasionally, a victim is able to suggest the names of suspects and their motives. When several individuals have motive, their number can be pared down by ascertaining who had the opportunity of time and place, and who among them had the enterprise. This sifting process allows investigators to channel their efforts into those aspects of the inquiry most likely to produce evidence of the offender's involvement.*
>
> *There is another important reason for establishing motive- even when it is not helpful in suggesting possible suspects. Though not an element of any crime, a jury is more likely to be convinced of a defendant's guilt if a motive for committing the crime can be shown.*

The determination of motive is more than a simple academic exercise. Motive refers to *why*; therefore, for investigative purposes, establishing *why* can often lead directly to questions relating to *who*: who presented most risk to this victim; who is within the suspect pool; who should be interviewed or reinterviewed with a view to providing new direction? But this is not the limit of its utility, with motive speaking to many other issues: whether the crime has aggravating circumstances that should elevate the charge or sentence; predictions of risk where these may be elevated for revenge or sadistic offenders and reduced for a reassurance oriented offender; treatment options; crime prevention efforts targeted at determined (motivated) offenders; and premises liability where deferability and foreseeability relate specifically to offender motive.

8.1.2 **Motives, theories, and explanations**

Motives can operate at the broad level of criminal activity, or represent the general or typical motivational type involved in specific crime categories. For example, with specific exceptions, the motive for robbery and burglary offenses will be some form of financial gain. However, a particular theft intended to deprive an individual of a specific item may be better described as retaliatory or revenge in nature. The applied crime analyst is less likely to be concerned with motives at this broad level, unless this involves an attempt to understand the array of possible motives with a view to narrowing these down to only those evident in the current case.

Before delving further into the motivational types, it is necessary to discuss another common source of confusion in this area. That of the difference between motives, theories, and explanations.

Theories represent an attempt to make sense of the world, or explain some aspect of it (Brown, Esbense, & Geis, 2010). Ideally, they provide of-this-world explanations and can be tested against the known body of evidence with a view to providing an understanding of how something works. In criminology, theories of crime are many and varied, including strain, classical, and positivist theory, among a host of others. Strain theory suggests that crime is a result of frustration and anger among lower socioeconomic classes, where similar values are shared but legitimate avenues for acquisition of goods are closed (Siegel, 2011). This theory works for some cases and not others, and by definition would be more suited to crimes involving profit or materialism, such as theft and burglary. Classical theory sees individuals as calculating the costs and benefits of certain behaviors (Andrews & Bonta, 2010). This theory posits that offenders are less driven by base or evolutionary drives and more by rational decision making. Positivism, by contrast, argues that criminals are largely biological throwbacks to earlier stages of evolution, or *atavists* (Brown et al., 2010).

Although these are widely accepted theories, from a motivational perspective they are often less than useful. Under strain theory, because of the dissonance created by differential access to goods and wealth, individuals turn to crime in an attempt to alleviate the "strain." However, does the strain result from the desire to look good amongst one's friends because of a low self-esteem (reassurance oriented, discussed later) or is it owing to a long-standing and pervasive dislike of the upper class and the resulting desire to deprive them of their hard earned goods (revenge or retaliatory, in the language of motivations)? Similarly, how does classical theory adequately inform attempts to understand what truly motivates someone? Any given rapist may make a rational choice to commit an offense, but what is the real underlying motive? An assertive offense is that in which the offender is expressing dominance and control over a victim as a result of low self-esteem, while a sadist receives sexual satisfaction from the pain and suffering of their victims. Both have made a rational choice, yet both have executed their offenses in different ways and for different underlying reasons. As such, theories may be of less use in understanding the core aspects of motivation in a given case.

Explanations may form a broader part of our understanding of criminal behavior, but not necessarily. Explanations are likely to be casual and therefore nonscientific, and thus may be more representative of popular conceptions of crime and criminality. The reader will undoubtedly be familiar with such claims as "young people are just bad," "drugs are the problem," or "the problem with society is teen pregnancy." These claims tend to be sweeping generalizations relating to a problem, and as such may be heavily reliant on bias or stereotyping. Explanations are often typically not open to testing—at least, not any testing that would yield valid results. It is undoubtedly true that some young people are bad (as are some adults), but are these individuals inherently bad, the result of some flawed genetics? Are they bad because they have made bad decisions? Or because of inhibited opportunities? Could it be a result of all three?

Explanations generally come in one of three forms: spiritual, scientific, and popular. Spiritual explanations explain crime as a result of "other-worldy" forces such as demonic possession or classify criminal behavior in terms of the degree to which it is an affront to God. Consider the Westboro Baptist Church, who are outspoken about a variety of alleged "crimes" against God (The Post and Courier, 2013):

Members of the anti-gay, anti-Semitic Westboro Baptist Church have threatened to protest tonight outside of the College of Charleston, where an activist against hate crime is expected to speak. College of Charleston officials said the school will take the appropriate safety precautions and extra officers will be there.

"But this is a public event so we don't anticipate any problems," said Mike Robertson, the college's spokesman.

Judy Shepard is scheduled to speak at 7 p.m. at the Stern Student Center Ballroom about the loss of her son, Matthew.

It's been 15 years since Matthew Shepard, who was gay, was abducted and beaten to death in Laramie, Wyo.

His mother and family then formed the Matthew Shepherd Foundation. Their work at raising awareness helped expand the federal hate crimes laws to include crimes against people based on their sexual orientation.

This is not Westboro's first visit to Charleston. The group picketed outside of Joint Base Charleston and James Island Charter High School in 2010.

The group is monitored as a hate group by the Anti-Defamation League and the Southern Poverty Law Center.

Although these are among the oldest (and were initially the only) attempts to explain criminality (Samaha, 2005), they are perhaps less popular in modern society. Despite a decline in usage, this does not mean that such popular attempts to understand crime and its causes are not relied on at all. A spiritual explanation today may be that a rising crime rate is the result of less people attending church (despite the fact that certain types of crimes are in decline). In times past, crimes were classified according to the degree to which they represented sin, and early inquisitions revolved around witch hunting and demonic possession. This attempt to understand crime is not open to testing because the premise of the malfeasance is faith. Put another way, explanations of this type cannot be scientific nor subjected to the principles of scientific inquiry.

The other two types of explanation are naturalistic (Samaha, 2005) and refer to of this world phenomena in an attempt to explain crime. The first is scientific, which is a casual explanation grounded in some way on past, current, or accepted scientific understanding. Some examples include assertions that the physical presentation of the skull reveals underlying brain structure based on phrenology, and assertions that criminals have smaller brains, based on the work of Cesare Lombroso—a finding later disproven by Goring. These tend to be distortions of scientific reality, misrepresenting the scientific aspect underpinning the particular explanation, or taking one small part of a complex whole and assuming it represents the entirety of the argument.

The second naturalistic explanation is popular, and refers to commonly accepted or stated reasons for crime and antisocial behavior. These may be the result of media influence and the way that certain crimes are reported, such as the degree to which the crime portrayed plays to public fears, or the repetition of crime stories. These may lead to perceptions that crime is more prevalent than it is or that factitious risks are in fact possible or probable. These are based on community perceptions of the causes of crime, such as roving bands of miscreant youth, the representation of certain ethnicities in crime statistics, or that drug abuse among teenagers is out of control (for an excellent discussion of the misplaced nature of fear and panic, see Glassner's (1999) *A Culture of Fear*.

8.1.3 Motive, planning, and intent

Planning refers to preparations made to execute a course of action. These preparations may include surveilling a target, choosing a crime site or location, becoming familiar with security routines, and acquiring weapons, ligatures, other tools, or disguises, among others. Evidence of planning may be found in the preparatory crime scene if this location is known or has been established; in addition, it can be inferred from other behaviors or may be openly stated to other parties. When an individual communicates to a third party an intent to do harm, this is referred to as leakage (see Meloy & O'Toole, 2011) and is commonly employed in the areas of threat and risk assessment. Leakage may be especially useful for the purpose of risk imminence. According to O'Toole (2000, p. 14), with specific reference to school shootings:

> *A student intentionally or unintentionally reveals clues to feelings, thoughts, fantasies, attitudes, or intentions that may signal an impending violent act. These clues could take the form of subtle threats, boasts, innuendos, predictions, or ultimatums. Clues could be spoken or conveyed in stories, diaries, essays, poems, letters, songs, drawings, doodles, tattoos, or videos.*

Evidence of leakage can be used to understand what planning was involved. It can also speak to intent when statements made prior to the event can be mapped directly to behaviors executed during the event.

To better understand the degree of planning involved, the analyst needs to establish what the offender already had, what they acquired prior to the event, what they brought with them, and what they used from the crime scene or victim. Prior preparation of the crime scene will help understand motive, such as placing a blanket on the ground before taking the victim to the location for sexual assault in order to ensure comfort or leaving pretorn sections of duct tape affixed along with instruments of torture at the crime scene.

Intent is the specific aim or purpose the individual has. It may include the offender's state of mind leading up to, or at the time of, the offense. Intent is critical in criminal law and a necessary component of charging an individual with any given number of crimes. In homicide, being able to establish motive will be the difference between murder and the lesser offense of manslaughter. According to de Jong (2011, p. 1):

> *The concept of intent constitutes one of the central concepts in substantive criminal law. Before the court can convict a person of a certain intent crime, it must have been declared proven that the offender committed an unlawful act under consideration* intentionally, *that is: knowingly and willingly.*

Although intent may be apparent from the act itself, there may also be instances when interpretation and explanation are necessary. Once the evidence has been fully examined, the applied analyst will be well placed to provide interpretations of not only motive, but planning and intent also.

8.1.4 The evolution of motive

Any given person is a culmination of various biological and life influences (Petherick & Sinnamon, 2014). This includes any number of genetic predispositions that represent vulnerabilities or risks, which are mediated and moderated through socialization, interactions with our family and friends, and other environmental impacts. Added to this would be any number or type of traumas, whether psychological or physical, such as relationship problems, work hardships, motor vehicle accidents, or illnesses. Trauma may compound existing problems or create new ones, such as social withdrawal and anxiety.

Before moving onto the motivational typologies, this section will present the pathways model to motivation of Petherick and Sinnamon (2014). Unless otherwise stated, the following information will be drawn from this source. In this model, emotional development precedes self-esteem formation, which is the basis for the development of personality. In some cases where trauma is present, personality aberrations may surface as personality disorder. Despite the movement from one stage to another, the model as presented is not rigid and inflexible; each stage is connected to others via feedback loops.

Emotion is a relatively elusive concept, and is the subject of much debate in personality psychology. According to Rockelein (2006, p. 187) "the term *emotion* derives from the Latin *emovere*, meaning to excite, to move, to agitate, or to stir up." It is for this reason that we refer to emotional connections as moving or stirring. Emotions are the result of complex interactions between subjective and objective factors, including neural and hormonal system and cognitive processes (Kleinginna & Kleinginna, 1981).

Our emotional experiences start from an early age. In the hyperplastic brain of the child, they set up a number of predispositions; some of these will be positive and some negative. The amygdala, the brain's fear and threat center, and the frontal cortex, where complex decisions and reasoning occur, usually maintain a harmonious relationship (Arden, 2010). In some cases, however, trauma can create a strong and enduring effect whereby the amygdala overrides rational cognitive processes; as a result, a person fears (an emotion) situations simply because they *may* represent a threat. This is a protective process gone awry. A child who encounters a potentially fearful situation (such as their first roller coaster ride) may come to associate this with fun or pleasant feelings if the parent demonstrates happy or positive emotions such as excitement, thereby positively reinforcing the act. A parent who reacts negatively may reinforce to the child that this event is potentially dangerous and is to be feared. In short, emotional states start early in life, are reinforced continually, and are powerful predictors of behavior.

Self-esteem development follows, with both implicit and explicit links between emotion and self-esteem. If we consider a very basic definition of emotion as *how we feel*, then it becomes a necessary component for the very basic definition of self-esteem, which is *how we feel about ourselves*. Once some foundational emotional experiences have formed, we begin to develop an understanding of our place and position within society and relative to others, known as self schemas. There is a bidirectional relationship where emotional experiences shape our self-esteem; once formed, our self-esteem becomes a determinant of our emotional bias. Although self-esteem is a purely cognitive component of being, it can be expressed behaviorally for interpretation by the applied crime analyst. These behavioral expressions include "attention seeking, sexual promiscuity, and addictive or destructive behaviors" (Petherick & Sinnamon, 2014, p. 397).

Emotions and self-esteem then shape our personality, which could be described as a coping mechanism for our environment. This is a complex and multifaceted aspect of the self, and it develops from biological factors moderated by emotional states and cognitive representations of ourselves. According to Millon, Grossman, Millon, Meagher, and Ramnath (2004, p. 2):

> *The word personality is derived from the Latin term persona, originally representing the theatrical mask used by ancient dramatic players. As a mask assumed by an actor, persona suggests a pretence of appearance, that is, the possession of traits other than those that actually characterise the individual behind the mask. Over time, the term persona lost its connotation of pretence and illusion and began to represent not the mask, but the person's observable or explicit features.*

Cohen (2008, p. 479) provides the following on factors that interact and shape the personality:

> *Over time, the basic positive and negative emotions are gradually replaced by emotional schemas in which cognitive frames, appraisals, and attributions develop out of the individual's emotional experience and replace the basic emotions as predominant motivators. These "motivators" may be seen as temperament in early childhood, in a period in which it may be normative that biological differences may have a dominant influence. With increasing age, the schemas combine these emotional states in relatively common/correlated patterns but more-or-less uniquely across individuals, based on genetic and experiential combinations. Thus, personality differences develop from combinations of individual genetic-based differences in the relative strength of these emotions and life experiences that shape the nature of the individual's schema regarding self, others, and the world they live in.*

Emotion and self-esteem are directly implicated in the development of extreme personality features such as narcissism. From about 8 years of age, developmental processes of self reflection and abstract reasoning allow for self-esteem formation; from this age, children become motivated to create and maintain a favorable self-view (Thomaes, Bushman, de Castro, & Stegge, 2009). In some individuals, excessive and undeserved praise may lead some to believe their own press; others who are never praised may seek affirmation from their environment. In either instance, these individuals may be especially prone to insult or slight (American Psychiatric Association, 2013, p. 671):

> *Vulnerability in self-esteem makes individuals with narcissistic personality disorder very sensitive to "injury" from criticism or defeat. Although they may not show it outwardly, criticism may haunt these individuals and may leave them feeling humiliated, degraded, hollow, and empty. They may react with disdain, rage, or defiant counterattack. Such experiences often lead to social withdrawal or an appearance of humility that may mask and protect the grandiosity. Interpersonal relations are typically impaired because of problems derived from entitlement, the need for admiration, and the relative disregard for the sensitivities of others. Though overweening ambition and confidence may lead to high achievement, performance may be disrupted because of intolerance of criticism or defeat.*

When personality characteristics develop in such a way that they become extreme representations of the self and affect the individual's interactions with others, they may become personality disorders. According to the American Psychiatric Association (2013, p. 647):

> *Personality traits are enduring patterns of perceiving, relating to, and thinking about the environment and oneself that are exhibited in a wide range of social and personal contexts. Only when personality traits are inflexible and maladaptive and cause significant functional impairment or subjective distress do they constitute personality disorders. The essential feature of a personality disorder is an enduring pattern of inner experience and behavior that deviates markedly from the expectations of the individual's culture.*

At each and every stage of this model, motivations develop to enhance or suppress aspects of the life experience, which will be more or less achievable through the degree to which introspection and insight play a role in emotional ruminations. Introspection is the degree to which one can identify, understand, and appreciate their emotional state (Petherick & Sinnamon, 2014, p. 396); this is the process through which change is achieved and dysfunctional aspects of personality abated. For example, feelings of abandonment may lead to a sad state (an emotion) in children, causing feelings that they are not loved or wanted and lowering their self-worth (self-esteem). From this, they become introverted and lack self confidence (a feature of personality). As an adult, they cling to relationships, fear abandonment, have an unstable self-image, and repeatedly contemplate suicide (all characteristics of borderline personality disorder).

Their motivation would be characterized as reassurance-oriented victim behavior, as discussed in Petherick and Sinnamon (2014, p. 415):

> *Reassurance-Oriented victims have low self-esteem and attempt to restore this by establishing relationships and engaging in behaviours that are intended to restore their self worth...They tend to feel inadequate and may perform poorly in social interactions. These individuals have sometimes been victimised, sometimes repeatedly, because they feel that is their lot in life, or that they are somehow deserving of emotional or physical abuse. By extension, their need for companionship and to feel a sense of worth through being in relationships may lead them to place more emphasis on their partner's needs, neglecting their own as a result.*

The above illustrates the complex interplay between a variety of factors that collectively form our personality, and in some cases lead to aberrations known as personality disorder. Emotions, self-esteem, and personality all work to form the core reason that underpins our behavior: motive. It should be noted that the present discourse relates primarily to criminal and antisocial behaviors, but the same general principles apply to noncriminal behaviors. For example, individuals who impose their will on others in the workplace or assert their dominance in a relationship would be identified as "power assertive" in motivational parlance.

8.2 The Groth typologies

In 1977, Groth, Burgess, and Holmstrom published *Rape: Power, Anger, and Sexuality*. The data for this paper was a random sample of 133 convicted rapists referred for treatment to the Massachusetts Treatment Center (MTC) for the Diagnosis and Treatment of Sexually Dangerous Persons. A further 146 victims presented to the Boston City Hospital with the complaint "I've been raped," with 92 adult rape victims from that sample being used in this study. None of the women studied in the victim sample were victimized by those in the offender sample. This study resulted in one of the first typologies of rape, with four categories: power rape (with both assertive and reassurance subtypes) and anger rape (with both excitation and retaliatory subtypes).

The authors state that, according to their clinical observations, all rape involves three dimensions: power, anger, and sexuality. The specific type of rape will be dictated by the degree to which these three components are present and how they interact within, and are expressed by, an individual: "We have found that either power or anger dominates and that rape, rather than being primarily an expression of sexual desire, is, in fact, the use of sexuality to express issues of power and anger" (p. 1240). Although this publication presents four types along two primary axes, it should be noted that *Men Who Rape*, published by Groth in 1979, presented three types: power rape, anger rape, and sadistic rape. No explanation is provided in either work for the difference, although one possible explanation is the amount of time it takes to publish a book compared to a journal article.

It is the latter work which will be discussed first. According to Groth (1979, p. 13):

In every act of rape, both aggression and sexuality are involved, but it is clear that sexuality becomes the means of expressing the aggressive needs and feelings that operate in the offender and underlie his assault. Three basic patterns or rape can be distinguished in this regard: (1) the anger rape *in which sexuality becomes a hostile act; (2) the* power rape, *in which sexuality becomes an expression of conquest; and (3) the* sadistic rape, *in which anger and power somehow become eroticised.*

Rape is complex and multi determined. It serves a number of psychological aims and purposes. Whatever other needs and factors operate in the commission of such an offence, however, we have found the components of anger, power, and sexuality always present and prominent. Moreover, in our experience, we find that rape, rather than being primarily an expression of sexual desire, is, in fact, the use of sexuality to express these issues of power and anger. Rape, then, is a pseudosexual act, a pattern of sexual behaviour that is concerned much more with status, hostility, control, and dominance than with sensual pleasure or sexual satisfaction. It is sexual behaviour in the primary service of non-sexual needs.

In anger rape, sexuality is a means through which anger or rage is discharged. The assault involves physical violence, with more used than necessary to simply overpower the victim and achieve sexual penetration. "This offender attacks his victim, grabbing her, striking her, knocking her to the ground, beating her, tearing her clothes, and raping her" (p. 14). The aim of this offender is to hurt, humiliate, and debase the victim. The offender may use profanity and may force the victim to perform degrading or humiliating sexual acts. These offenders may not anticipate committing the offense, but they may acknowledge that something was going to happen. The acts tend to be short in duration.

In power rape, the offender desires not to hurt the victim but to possess them physically. Sexuality is a mechanism through which underlying feelings of inadequacy are addressed and the goal is sexual conquest, using only the force necessary to complete the offense: "The intent of the offender usually is to achieve sexual intercourse…and to accomplish this, he resorts to whatever force he finds necessary to overcome his victim's resistance and to render her helpless" (p. 26). Due to feelings of inadequacy, he is not reassured by his performance nor the victim's response to it, and more victims will be sought out. Victims will be the same general age as the offender and will be chosen based on availability, accessibility, and vulnerability. These offenders tend to converse with their victims and will ask about sexual performance to reassure themselves.

The third type is sadistic rape, which is discussed elsewhere in this chapter as a subtype of anger. Here, sexuality and aggression are fused into sadism, where anger and power become eroticized. Assaults of this nature tend to involve bondage and torture. The intentional mistreatment of the victim is gratifying to the offender. The sadistic offender may target individuals they perceive as promiscuous or prostitutes, and victims are usually strangers who share similar features: "They are symbols of something he wants to punish or destroy. The assault is deliberate, calculated, and preplanned. The offender takes precautions against discovery, such as wearing a disguise or blindfolding his victim. The victim is stalked, abducted, abused" (p. 45). These offenses are characterized by sexual satisfaction from a victim's pain and suffering, and arousal is a function of aggression.

Groth, Burgess, and Holmstrom (1977) presented four types that reflect largely the same psychological and behavioral considerations as that of Groth (1979). This typology is more refined, with the distinction made on two main axes, Power and Anger, with two subtypes each: Assertive and Reassurance, and Retaliatory and Excitation. In Power-Assertive rape, the rape is an expression of virility and mastery. The rapist feels entitled to the sex, and entitlement is another name for this classification. The offense is a reflection of inadequacy. In Power-Reassurance rape, the rapist is resolving doubts about his sexual adequacy and masculinity. He wishes to place a woman in a position where she cannot refuse him. In Anger-Retaliation rapes, the offense is an expression of hostility and rage toward the victims. The underlying motive is revenge. The final type, Anger-Excitation rapists, find pleasure and thrill in victim suffering. Aggression is eroticized and the offender is a sadist.

8.3 The Massachusetts Treatment Center typology

The original version of the MTC was based on the same four elements of the typology presented by Groth et al. (1977) and from the dimensions of Cohen and colleagues (Gannon, Collie, Ward, & Thakker, 2008; Prentky, Cohen, & Seghorn, 1985). The MTC typology is currently in its third edition, the MTC:R3 (see Knight, 1999; Knight et al., 1998). The typology was developed for use in a treatment and not investigative setting, although the axes on which the types are divided are motivational in nature. Four motivations are represented, some with subtypes: Opportunistic, Pervasive Anger, Sexual Gratification, and Vindictiveness.

For the Opportunistic type, there are both High Social Competence (Type 1) and Low Social Competence (Type 2). The sexual assault is an impulsive and predatory act dictated by situational and contextual factors. Type 1 offenders are higher in social competence and the impulsivity is manifested in adulthood, whereas Type 2 are lower in social competence and manifest in adolescence (Knight et al., 1998). For both types, adult unsocial behavior is high, with low expressive aggression evidence in the offenses, with the offenses also being poorly planned (Knight, 1993).

Type 3 is Pervasive Anger, with both adult unsocial and juvenile unsocial behavior being high, and high expressive aggression and low offense planning (Knight, 1993). The main motivation is generalized anger that runs throughout every aspect of the offender's life. Their unmanageable aggression is equally likely to be directed at men as women (Knight et al., 1998). "These offenders express this anger and aggression in their sexual assaults, and they cause their victims high levels of physical injury" (Knight, 1993, p. 312).

Type 4 and Type 5 are both sadistic types, with the former reflecting overt sadism and the latter muted sadism. Both types are considered to be primarily sexual in their motivation. For the overt sadist, the aggressive sexual fantasies are actualized and are expressed in violent acts (Knight, 1999). For the muted type, the fantasies exist but are not acted upon. As such, the muted sadist may be common in a clinical setting, where they present because they find the violent fantasies disturbing, but they may be rare in crime analysis work. One exception may be where their violent fantasies are used to support or determine some other aspect of their behavior or support an assertion of a tendency or inclination toward violence, as may be the case with risk assessment. It would be safe to say also that not all muted sadists will remain muted.

Also under the rubric of sexual motivations are the nonsadistic types (Types 6 and 7), including low social competence and high social competence types, respectively. These are characterized by dominance needs and/or acute feelings of inadequacy (Knight, 1999). The nonsadistic types are subdivided on their level of social competence (Darcangelo, 1996).

The last type is based on the motive of vindictiveness, and also includes both low social competence (Type 8) and high social competence (Type 9). Their primary motivation is rage and anger directed toward women. Unlike the pervasively angry type who display generalized anger, this type shows little evidence of anger toward men Darcangelo (1996) and is "exclusively misogynistic" (Knight, 1999, p. 312).

Empirical data have highlighted potential problems with parts of this typology, including the sadistic and nonsadistic types and the vindictive types (Gannon et al., 2008). Other studies have shown support for the taxonomy (McCabe & Wauchope, 2005); however, it should be stated that this latter research included in their sample men who were only accused of rape. As stated in the article by McCabe and Wauchope (2005) and as outlined in the chapter on false reports regarding their prevalence, the inclusion of this sample in a study to validate the typology could be problematic (see Chapter 7).

8.4 The Hazelwood adaptation of Groth and colleagues

The first significant adaptation to the original Groth typology was that of Hazelwood in *Practical Aspects of Rape Investigation*, now in its fourth edition (Hazelwood & Burgess, 2009). In this work dedicated solely to rape, Hazelwood (2009) proposes that the first objective is to determine whether the offender intended the assault to be selfish or unselfish. In this context, "pseudo-unselfish behavior by the rapist evidences a belief that his exhibited concern for the victim's welfare will win her over and his hope that she will come to believe he is not a bad person" (p. 98). By contrast, selfish behavior indicates that "he does not want the victim to be involved in any way except as an object for his use" (p. 101). This initial classification may not be entirely useful because of the fact that all of the following types exhibit selfish behavior bar one. The following information is taken from Hazelwood (2009) unless otherwise stated.

In reassurance-oriented rape, the offender is driven by a relational component of complex and ritual fantasy, and he feels the victim is someone special. There is no intent to punish or degrade the victim, and this type is also the least likely to injure his victim. This offender proves himself to himself, attempting to reassure himself of his masculinity. In a way, the rapist is trying to prove to the victim that he is a "nice guy" and that the offense is somehow mutual or consensual. The victims are usually pre-selected, and many will be targeted in advance. This way, if the offense is thwarted for any reason, he can move on to the next victim.

This rapist will target victims of the same general age. These rapists shy away from forcing entry as this will shatter the illusion that the victim is somehow consenting. Following the offense, they will apologize to the victim. They may take souvenirs and will sometimes attempt to recontact the victim. In one case of this author, the offender had access to the victim's properties prior to the offense during his employment as a handyman. While performing maintenance, he would steal spare keys to the property, later returning and letting himself in, then attempting rape. Some of the victims fought back, at which point he would leave the residence and later attempt to assault another victim in the same general vicinity.

The *Assertive Offender* also has low self-esteem; this is bolstered by expressing mastery and control over the victims. While much of the reassurance-oriented offender's behavior and cognition is internally directed (they will often be self-deprecating and much of the offense behavior is fantasy related), the assertive offender typically expresses low self-worth externally toward the victim. They have no desire for the victim to perceive them in a positive light, and they may use violence in the execution of their offense.

This offender is low to moderately impulsive and uses rape as an expression of virility and dominance, and the victim is simply an object. The victim's clothing may be torn or ripped, and they will subject the victim to multiple assaults. In an assertive case from the author's files, an East Asian male sexually assaulted a number of women simply because they were women. It is important to note here that he was not punishing the victims for being female, as this would be more appropriately described by the following type, *Anger Retaliatory*. He assaulted them because he harbored a cultural belief that women were subservient to men, and that they were essentially there for his pleasure.

In revenge or *Anger Retaliatory* assaults, the offender acts impulsively; these are less common but more violent. Women are hated and the offense is a form of punishment, and an attempt to get even with women over real or imagined wrongs. As a result, the victim will be intentionally brutalized. The attack is an emotional outburst and may occur whenever the offender feels angry. These may be cyclical; following the offense, the anger subsides until something happens that triggers another assault. The victims may be the same age range or a little older, and may symbolize someone in the rapist's life, such as a past or current girlfriend. Victims are opportunistic, as are weapons.

The least common but most brutal is the *Anger Excitation offender*. This is the sadistic type. The offender has a complex fantasy, is highly ritual, and may employ complex mechanisms, tools, and scripting in the offense. The rapist is aroused by pain and suffering, which are the defining criteria of the sadist. Every aspect of this offense is preplanned, except that of victim selection. Victims in this offense are typically strangers, as it is easier to torture and humiliate a stranger than a person who is known or familiar. This offender is said to be polymorphous, in that they may not have any victim selection criteria.

The remaining two types in this taxonomy are *Opportunistic* and *Gang*. The opportunistic type is an impulsive offender who typically commits the rape during the course of another offense, and is simply presented an opportunity with a particular person at a fixed point in time. As a result, it could be said that this is not a motivation at all, but rather is a context in which other motivations will emerge. For example, during a burglary, the offender finds a female home alone. He is macho and believes it is his right as a male to take what is his. He assaults the lone occupant of the house before fleeing. This offense would be more accurately described as *assertive*.

For gang offenses, the milieu of the group provides the impetus for the assaultive behavior. These cases involve three or more individuals with a pack mentality, and the crimes are typically impulsive. Of this type, Hazelwood (2009, pp. 109–110) stated:

> *In almost all gang rapes, one person emerges as the leader and it is this individual upon whom the analyst should focus by eliciting detailed information from the victim. In many instances, there is also a reluctant participant involved in the gang rapes. This individual is relatively easy to identify because he physically or verbally indicates to the victim that he is not in favour of the attack. He may argue with the others or even attempt to help her escape. Obviously, this individual is the weak link in the group and if such a person is described by the victim, the analyst should attempt to profile him as well.*

What Hazelwood described are essentially different motivations that exist within individual members of the group. The group leader may be the assertive, retaliatory, or sadistic type, and the reluctant individual described above may be the reassurance-oriented group member. This latter type may be more likely to seek out group membership because of a lack of feeling of belonging to other social groups. As such, gang does not so much describe motive as it does the context under which individuals came together and act in concert. The case of Bilal Skaf is further illustrative of this (News.com, 2014):

> *Australia, August 4, 2000: A 14-year-old girl is approached by four men while traveling on a train and they begin to punch, slap, and tell her she will be forced to perform sex acts on the men and then she will be raped. The girl manages to escape, but the men begin a month-long rampage of attacks.*

August 10th, 2000: Two girls, aged 17 and 18, who are shopping at Chatswood Mall in Sydney at night are approached by eight men, including Bilal Skaf, and are persuaded with the promise of marijuana to get into a white van with four of the men, which is followed by a red car containing the other four men. Bilal introduces himself as Adam, and says he is 22 years old. The men coordinate on their phones and speak to each other in Arabic as they drive to Northcote Park in Greenacre, where the teens are beaten and forced to repeatedly perform oral sex on the attackers.

After the attacks, the girls are left in the park at close to midnight with no phones and no money and are rescued by a couple walking by.

August 12, 2000: A 16-year-old girl is taken to Gosling Park in Greenacre by her friend of six months, Mohammed Skaf, 17, where she is raped by Mohammed's brother Bilal Skaf and another man while 12 men watched. They then hold a gun to her head and kick her in the stomach before she escapes to a phone booth.

August 30, 2000: A woman is approached by the men at Bankstown Railway Station and offered marijuana, which she accepts. The woman is then taken to three separate locations and raped 25 times by 14 men over six hour, while being called an "Aussie pig" and being asked if "Leb c*** tasted better than Aussie c***". The attacks began at toilets in Bankstown with one group of men, then the woman was passed to another group at Bansktown Trotting Club, and finally she was taken to an industrial estate in Chullora where the attacks culminated and she was hosed down by the attackers.

September 4, 2000: Two 16-year-old girls, are lured from Beverly Hills train station and taken to a house in Lakemb, where they are were raped by three men over a period of four hour and told "You deserve it because you're an Australian."

Subsequently, the Skaf brothers and several of their accomplices are arrested and tried, then retried after an initially lenient sentence. Justice Finnane points out during the trial of Bilal Skaf that the crimes were "carefully planned and coordinated". Bilal Skaf is sentenced to 55 years in jail, which is later reduced to 28 years. Four of the other men convicted of the gang rapes have been granted parole.

The behavior in this case is indicative of a retaliatory-oriented motive exhibited by many members of the group. As such, the mere coming together of the group is not motivational in nature, but the punishing nature of the crimes most certainly is.

While this and the previous typologies relate strictly to the crime of rape, in the first edition of *Criminal Profiling: An Introduction to Behavioural Evidence Analysis* (this work is now in its fourth edition), Turvey proposed that the motivations put forth by Hazelwood were not limited to rape; they were indeed general motivational types that could be applied to a variety of crimes. Freeman and Turvey (2012) suggest that, as adapted, the behavior-motivation typology moves from classifying offender behavior to classifying crime scene behavior. They suggest that (p. 318):

This changes the typology from a nomothetic offender labelling system to an idiographic tool from crime scene analysis. This typology is constructed as a guide to help investigators and criminal profilers classify behaviour, in context, in relationship to the crime scene behaviour evidenced and the offender is serves. It is not intended for use as a diagnostic tool, where offenders are crammed into one classification or another and conclusively labeled. Therefore, it is not investigatively helpful to think of this as an offender typology, but rather as a crime scene-oriented behaviour motivational typology.

From here, they map out the verbal, sexual, physical, modus operandi, and signature behavior of each type. It should be noted that while the authors state this is a general motivational typology, they still relate each of the above to the crime of rape. They also add *Administrative Behavior*, which is elsewhere identified as instrumental behavior. These administrative behaviors do not necessarily meet the rule of motivation by satisfying psychological or emotional needs.

Taking this one step further, Petherick and Sinnamon (2014) propose that this typology is useful from a thematic perspective in explaining a far more broad range of behaviors. Here, the classification of individual behaviors may not be possible (there may be no sexual behavior in stalking, for example), but at a general level, the emotional or psychological need that is satisfied through the offending behavior will be representative of one of the types. Indeed, we proposed therein that these same classifications can be used to describe a host of victim and other nonoffending behaviors, and that they may be seen in many relationship and workplace behaviors. Further types were added.

The *Materially Oriented* victim is seeking some personal gain. This may be monetary or material. The previously used *Profit* label may be misleading, as the heroin addict who steals to support his habit may only seek enough money to acquire his next fix. For victims who stay in an abusive relationship because they could not support themselves financially, their entire lives may be spent at a subsistence level with the abuser maintaining tight control of the finances. As such, there is no specific profit to be had. The other type is the *Preservation-Oriented* individual. Self-preservation or the preservation of another is a purely survival-based drive and usually reflects a "me/us-or-them" dynamic in which a person is (typically) killed to stop that person from killing others. While many offenders, such as those involved in white collar and other financial crimes, will undoubtedly net considerable gains, it may be all others can do simply to get by on a day-to-day basis.

Petherick and Sinnamon (2014) also propose the reintroduction of a *Pervasively Angry* type in line with that of the MTC:R3. This accounts for those cases in which anger is an underlying issue, though there is no real or imagined wrong, and they are not targeting a representative group or individuals.

A secondary motivation is one which comes to the fore after another offense has been committed, usually where a cover-up is required to hide a more serious primary offense. Staging may be involved (see Chapter 5); in general terms, this could be considered a type of false report. This motivational type is referred to as *Crime Concealment* and may involve murder or arson as examples. In one case from the author's files, a family business was set fire to in an attempt to claim the insurance money to cover fraud and embezzlement of funds from the business. Murder may also be used to conceal sexual assault by eliminating the sole witness.

8.5 Other typologies

Other typologies have been created in an attempt to explain motives in other crimes. One more will be examined to illustrate similar motivational considerations in other crimes—a multiaxial classification for stalking developed by Mullen, Pathé, and Purcell, with motive being just one of the axes covered. This was developed largely for clinical use and also serves as the basis for the *Stalking Risk Profile*, a risk assessment instrument for stalking cases.

8.5.1 The multiaxial stalking typology

Mullen, Pathé, and Purcell (2009) present a typology in *Stalkers and Their Victims* which they note is "primarily motivational, though acknowledging the relational and psychopathological dimensions" (p. 58) of stalking. They propose five types, described in the following paragraphs.

The *rejected* is one of the most common types and results from a relationship breakdown. The stalking begins at the dissolution of the relationship or where there are indicators that it may come to an end. The aim is an attempt to reconcile that which has broken down or to exact revenge for the rejection. Stalking can also be seen as a proxy to an actual relationship where the pursuit maintains a facade of closeness. This type can occur between friends, business relationships, parents and children, or romantic partners. The rejected can be seen as among the most persistent. This type has the lowest level of mental disorder, although personality problems such as egocentricity and externalizing blame are common.

In the *Resentful* and *Retaliatory* types, the main goal is retribution, and the stalking is intended to frighten or distress the victim. The stalking persists because of the power and control the stalker feels, and they may indeed present themselves as the aggrieved victim. The retaliatory may confine themselves "to a brief episode of insulting phone calls and anonymous letters of the poisoned pen variety or repeated approaches and following lasting only a few days" (p. 76). Victims may be specific, such as a former partner, or they may be representative of a significant person in the life of the stalker.

The *intimacy seeker* and *incompetent suitor* account for the majority of stalking situations. Although they share common features or characteristics, these authors note that there are sufficient differences to warrant separate classification. It is further noted that "there is considerable merit in the notion that some stalking behaviors are the product of the insensitive, inept, and grossly overconfident actions of individuals who cannot conceive that their approaches would generate anything less than reciprocal interest" (p. 82).

In line with this, intimacy seekers are attempting to establish a relationship with the object of their unwanted affectations. A subgroup of this type are those who pursue public figures, the difference between the romantic type and the public figure pursuer is that "the first subgroup are seeking romantic intimacy, the second the intimacy of amity" (p. 83). The incompetent suitor is similarly bereft of the socially appropriate ways to go about establishing a relationship, being impaired in social skills and courting behaviors. These may be common but the duration of their offenses is comparatively short (p. 86):

> *The incompetent suitor usually harasses any particular victim for relatively brief periods, having the lowest average duration for stalking of any type. They are, however, the group with the highest recidivism rate in terms of embarking on a course of stalking targeting a new victim.*

As such, this type is the most likely to be a serial stalker (see Petherick (2014) for more information on serial stalkers).

The final type is the *predatory stalker*, whose behavior is a means to an end that is usually sexual. The stalking is therefore a prelude to another attack. In contrast to the resentful stalker, whose intrusion revolves around fear and distress, the predatory "derive excitement from the surreptitious observation of their victim" (p. 110). Victims may not be aware of the fact they are under surveillance, and the offender may receive erotic gratification from the sense of power they have over their target.

8.6 Considerations in determining motive

As most of the typologies developed based on Groth relate to sexual crimes, there is more guidance available in crimes of this type than of other crimes for determining motive. The result is more theory and specific characteristics on which to base an assessment. In others, the specific behaviors included in various works may not apply in each and every facet of the behavior under study. In these instances, motivations may need to be inferred at a thematic level represented by the general pattern of emotional and psychological needs that are being met in a given situation.

The following are important considerations in determining motive. It is suggested that the reader follow the general patterns of emotional and psychological needs being fulfilled by offense behaviors, as given in Petherick and Sinnamon (2014). This includes the reassurance, assertive, retaliatory, pervasively angry, excitation, materialism, and preservation types.

With all cases, the main question that is being asked is *what needs are being met?* If the needs can be identified through multiple behaviors or evidentiary items, then a good argument can be made for particular motives. While there is the possibility of encountering signature behaviors that are better indicators of motive than modus operandi, it should be noted that these are rare in practice. As such, a further important consideration for the determination of motive is *whether the behavior(s) under examination are functional or related to fantasy*. A good general rule is to always consider whether component behaviors have a functional purpose and to *rule out function before you consider fantasy*.

8.6.1 Examine the available evidence

The evidence should first be compiled and studied en masse to determine the nature and quantity of the materials. This will allow the analyst to form a big picture of what they have and what they do not have. For each evidence item, validity, reliability, and sufficiency must be established before any determinations are made from them. Greater weight should be given primary evidence over secondary evidence, such as statement or hearsay evidence, especially where the veracity of this has not been established.

Wound patterns should be established to determine the difference between that which was instrumental in nature, where the attempt is simply to kill an individual, as may be the case in preservation-oriented actions. Evidence of wounds beyond that simply necessary to kill the victim must be examined (called overkill), and pre-, peri-, and postmortem wounds must be identified and correctly assigned their place relative to the time of death. This may help establish those that are sadistic in nature versus other wound constellations. Different types of wounds must be identified such as abrasions, contusions, and sharp and blunt force injuries, along with the weapon that created them. Try to establish, where possible, whether the items were brought to the scene by the offender and which may have been taken from the scene or victims. In serial cases, any change of weapon or wound pattern may be used to understand a change or escalation in motive.

The analyst will need to give serious consideration as to whether any of the evidence indicates an attempt to conceal or cover up other behavior, such as may be the case where attempts to burn the victim or other item are evident. If the offender took something away from the scene, the precise nature of items taken must be established. Any or all of these things may indicate an attempt to hide another crime, in which case the analyst knows they are dealing with a secondary motive and must make every effort to establish what the original motive was.

Some cases will involve a written script or other types of written communications between the victim and offender. Other cases will involve verbal behavior, or there may be other records of verbal behavior available such as answering machine messages left on the victim's home, work, or mobile phones.

E-mails and text messages will also be instructive in this regard. The specific types of language used to communicate, instruct, compliment, or demean the victim will provide vital information as to motive. The precise timing, wording, and tone of each and every individual comment, statement, or claim must be determined by the analyst. If from the offender, was something they said volunteered of their own free will, or was it in response to something the victim did or said? If from the victim, was something they said volunteered of their own free will, or was it in response to something the offender did or said? If the victim resisted, how did the offender respond? Did they reply in kind, or did they simply warn the victim not to do that thing again? Did the offender specifically ask the victim to do something and in what way did they ask? Did the victim ask the offender to do something ("can you please stop, that hurts") and what type of reception did this receive? Is there any profanity used, and if so how does it portray the victim or offender? Does the use of any language indicate a belief on the part of the offender that he and the victim are in some type of relationship, or is the language simply objectifying the victim or self-deprecating to the offender?

When examining the evidence, the analyst will always need to consider whether there is any evidence of drug and/or alcohol use along with potential psychopathology. The presence of any or all of these conditions can seriously impact mental functioning, decision-making, and perceptions of reality. As such this can grossly distort the presentation of the crime and the interactions with the victim. This means that the behaviors on which motives are based may not be the result of rational thought or planned action, and may instead be the result of distorted cognitions or drug-fueled rage and paranoia.

When examining a number of cases that have been linked, any discordant behavior between offenses needs to be identified and accounted for. This change needs to be viewed in the context of the evolution or de-evolution of motivation. This may be the result of a change in the offender's emotional needs, mental stability, alcohol or drug use, or may even indicate that an offender has stopped taking prescription medication. Victim behavior may also bring about a change, so this needs to be factored into the overall analysis. In one case from the author's files, a reassurance-oriented stalker grew increasingly frustrated when, over time, the victims did not respond to his advances with positive affirmation. He was under the false impression that one of the victims would eventually succumb to his romantic endeavors (calling on the phone and heavy breathing, leaving letters complimenting them on their dress, etc.). When the reality did not comply with his fantasy, he become angry and started to abuse his later victims, threatening some with rape and assault.

8.6.2 Perform a thorough victimology

As with all other types of examination, victimology is critical in helping to understand the offense in toto. This may be especially true where the victim is chosen because they have certain features or characteristics that are important to the offender.

Where there are commonalities between victims in serial cases, the common features must be determined to see if these constitute victim selection criteria. Hair color, weight, employment, or personality and general disposition may all be defining criteria. For example, were all victims generally shy and withdrawn? Were they all of a particular body type? Did they all work for the same company or business, or were they employed in the same general industry?

In a serial case, all aspects of the victim's personal, social, and professional life need to be established. Hobbies, habits, and routines must be determined, as will memberships to clubs or societies. Schools or universities attended must also be established. Where victim selection criteria are not important or where offenses are impulsive or opportunistic, this still needs to be established and cannot simply be assumed.

8.6.3 Crime scene considerations

Where crimes scenes have been identified and information from them available, this needs to be studied. What does the preparatory scene tell you about planning and intent? Are there photographs or other information about the victim? In multiple victim cases, this may provide insight into victim selection and targeting. Does the primary or secondary crime scene provide any valuable insight into actual behavior? The analyst should look for preparations that are made in these areas, such as providing materials for the victim's comfort, or items that were placed in the scene prior to the offense to facilitate execution, such as rope or duct tape.

It must be established whether the victim was at the crime scene when they were selected or when the offense occurred. By extension, what does the crime scene demonstrate about the movement of parties involved? For example, if this was a location frequently used by the victim, it may tell you something different to them being at a location they were unfamiliar with or had not previously visited. The same considerations apply for the offender. Was this a location they displayed some familiarity with, or was it perhaps selected at the last moment in an impulsive or spontaneous act? Where there are multiple crime sites, the relationship of each to others needs to be established. Other factors for consideration are the precise nature of the scene (business, commercial, residential); how far each crime scene is from other crime scenes; whether the offender walked there or used transport; and whether the victim walked there or used transport. Obviously, these are but a few of the elements that need to be considered.

With regards to time, it needs to be established how long was spent at each of the crime scenes, and what, if anything, caused a change in the time spent at each. Was the offense interrupted by a third party? Was the offense interrupted by the victim? Did the offender willingly cease the interaction? Another important element of timing is how much and how long the offender interacted with the victim. Did they spend time talking to them, or were they not interested in discourse? Does it appear that the offender was only interested in executing any sexual or other behavior, such that the plan appeared to be execution of the act only?

Privacy and openness are also important. This can be answered by determining the type of location, whether there is general access for the public, or whether it was a private residence or other generally inaccessible location. If public, why was no one else present at the time of the offense? If private, was the residence shared and were any of those present at the time? If not, when were they planning on returning, and does this appear to be known to the offender?

In conclusion, and as a general rule, the best way to establish motive is to apply the scientific method to rule out those motives that are not in play. If this can be done to the exclusion of all but one, and that one consistently fails to be disproved, then the analyst is subsequently left with the strongest basis on which to put forth their determination of motive. Should they be left with more than one option, the process of analysis of competing hypotheses (ACH) presented in Chapter 2 may help establish which of the remaining is most likely.

Conclusion

The determination of motive may not be necessary from a legal point of view, but it is critical from an investigative or analytic perspective. As it constitutes the reason the offender engaged in the offense in the first instance, attempts must be made to understand the motive whenever possible. Any failure to do so represents a significant void in the analyst's attempt to understand the offense in its totality. In cases where the offender is known, knowing why the offense is occurring will be pivotal not only in determining the risk present but also in the best way to manage any threat that exists. Where the offender is unknown, establishing motive can help understand what the suspect pool may be—that is, which collective of individuals (known to the victim, not known to the victim, etc.) will help to narrow down the possible number of individuals that investigators have to sift through to find the offender.

Despite being such an important consideration and being fairly prevalent in the literature, motive is not well defined, with only a few works providing a base definition from which they are working. Most other sources use the term without operationalizing its meaning, apparently working from the assumption that the meaning is self-evident.

There have thus far been a few attempts to categorize and classify offenses and offenders on the basis of the motive for the offense. These include the Groth typologies and the later attempt of Hazelwood with specific regards to rape. The MTC:R3 is largely a clinical typology for rapists whereby offenders are grouped by virtue of four main motives: opportunistic, pervasively angry, sexual, and vindictive. However, as discussed elsewhere, opportunistic is not a motive in itself; it more accurately describes the context in which the offense occurs. Sexual may be similarly problematic in that it describes a type of behavior rather than the underlying reasons for that behavior. In this instance, the subtypes of sadistic and nonsadistic may be more useful.

For stalking, Mullen and colleagues provide a multiaxial typology that is based primarily on the motive, but also includes relational elements and considerations of psychopathology. This typology also grew from clinical observations with the type's authors all being mental health professionals who have worked with stalkers and their victims. While discrete types exist, these authors acknowledge that there is some overlap and sometimes only subtle differences among them.

There will be obvious overlap among some taxons, especially given that Groth's typology served as the basis for Hazelwood's and that proposed by Turvey, as well as forming the core of the first series of MTC typologies for rapists. The same applies to that proposed by Petherick and Sinnamon as both use a general motivational scheme and a specific approach for understanding victim behavior. As such, it would be unfair to claim that these other typologies essentially validate the original types proposed by Groth. However, given that the typology proposed by Mullen, Pathé, and Purcell was developed based on a different crime and that they do not cite Groth in their work, the inferred similarity between their types and those of Groth and others may be taken as by-proxy validation—at least to the degree to which the core or patterns are consistent between the two types.

For example, the core cause of behavior in reassurance-oriented cases closely parallels the poor social and courting skills of the incompetent suitor. The predatory stalker who surveils a victim in preparation for an attack, especially where there may be sexual gratification from the humiliation and suffering of the victim, is similar to that of the anger excitation offender of Groth. Both systems employ retaliatory terminology to describe the offender who is striking back for real or perceived wrongs.

This chapter has discussed motive from a definition and operational perspectives. Motivation was distinguished at a theoretical level from criminological theories and popular explanations, and the difference between motive and planning and intent was explicated before providing an overview of various typologies that describe motivation, among other elements of offense-related behaviors. The chapter closed with some suggestions for the crime analyst who is attempting to determine motive from offense-related behavior.

REVIEW QUESTIONS

1. Motive has its roots in the Latin, meaning "to move." True or false?

2. Motive affects behavior. True or false?

3. Which of the following is not one of the motives in the Hazelwood adaptation of the Groth typology?

    a. Reassurance
    b. Assertive
    c. Retaliatory
    d. Pervasive
    e. Gang

4. What is the problem with using opportunity and gang as motives?

5. What is the difference between a motive, an explanation, and a theory?

6. What are the three types of explanations?

References

American Psychiatric Association.(2013). *Diagnostic and statistical manual of mental disorders* (5th ed.). Washington: American Psychiatric Association.

Andrews, D. A., & Bonta, J. (2010). *The psychology of criminal conduct* (5th ed.). New Jersey: Andersen Publishing.

League, Anti-defamation (2011). *Anti-defamation league state hate crime statutory provisions.* http://www.adl.org/assets/pdf/combating-hate/state_hate_crime_laws.pdf. Accessed 30.12.13.

Arden, J. B. (2010). *Rewire your brain: Think your way to a better life.* New Jersey: John Wiley and Sons.

Brown, S. E., Esbense, F. A., & Geis, G. (2010). *Criminology: Explaining crime and its causes* (7th ed.). New Jersey: Andersen Publishing.

Burgess, A. W., Commons, M. L., Safarik, M. E., Looper, R. R., & Ross, S. N. (June 2007). Sex offenders of the elderly: typology and predictors of severity of crime. *Aggression and Violent Behavior, 12*, 582–597.

Cohen, P. (2008). Child development and personality disorder. *Psychiatric Clinics of North America, 31*(3), 477–493.

Darcangelo, S. M. (1996). *Psychological and personality correlates of the Massachusetts treatment centre classification system for rapists.* Simon Fraser University. (Unpublished Doctoral Thesis).

Felson, M. B., & Krohn, M. (1990). Motives for rape. *Journal of Research in Crime and Delinquency, 27*(3), 222–242.

Freeman, J., & Turvey, B. E. (2012). Interpreting motive. In B. E. Turvey (Ed.), *Criminal profiling: an introduction to behavioural evidence analysis* (4th ed.). San Diego: Elsevier Science.

Gannon, T. A., Collie, R. M., Ward, T., & Thakker, J. (2008). *Rape: psychopathology, theory, and treatment. Clinical Psychology Review* (28). 982–1008.

Glassner, B. (1999). *The culture of fear: Why americans are afraid of the wrong things*. New York: Basic Books.

Groth, A. N. (1979). *Men who rape: The psychology of the offender*. New York: Plenum Press.

Groth, A. N., Burgess, A. W., & Holmstrom, L. L. (1977). Rape: power, anger and sexuality. *American Journal of Psychiatry, 134*(11), 1239–1243.

Hazelwood, R. R. (2009). Analysing the rape and profiling the offender. In R. R. Hazelwood & A. W. Burgess (Eds.), *Practical aspects of rape investigation: an investigative approach* (4th ed.). Boca Raton: CRC Press.

Hazelwood, R. R., & Burgess, A. W. (2009). *Practical aspects of rape investigation. A multidisciplinary approach* (4th ed.). Boca Raton: CRC Press.

de Jong, F. (2011). Theorising criminal intent. *Utrecht Law Review, 7*(1), 1–33.

Katz, J. Seductions of crime: Moral and sensual attractions in doing evil. Jackson: Perseus Book Group.

Kleinginna, P. R., & Kleinginna, A. M. (1981). A categorized list of emotion definitions, with suggestions for a consensual definition. *Motivation and Emotion, 5*(4), 345–379.

Knight, R. A. (1993). Validation of a typology for rapists. *Journal of Interpersonal Violence, 14*(3), 303–330.

Knight, R. A., Warren, J. I., Reboussin, R., & Soley, B. J. (1998). Predicting rapist type from crime-scene variables. *Criminal Justice and Behaviour, 25*, 46–80.

Leonard, D. P. (2001). Character and motive in evidence law. *Loyola of Los Angeles Law Review, 34*, 439–536.

McCabe, M. P., & Wauchope, M. (2005). Behavioural characteristics of men accused of rape: evidence for different types of rapists. *Archives of Sexual Behaviour, 34*(2), 241–253.

McDevitt, J., Levin, J., & Bennett, S. (2002). Hate crime offenders: an expanded typology. *Journal of Social Issues, 58*(2), 303–317.

Meloy, J. R., & O'Toole, M. E. (2011). The concept of leakage in threat assessment. *Behavioural Sciences and the Law, 29*, 513–527.

Millon, T., Grossman, S., Millon, C., Meagher, S., & Ramnath, R. (2004). *Personality disorders in modern life* (2nd ed.). New Jersey: John Wiley and Sons.

Mullen, P. E., Pathé, M., & Purcell, R. (2009). *Stalkers and their victims* (2nd ed.). Cambridge: Cambridge University Press.

News.com. (2014). *Rapist may soon be free, 13 years after Skaf gang rapes*. Accessed 25.01.14 . http://www.news.com.au/national/rapist-may-soon-be-free-13-years-after-skaf-gang-rapes/story-fncynjr2-1226807690189.

Osterburg, J. W., & Ward, R. H. (2010). *Criminal investigation: A method for reconstructing the past* (6th ed.). New Jersey: Andersen Publishing.

O'Toole, M. E. (2000). *The school shooter: A threat assessment perspective*. Quantico, Virginia: FBI Academy Critical Incident Response Group, National Centre for the Analysis of Violent Crime.

Petherick, W. A. (2014). Serial stalking: looking for love in all the wrong places? In W. A. Petherick (Ed.), *Profiling and serial crime: theoretical and practical considerations* (3rd ed.). Boston: Andersen Publishing.

Petherick, W. A., & Sinnamon, G. (2014). Motivations: offender and victim perspectives. In W. A. Petherick (Ed.), *Profiling and serial crime: Theoretical and practical considerations* (3rd ed.). Boston: Andersen Publishing.

Prentky, R., Cohen, M., & Seghorn, T. (1985). Development of a rational taxonomy for the classification of rapists: the Massachusetts treatment centre system. *Bulletin of the American Academy of Psychiatry and the Law, 13*(1), 39–70.

Purcell, R., Powell, M. B., & Mullen, P. E. (2005). Clients who stalk psychologists: prevalence, methods, and motives. *Professional Psychology, Research and Practice, 36*(5), 537–543.

R v. Hillier (2007) HCA 13 (2007). 233 ALR 634; 81 ALJR 886 (March 22, 2007).

Rockelein, J. E. (2006). *Elsevier's dictionary of psychological theories*. San Diego: Elsevier Science.

Samaha, J. (2005). *Criminal justice* (5th ed.). Belmont: Wadsworth Cengage.

Siegel, L. J. (2011). *Criminology: The core* (4th ed.). Belmont: Wadsworth Cengage.

Steven Wayne Hillier v. The Queen. (2005). *ACTCA 48 (December 15, 2005)*.

Steven Wayne Hillier v. The Queen. (2008). *ACTCA 3 (March 6, 2008)*.

Steven Wayne Hillier v. The Queen. (2010). *ACTSC 33 (April 16, 2010)*.

The Post and Courier. (2013). *Westboro Baptist Church threatens to protest gat hate crimes activists' speech at CofC*. Available from. http://www.postandcourier.com/article/20131021/PC16/131029916. Accessed 08.01.14.

Thomaes, S., Bushman, B. J., de Castro, B. O., & Stegge, H. (2009). What makes narcissists bloom? A framework for research on the aetiology and development of narcissism. *Development and Psychopathology, 21*, 1233–1247.

Risk Assessment

9

Jessica Gormley, Wayne Petherick
Bond University, Robina, QLD, Australia

Key Terms

Risk-assessment
Actuarial risk assessment
Structured professional judgment
Risk of recidivism
Static-99
Base-rates

INTRODUCTION

Risk assessment is practiced in many disciplines, including those not related to forensic disciplines. For instance, clinical psychologists are often required to make an assessment of the likelihood of a client self-harming or committing suicide. Physicians are often asked to consider the likelihood of a patient surviving surgery. Similarly, risk assessments are common practice in legal settings in civil and criminal domains, and are often context specific. In the fields of forensic psychology and psychiatry, for instance, clinicians are often required to evaluate the likelihood of danger an individual poses to society, particularly during the court proceedings of sentencing, placement in prisons, and release from custody. At these stages, clinicians are often required to make predictions regarding the likelihood that the offender will re-offend, and provide some opinion regarding his or her prospect of rehabilitation. Within criminology, risk assessments are often performed on public spaces for crime reduction or prevention purposes, and assessments of risk within education settings for bullying and playground violence are also conducted.

Forensic risk assessments are of particular importance when considering whether a sexual or violent offender should be released from custody, and clinicians are often asked to determine the likelihood the individual will re-offend, should he or she be allowed back into the community. In fact, the introduction of legislation, such as the *Dangerous Prisoners (Sexual Offenders) Act 2003 (Qld)* and similar legislation across various other jurisdictions, means that "dangerous" sexual offenders are able to be sentenced to further conditions after the completion of their original sentence (a type of civil commitment). These types of orders affect both the offender and society given the expense of continuing parole-like conditions for significant periods of time. Similarly, these assessments are also of particular importance when considering the risk a violent or sexual offender may pose to prior and potential victims in the future. The implications of these assessments are vast, given the stakeholders who can be affected by these predictions. In fact, the consequences of inaccurate prediction can be devastating for the offender, the victims, and the community as a whole.

9.1 What is risk assessment?

To define risk assessment, it is useful to first consider what is meant by the word *risk*. Despite the simplicity of the widely used term, the wide variety of definitions appears somewhat convoluted. For instance, researchers have defined risk as "a combination of an estimate of the probability of a target behavior occurring with a consideration of the consequences of such occurrences" (Towl & Crighton, 1996, p. 55). This definition highlights that the behavior of interest is neutral; it could be either positive or negative. However, the word risk lends itself to some negative connotation: we do not tend to consider risk an appropriate term when discussing a positive outcome (i.e., a reduced risk of re-offending, as opposed to risk of re-offending). The term risk has been more specifically defined during the past few decades to encapsulate the underlying negative outcome that is present in the context of recidivism (habitual, or repetitive, criminal behavior).

For instance, the emphasis of the definition shifted to conceptualize prediction of future *dangerousness*, as opposed to a more broad, target behavior (Harris, Rice, & Quinsey, 1993). The concept of *dangerousness* has become an important part of the definition of risk largely because of legislative, as opposed to clinical, changes, which allow for civil commitment as a result of how dangerous someone could be, as opposed to how much the person requires treatment (Cooper, Griesel, & Yuille, 2008). The issue with the term dangerousness in a clinical setting is that the word itself evokes dichotomous thinking, implying that an individual is either dangerous or not dangerous, without considering dependent factors. In line with this, Prentky and Burgess (2000) concede that the term risk is more accurately a depiction of likelihood of future harm, given that it suggests continuity and allows for calculation and consideration of several variables. In addition, contemporary definitions consider risk as occurring along a continuum or range (Steadman, Silver, Monahan, Appelbaum, Robbins, Mulvey, Grisso, Roth, & Banks, 2000).

There is consensus in the literature that risk refers to some kind of probability of harm. Consider the most basic definition of the word itself: "1. the possibility of something bad happening. 2. a person or thing causing this" (Oxford Dictionary and Thesaurus, 2004). When the term risk is used in its simplest form, particularly in disciplines such as criminology and psychology, this is essentially what is being discussed (i.e., the chance that someone will do something bad). Risk assessment, therefore, refers to the evaluation of the probability that an individual will engage in a particular behavior in the future. More specifically to dangerousness, clinicians often make predictions about the level of potential harm persons may cause to themselves or others under certain circumstances (i.e., if released from prison).

Risk assessments should involve a balance of a prediction of risk, through systematic evaluation, and recommendations on what should be done to ameliorate the particular level of risk (Heilbrun, 1997). The prediction element of risk assessment involves clinicians making an estimate of the probability of individuals causing future harm to themselves or others, through the identification and consideration of factors known to contribute to risk. The management element of risk assessment involves clinicians making some recommendations toward how that risk should be managed with regard to the context the individual is in (in the community or in custody), and, ideally, how treatment may ameliorate that level of risk. Assessment of future risk can be a challenging and demanding task. However, as long as assessments are conducted thoroughly, and with transparency regarding limitations, risk assessments can be helpful for legal decision makers (Cooper et al., 2008), and crucial for the deterrence of future serious crime (Douglas, Ogloff, & Hart, 2003).

Table 9.1 The Four Possible Outcomes from Predicting Risk

	Outcome	
Decision	**Re-offends**	**Does Not Re-offend**
Predicted to re-offend	True positive	False positive
Predicted not to re-offend	False negative	True negative

9.1.1 Prediction outcomes

In conducting a risk assessment, there are four possible prediction outcomes (Table 9.1). The first two, which are correct predictions, are known as true positives and true negatives. A true-positive outcome occurs when a prediction is made that an individual has a high likelihood of committing a crime, and that individual does commit a crime at some point in the future. Similarly, a true-negative outcome occurs when it is deemed that an individual is unlikely to commit an offense in the future, and the individual does not go on to offend. The second two predictions, which are inaccurate, are known as false positives and false negatives. A false-positive outcome occurs when a prediction is made that an individual has a high likelihood of committing an offense in the future, and that individual does not actually go on to offend. A false-negative outcome occurs when a prediction is made that an individual is unlikely to offend, and the individual does actually commit an offense at some point in the future.

The two incorrect predictions can each have different, but nonetheless devastating, consequences. For instance, a false-positive outcome can result in an individual being sentenced to a longer period in prison, or being subject to a civil commitment order at the expense of his or her freedom, in addition to adding unnecessary financial burden to the community. Similarly, a false-negative outcome can result in an offender who is likely to re-offend being released on parole into the community, or being denied necessary treatment that could ameliorate his or her risk.

9.2 Types of risk assessment

Insofar as assessments of risk go, are we just glancing through an educated crystal ball? Because there are consequences for the inaccurate prediction of risk, the question remains, how can accurate assessments of risk be increased? During the past few decades, researchers have attempted to develop the latter and subsequently reduce the former, and extensive development has been made in the field. Broadly speaking, there have been three generations of forensic risk assessment, with varying levels of success, all having advantages and disadvantages.

9.2.1 Unstructured clinical judgment

Before the more structured types of risk assessment, clinicians collected and collated extensive information, and were required to interpret and communicate this information into some kind of meaningful risk prediction. This type of clinical approach to risk assessment is known as unstructured clinical judgment (Falzer, 2013). Unstructured clinical judgment is a form of professional opinion, and relies solely on the discretion of the clinician, without any standard on how these assessments should be made or what factors should be assessed or deemed important (Pedersen, Rasmussen, & Elsass, 2010). One advantage to these types of assessments is that they utilize the expertise of the clinician conducting them. However, unstructured risk assessments have been considered somewhat informal and research has yet to demonstrate support for their predictive validity (Grove & Meehl, 1996), partly because of the lack of uniformity.

Despite the expertise of clinicians conducting these assessments, without a set of guidelines to follow, the opinions of experts vary, likely owing to varying levels of experience. Monahan (1981) for instance, found that clinicians were not able to accurately make predictions about future violent behavior when exclusively relying on unstructured clinical judgment. In fact, several researchers have highlighted the accuracy (or lack thereof) of such unstructured predictions (Andrews, Bonta, & Wormith, 2006; Grove & Meehl, 1996; Quinsey, Harris, Rice, & Cormier, 1998). As researchers began to acknowledge the pitfalls of unstructured clinical judgment, a more standardized structured approach was developing, and this came to be known as actuarial risk assessment (Cooper et al., 2008).

9.2.2 Actuarial risk assessments

A more statistical approach was then taken to risk assessment in an effort to reduce inaccurate risk prediction, by providing clinicians with a set of standardized risk factors associated with different crimes (Scurich & John, 2012). These assessments, known as actuarial risk assessments, involve the evaluation of risk factors (a measurable attribute found to predict behavior), usually determined through statistical or empirical analysis, to determine the level of risk an individual poses (Cooper et al., 2008). Actuarial risk assessments are used across both civil and criminal domains, and are probably most well known in the cases of violent and sexual offenders, particularly regarding questions about recidivism.

The development of the widely used Static-99 (Hanson & Thornton, 2000) provided clinicians with a set of salient factors, which have contributed to risk of sexual, and general, recidivism. The 10-item clinician-rated instrument assesses risk through a focus on static factors (historic factors not amenable to change), including whether the individual has been convicted of prior sexual offenses or nonsexual violence, his or her age, and prior relationship history. In addition, the Static-99 also assesses the type of victim involved (i.e., whether the victim is unrelated, known to the offender, and his or her sex).

The Static-99 has had mixed reviews regarding accuracy, with Hanson and Thornton (2000) reporting the instrument to have moderate predictive validity (area under the curve (AUC)=0.71), although other researchers report much higher predictive validity (AUC=0.91; Thornton & Beech, 2002). In addition, the Static-99 has had more accurate predictive validity than other actuarial sex offender risk assessments, such as the Rapid Risk Assessment of Sexual Offense Recidivism, which the Static-99 was based on Hanson (1997) and Craig, Beech, and Browne (2006). However, developers and their colleagues have since reported that, although the Static-99 yields good predictive validity, estimates of risk vary significantly, meaning that conclusions about likelihood of re-offending are likely to be "meaningfully different" (Helmus, Hanson, Thornton, Babchishin, & Harris, 2012). This assertion loosely reflects similar concerns held regarding the use of unstructured clinical judgment, often criticized for an inability to produce standardized conclusions regarding re-offending risk. Furthermore, research has indicated that the Static-99 tends to overestimate risk as a result of issues with base rates (for discussion on base rates, see issues with risk assessment; Hood, Shute, Feilzer, & Wilcox, 2002). The Static-99 has also been criticized, along with other actuarial risk assessments such as the Violence Risk Appraisal Guide (VRAG; Quinsey, Harris, Rice, & Cormier, 2006), for reliance on static risk factors. For instance, if the results of a Static-99 indicate that a sexual offender is a *moderate* risk of re-offending, even after successful treatment, this score can only increase, not decrease, to reflect his or her engagement in therapy.

Similarly, the 12-item clinician-rated VRAG is considered one of the most well-known actuarial risk assessments for prediction of future violence (Scurich & John, 2012). The VRAG also focuses on static risk factors, such as those related to maladjustment at primary school and home, previous marital status, criminal history (including sex of victim, age at index offense, whether the victim sustained injury, and supervision failure on conditional release), and diagnostic features (related to alcohol and personality disorders). The VRAG assesses the level of psychopathy the individual demonstrates through the suggested use of the Psychopathy Checklist, or the updated version, the Psychopathy Checklist-Revised (Hare, 2003). Part of the reason why the level of psychopathy is considered in violence risk assessments is because it has been documented to be a fairly salient predictor of future violence (Leistico, Salekin, DeCoster, & Rogers, 2008).

Although some researchers have found lower, but still adequate, results (AUC=0.65; Coid, Yang, Ullrich, Xhang, Sizmur, Roberts, Farrington, & Rogers, 2009), the VRAG has the benefit of having fairly well-documented psychometric properties among violent and psychologically impaired males (Glover, Nicholson, Hemmati, Bernfeld, & Quinsey, 2002; Kröner, Stadtland, Eidt, & Nedopil, 2007), with estimates of predictive validity being well supported and generally stable across studies (AUC=0.71; Harris, Rice, & Camilleri, 2004; AUC=0.76; Quinsey, Harris, Rice, & Cormier, 1998; See Harris, Rice, and Cormier (2002) for further review). However, as with the Static-99, the VRAG is confounded by reliance on historic, unchangeable risk factors, which only allows for prediction, rather than management, of risk. Some researchers assert that the VRAG is slightly superior to other actuarial instruments, in that it can be considered an *adjusted actuarial* approach (Cooper et al., 2008). More specifically, adjusted actuarial assessments allow for predictions to be anchored in statistical estimates, while also allowing the clinician conducting the assessment to override the outcome based on other important factors. Despite this difference, the VRAG itself does only measure static risk factors.

Actuarial risk assessments led the way for a more structured approach to risk assessment away from the subjective approach of unstructured clinical judgment. However, these risk assessments are often used in high-stakes situations, such as in cases of civil commitment (Babchishin, Hanson, & Helmus, 2002). Clinicians using these tools need to be aware of the limitations of those that are likely to overestimate risk, or have significant variability in results across raters. Static risk assessments certainly are useful in providing clinicians with guidelines for conducting risk assessments; however, they should not be used alone to determine the risk offenders pose to themselves or others. In addition, overreliance on static risk factors, whether the outcome can be altered or not, does not inform the second and equally significant part of assessment: risk management.

9.2.3 Structured professional judgment

As a result of issues with the previously mentioned approaches to risk, researchers have attempted to integrate the scientific basis of actuarial approaches with the benefits of clinical opinion, known as *structured professional judgment* (SPJ; Cooper et al., 2008). Although based on past approaches, SPJ differs from both actuarial and unstructured risk assessments in that it focuses on a combination of static factors, while giving consideration to risk factors that are not predefined, known as *dynamic risk factors* (Guy, Packer, & Warnken, 2012). Dynamic risk factors are factors that contribute to, or ameliorate, the level of risk that an individual will re-offend. Dynamic risk factors fluctuate, meaning that they are amenable to therapy and change in general (Chu, Thomas, Ogloff, & Daffern, 2013). For instance, should an offender be deemed high risk based on static factors, although having adequate social support, being engaged in psychological therapy, and feeling motivated to change, clinicians can incorporate this information into the prediction.

An SPJ instrument requires clinicians to code empirically derived risk factors, and then interpret these risk factors through the integration of collateral information, allowing the conclusion to be derived from consideration of all risk factors present (Pedersen et al., 2010). Another benefit to SPJ tools is that they allow for the second element of risk assessment, risk management, to be considered, along with risk prediction. More specifically, clinicians are able to make recommendations about how to treat and manage dynamic risk factors that may contribute to an individual's risk of re-offending (de Vries Robbé, de Vogel, & Stam, 2012).

One of the most well-known SPJ risk assessment tools, the Historical Clinical Risk Management-20 (HCR-20; Webster, Douglas, Eaves, & Hart, 1997), is used to assess the risk of future violence. The 20-item clinician-rated instrument assesses 10 static risk factors, past criminal and antisocial behavior, past psychosocial difficulties (such as with employment and relationships), substance abuse, major mental illnesses, and level of psychopathy. The HCR-20 also assesses risk across five clinical items (insight, attitude, active symptoms, impulsivity, and responsivity to treatment) and five risk management items (realistic future plans, exposure to psychosocial stressors and pressure, and personal support and general level of compliance with remediation attempts).

Tools, such as the HCR-20, are considered *best practice* in the field of risk assessment (Guy et al., 2012), partly because of the comprehensive and holistic nature of the process involved in completing such assessments. The HCR-20 has the added benefit of having reflected comparable psychometric results to those demonstrated by actuarial risk assessments (AUC=0.78 for violent crime; Douglas, 1996). However, further studies have indicated that the clinical and risk management items, which set the HCR-20 apart from actuarial assessments, such as the VRAG, have not yet been found to correlate as well with violent outcomes. More specifically, research indicates that the historic risk factors demonstrate more accurate predictive validity (Cooper et al., 2008). Other issues that are relevant to SPJ risk assessments are those highlighted in the section on unstructured clinical judgment. For instance, although SPJs provide clinicians with a more standardized guideline on how to conduct risk assessment, these approaches still reflect a level of human error present in interpretation and application of those guidelines.

9.2.4 **Risk and the power of perception**

Questions of risk are common among both the public and victims of crime and other antisocial behaviors. During any given time period, there are natural fluctuations in the frequency of any given action or behavior, and we may be influenced by our perceptions during these periods of flux. These perceptions may be personal, such as experiencing an event first hand, or they may be communicated vicariously through others, such as family, friends, colleagues, and even strangers. Perhaps one of the largest sources of influence is the media, which serves as a potent force in shaping public perception on important issues, such as crime (Chadee, 2001). As a result, media coverage of events may be an important factor in how often we *believe* something occurs, or, more specifically, the likelihood that something may occur to us. Consider the following example from a major Australian holiday destination, reproduced in its entirety for illustrative purposes (Bartlett & Ferrier, 2011):

> *The Gold Coast's "Underbelly" image is demonizing the area and could cost the tourist industry big dollars, a resort operator warns.*
>
> *Garry McKenzie, whose company runs nine resorts on the Gold Coast, says recent shootings and wide coverage of claims about organized crime and a runaway drug trade could cost the glitter strip its family-friendly image.*

"Nobody wants to holiday in a war zone and that's the image the Gold Coast is getting with all this crime out of control talk," he said.

Queensland Treasurer Andrew Fraser says the coast's reputation is at risk and Gold Coast Tourism (GCT) chief executive Martin Winter agrees the bad publicity has the potential to do serious damage.

Mr Winter said the Gold Coast had no more crime than any other city of a similar size, but per-ception was everything.

"We definitely don't believe it's damaged the image of the Gold Coast at this stage, however we're very concerned that if it was to continue or escalate, there would be damage to the reputation of the destination," he told AAP.

The Queensland Police Union on Thursday warned that innocent lives could be lost as organised crime spilled onto the streets.

Union president Ian Leavers also said the Gold Coast had cemented its position as Australia's crime capital.

Mr Fraser on Friday said some perspective must be kept on the situation.

"And that is we see overall crime rates down," he told ABC radio.

"Of course what's occurred in the last little while is unacceptable and there is a reputational risk here for the Gold Coast."

"But I think before we rush to engage in moral panic about these things, and seek to use labels that will do long-term damage to the Gold Coast, we should be mindful about the facts."

He said that by this he meant the state had the police resources to deal with the matter, a point on which the union disagreed.

Mr Leavers said while the union did not want to see long-term damage to the Gold Coast either, the government's state of denial could not continue.

"I have said many times that I am also worried about the reputational and long-term damage that the Gold Coast's crime epidemic is causing this once proud tourist, party and holiday mecca," he said.

"However, the only solution has to be more police numbers, not to simply hope the problem will just go away."

"The Gold Coast community is hurting. The government cannot pretend any longer that the num-bers of Gold Coast police are adequate."

Mr Fraser said about a quarter of the 203 extra officers added to the police force this year had gone into the fast-growing region.

He also pointed to a new serious and violent crime squad set up on the Gold Coast after the fatal shooting of detective Damian Leeding in late May.

But opposition treasury spokesman Tim Nicholls said the government came "kicking and scream-ing" to the party, long after the Liberal National Party called for such a squad.

Queensland Party spokesman on police Darren Hunt said recent incidents showed a need for urgent law and order reform.

"Armed robberies are through the roof, 94-year-old grandmothers are getting raped in their homes and repeat offenders who should clearly be kept in custody are getting bail and let loose on the innocent community to continue their crime sprees," he said.

"We need to send a message loud and clear. If you can't live by the rules of our society, then you will be removed from it."

But Mr McKenzie fears the whole issue is being used to push an agenda.

"The Gold Coast is no different to any other major city in Australia, but the Underbelly image is demonising us and could cost the city and district some major tourism dollars."

"What we desperately need is some balance in the discussion – the image being portrayed now suggests innocent holiday-makers could be gunned down in the streets."
"That's irresponsible and untrue."

As a media commentator, the second author had the opportunity to take part in this discussion at the time of the public panic, and to extensively review the local and regional crime figures. Even a cursory examination of these revealed that the current situation was not as bleak as the press, or representatives of government agencies, indicated. The state police service's own figures revealed that many types of crimes of violence were either stable or declining, and that there was no real evidence to indicate that this was a crime wave, as widely promulgated. If nothing else, it was far too early in the midst of several seemingly random shootings to tell if these were, in fact, increasing. To make matters worse, there was a rampant misuse or misunderstanding of the statistics occurring on many fronts; when told that recent shootings represented nothing more than a spike in the numbers at this point in time, an exasperated journalist proclaimed: "How can you say that? In one northern region there has been a 100% increase in the number of homicides this quarter?" On reviewing this figure, it did, in fact, turn out that there had been a 100% increase: the base figures had gone from one to two homicides. To gauge the level of public fear, one intrepid journalist set about conducting a poll of mothers who lived in the region, to determine their fear of letting children play outside. An overwhelming number were scared to let their children play in open spaces (e.g., yards and parks), for fear of "catching a stray bullet." Given the link between existing issues among shooters and their victims (put another way, an almost total lack of motiveless stranger shootings), and the fact that we are far more likely to be victimized by someone we know in a residential or other similar dwelling (especially among children), this concern over random, unpredictable street violence seems grossly misplaced. In fact, the opposite would appear to be true; children may be safer in front yards, parks, and other open spaces than in their own home.

Whether societal risk or individual risk, the expert is cautioned to exercise restraint. As a professional, his or her deliberations may carry great weight and can have a lasting impact on the party for whom the assessment has been undertaken. Work and social activities may be curtailed, familial obligations may not be met, and stress and anxieties may be heightened as a result of an assessment's being positive for risk. This applies when determining whether a domestic violence or stalking victim is at risk of future or elevated violence, or in determining the level of risk for a victim of violence at the time of the event (as discussed in Chapter 4, Forensic Victimology, this volume). The public may also find itself in a heightened state of alert after careless comments about crime rates, victimization, or the general risk of harm.

The important point to remember is that any discussion of risk, whether social or individual, must be undertaken from a balanced perspective that considers and incorporates all that is known about the risk under consideration, and that this must temper those unknown factors that are yet to be established. We must be measured, responsible, and cautious in any and all communications that arise from this type of analysis.

9.3 **The problems with risk assessment**

Forensic risk assessments have both advantages and disadvantages, and decisions made by clinicians in relation to risk must be conservative and balanced. There is still an overwhelming demand for legal decision makers to be advised on the level of risk of offenders. Even the recently developed "gold standard" (structured professional judgment tools) is not immune to limitations. One way to improve the overall accuracy of a risk assessment being conducted is to be transparent about the problems that occur with regard to all forensic risk assessments.

9.3.1 **Base rate problem**

The accuracy of risk prediction is confounded by the proportion of people who commit a particular offense. This percentage makes up what is known as a base rate, from which clinicians attempt to predict risk. This percentage becomes particularly problematic for prediction of risk when too low, because offenses that occur infrequently are much harder to predict. The offenses that are of most concern to society (e.g., sexual offenses) happen much less often than portrayed, and recidivism in those offenses occurs even less often than most would believe. For instance, the recidivism rate for sexual offenders has been fairly stable, with an average of 13% (Hanson & Bussière, 1998). Although this is great news for the community and clinicians treating sexual offenders, it makes the process difficult for determining which sexual offenders are likely to fall within that small percentage of those who re-offend. Consider this example: If 13 in 100 sexual offenders re-offend, and a risk assessment tool could accurately predict sexual recidivism 95% of the time, that would leave at least one person for whom the clinician could not predict the re-offense probability. Now consider that prediction of sexual re-offense is conducted on a much larger scale, and that risk assessment tools tend to have much lower predictive accuracy. When offenses have a low base rate, a false-positive prediction is more likely to occur.

9.3.2 **Bias and error**

Research in cognitive psychology has long documented the issues surrounding decision making. For instance, Tversky and Kahneman (1981) demonstrated, in their innovative research, that we often use *heuristics*, or cognitive shortcuts, as an easier way of decision making. Although this allows for computation of information in a more efficient manner, the cost of heuristics is that we bypass several other cognitive processes. For instance, the anchoring heuristic (the tendency to rely on the first piece of information provided to come to a final decision) means that humans can form a decision within the short time that it takes to obtain initial information. The problem with this is that we bypass the rest of the information provided. In a forensic context, heuristics can have detrimental effects. For instance, if the first information provided relates to offenses an individual committed some time ago, a clinician may be inclined to anchor his or her final decision about the level of risk the individual poses on these first pieces of information.

Also, consider the availability heuristic (the tendency to determine the probability of an event based on the cognitive availability of an example). In a forensic context, this means that clinicians may make decisions regarding the risk of an offender based on readily available information. For instance, the risk for sexual and violent offenses is often perpetuated by the media, as are those cases in which repeat offenders have re-offended while on parole or probation, because they highlight a failure of the justice system as a whole. If these cases are prominent in the media and in the minds of the general public, when another offender requires a risk assessment, he or she may be considered to be at a higher risk of re-offending than he or she actually is, because of the availability heuristic. Early research into this phenomenon found that when asked to determine a sentence, mock jurors who were exposed to intentionally biased public opinion regarding leniency in sentencing tended to indicate that judges had been too lenient in the sentences handed down to offenders. However, when in the position of the judge, participants actually tended to give sentences equal to, or even less severe than, judges had originally given (Diamond & Stalans, 1989). Similarly, Kahneman and Tversky (1982) also highlighted that people generally tend to be overconfident in their judgments, an innate tendency toward inflating self-assessments of competency, known as the better than average effect (see Kruger & Dunning, 1999).

Similarly, another potential bias in forensic risk assessment is that of the *illusory correlation*, a belief held that two events are related when they are not (or share a small correlation; Chapman & Chapman, 1967). People have this type of cognitive bias about many social issues (e.g., racial stereotypes can be formed through an illusory correlation, as can generational stereotypes, such as generation X being too lazy). However, although this is a more general societal issue, in a forensic context, these types of biases can affect the clinician's decision regarding risk. For instance, contrary to popular belief, schizophrenia is not actually directly related to an increase in violence (Walsh, Buchanan, & Fahy, 2002). However, some believe that mental illness is linked to a higher risk of violence, leading to a biased decision when conducting a risk assessment on the future violence of an offender with a diagnosed psychological condition. Fortunately, research indicates that education regarding the illusory correlation bias can lead to a reduction in such judgment errors (Murphy, Schmeer, Vallée-Tourangeau, Mondragón, & Hilton, 2011).

9.3.3 The importance of protective factors

The most commonly used forensic risk assessments rely heavily on risk factors. However, protective factors, or those factors that may reduce the likelihood of risk, have been valuable in the consideration of risk prediction and management (de Vries Robbé et al., 2012). The introduction of SPJ has led the way for protective factors to be considered. However, as stated, the static risk factors included in these assessments tend to be more salient in demonstrating predictive validity.

Treatment modalities for sexual offenders have adopted a strengths-based approach, such as the Good Lives Model (Ward & Brown, 2004), reflecting this emphasis on the importance of protective factors. For risk prediction to inform risk management, risk assessment must consider protective factors. As an example, suicide risk assessment, often conducted in both clinical and forensic settings, could be thought of as the balancing of risk and protective factors. Although individuals aged 18–24 years may present with fleeting suicidal ideation, previous suicide behavior, and feelings of hopelessness, should they also identify that they have no salient plan and no adequate social support, are engaging in treatment, and report strong religious faith, they may be deemed to have a mild risk of suicide (Bryan & Rudd, 2006). Now consider the same example, with the exception of the targeted behavior being assessed: Imagine that an 18- to 24-year-old individual was previously convicted of a sexual offense against an unrelated male victim not known to the individual, and presents with fleeting thoughts of sexually explicit images of children. However, the individual is engaged in treatment, has an adequate social support network, has strong religious faith, and has no salient plan to re-offend, would the individual still be deemed to have a mild (or low) risk of re-offending? The answer would depend on the risk assessment tool; however, as a rudimentary example, based on this information, the Static-99 would place the individual in the moderate-high risk of re-offending category. Consideration of protective factors in this example may significantly reduce the level of risk that a clinician may conclude the individual poses.

To date, there is considerably less research on protective factors; thus, it may be that, in any assessment, these factors will be assessed in the professional judgment part of SPJ. It is also important that the depth and complexity of a protective factor be incorporated into any assessment. It may not be sufficient to say "Person X has stable employment in the family business and lives with his parents," if the family business is producing methamphetamines in the home, or if they are involved in other illegal activities. This immersion in an atmosphere of antisocial behaviors may, therefore, serve to increase the risk of certain acts, and not reduce them, as may be implied on the face of it.

9.3.4 Atheoretical

Risk assessment research thus far has focused on improving the accuracy of prediction of re-offending. Risk assessment tools are developed, piloted, published, and revised based on statistical analysis. However, limited consideration has been given to the theoretical underpinning of crime, or behavior, in general. Consider the previously mentioned Static-99 and other actuarial risk assessments. All of the risk factors within these instruments focus on characteristics that have been found to be statistically significant in the prediction of sexual or violent re-offending, as opposed to considering the underlying theory related to what factors can precipitate or perpetuate these types of offenses. Research recommends that consideration is given to criminological theories of crime (Silver, 2006), which may further guide both risk prediction and especially risk management by providing insight into why people commit various crimes.

9.4 Communicating risk

One of the issues with conceptualizing risk as referring to dangerousness is that this type of thinking assumes risk is dichotomous: people are either dangerous or not dangerous. Although the focus has shifted to considering risk as existing on a continuum, it is unknown whether this shift in perspective is reflected within the justice system (Steadman et al., 2000). It is, therefore, paramount that communications concerning risk reflect an understanding that risk exists on a continuum in clinical practice, in addition to transparency about issues confounding risk assessment.

9.4.1 Levels of probability

There are two prominent methods of communicating risk to decision makers. The first involves reporting the percentage associated with the likelihood of recidivism. For instance, the actuarial Sex Offender Risk Appraisal Guide (SORAG; Quinsey et al., 2006) provides clinicians with a percentage of likelihood of recidivism over a period of 7 and 10 years. One of the advantages of providing a percentage is that it allows clinicians to provide an unbiased, objective assessment of the individual's risk level. This percentage total is not altered by the clinician on the basis of heuristics or biases.

The problem with communicating risk as a percentage is the potential for misinterpretation; also, it does not reflect a continuum paradigm (i.e., if it is reported that an individual whose score is between 26 and 31 on the SORAG has an 89% chance of sexually re-offending over a 10-year period, this does not consider dynamic factors in the individual's life that may reduce this risk considerably). The US Federal Bureau of Investigation reported in 2009 that the clearance rate of murder cases was approximately 62.1% (U.S Department of Justice, FBI, 2012), meaning that individuals have almost a 40% chance of getting away with murder. However, this is mediated by known factors, such as where the murderer lives, who the murderer kills, and even the murderer's intelligence. Therefore, if one was to report that an individual has an almost 40% chance of getting away with murder, this would not accurately represent who is likely to be part of that percentage. The same logic is applied to risk assessment: until dynamic factors are considered, we cannot be sure who will fall into that 11% of individuals with a SORAG score between 26 and 31 who do not re-offend.

It is not always possible for clinicians to avoid reporting risk as a percentage. For instance, when assessing populations that have limited follow-up research to assess recidivism, or do not have extensive information available to researchers, such as juvenile sexual offenders, clinicians are often forced to report some kind of probability. The Juvenile Sex Offender Assessment Protocol-II (J-SOAP-II; Prentky & Righthand, 2003) is one such instrument, in which the authors recommend the use of ratios because of the limited psychometric research. For instance, a ratio of 26 to 56 on the total score of a J-SOAP-II assessment indicates a risk of approximately 46%. One benefit to the J-SOAP-II is that beyond simply providing a total score, it provides a ratio for dynamic and static risks separately. These can then be translated more meaningfully in a report.

Clinicians who do report such percentages may consider the percentile rank (if available) of the normative data that estimates are based on, and then use sensitive language to ensure accurate communication of probability as opposed to statements of fact. Presenting risk in terms of a percentage can lead consumers to believe that the individual actually has that exact chance of re-offending, without consideration to limitations or base rates (Scurich & John, 2012). Therefore, because there is no way to predict risk accurately 100% of the time, clinicians conducting a risk assessment can ensure they convey this to their intended audience.

9.4.2 Thresholds and categories

The second main way of communicating risk is to report thresholds, or categorical information. The actuarial Static-99 (Hanson & Thornton 2000) provides suggested cutoff scores with descriptive categories. For instance, if individuals score between 0 and 1, they fall into the "low-risk" category, whereas if their score falls higher than six, they are considered to be in the "high-risk" category. Similarly, the VRAG provides a risk category descriptive that can be reported, as opposed to providing a percentage chance of recidivism. One of the benefits of providing a risk category with a risk assessment is that it can provide a more salient answer to the overall goal of the prediction of risk: "How likely is this person to re-offend?" With categorical information, clinicians can state the person has a "low risk" of re-offending or a "moderate risk" of re-offending. However, providing information about risk level in terms of risk categories may also provide the intended audience with a heuristic themselves. For instance, if legal decision makers are informed that on an actuarial risk assessment, an offender has scored within a "moderate-high" or "high" risk range, this may bias the decisions made, despite dynamic and protective factors ameliorating that risk level.

Although the categorical approaches do not cite a statistical probability of re-offending (e.g., 20% or 40%), another issue is that these can be "built into" the categories used to communicate risk. If we consider that risk occurs on a range of 0 (implying no risk at all) to 100 (implying that re-offending is a certainty, referred to as "imminent" risk assessment parlance), it could be suggested that the categories of low, medium, and high do exist on a fairly arbitrary place on the scale. Although these figures may not be so discrete because they relate to categories, consider the following for illustrative purposes: If low (or mild) risk occupies the position from 0 to 33, medium (or moderate) from 34 to 66, and high from 67 to 100, even assessment within the categories could represent different possibilities for future violence or harm. It goes without saying that an individual at the low end of the high-risk group (e.g., 66%, or more likely than not) poses a far lower risk than an individual at the upper end of the scale (e.g., 99% or 100%). Also, an individual at the bottom of the low-risk scale (e.g., 10%, or 1 chance in 10) is far less of a risk than an individual at the upper end of the low-risk group (e.g., 30%, or 1 chance in 3 of repeat offending).

9.4.3 Risk of what?

As discussed at the outset of this chapter, the term risk denotes several different possible outcomes or events. It is important, therefore, to be clear about the type of risk presented, or the type of prediction being made. Thankfully, these are usually domain specific (e.g., in the assessment of the likelihood of a sex offender committing another sexual offense). This relies largely on the assumption that offenders are criminal specialists and restrict themselves to a narrow range of criminal behaviors. It is also possible that a given offender group will be criminally versatile, availing themselves of a wide range of opportunities that arise in the course of their crime series. This may be particularly problematic when offenders have an extensive array of criminal behaviors in their past: The question then becomes, *which of the multitude of offenses are they at risk of repeating, if any?* The answer to this question may be far less than clear or obvious, and it is, therefore, incumbent on the risk assessor to determine the exact nature of the inquiry (general versus specific risk) and the quality of the information he or she has (both case based and theoretical or statistical) on which a given assessment will be based.

9.5 Risk assessment is not yet an exact science

Regardless of the method chosen to communicate risk, the information communicated should be fair and balanced. When conducting a risk assessment, it is important to consider all facets of risk through the use of all appropriate risk assessment tools (both actuarial and structured clinical judgment tools). Clinicians can read and consider the relevance of research in the area, because not all crimes lend themselves to a structured clinical approach to risk assessment and all are confounded by factors such as base rates. For instance, although clinicians can conduct a risk assessment for future violence on an offender who has committed murder, base rates of homicide are low. To compound the problem, there are no specific risk assessments for different types of homicide (e.g., infanticide, in which risk may be affected by number of children or post-natal depression, factors that are not specifically considered within a violence risk assessment).

In addition, it is important to give consideration to the individual being assessed to ensure that all risk factors, both static and dynamic, are identified and deliberated in the risk assessment process. More importantly, clinicians must communicate all of this information clearly and appropriately, in the context of the referral question. Often, people assume that the clinician providing the assessment is an expert and, therefore, his or her opinion can be held in high esteem. Clinicians must ensure that they are transparent and honest about the pitfalls of risk assessment, for several reasons, including that their opinion can contribute to an important decision-making process.

Consider the case of Dr James Grigson from Dallas, TX, better known as "Dr Death." He testified as a psychiatrist in death-penalty cases for almost 30 years, purporting that in murder cases he could predict with 100% certainty who would kill again. The problem with purporting that you can predict something 100% of the time, is that it only takes one incorrect prediction to prove you wrong. This is what occurred in the case of Randall Dale Adams, who was sentenced to death, partly on the basis of the testimony of Dr Grigson, who diagnosed Adams as being a severe sociopath who would kill again. After 12 years on death row, Adams' conviction was overturned because another prisoner confessed to committing the murder for which Adams was originally convicted. Dr Grigson was later expelled from the American Psychological Association for making false claims about the level of accuracy by which he could predict future homicide.

Compare this example with the expert witness testimony of Dr John Baron in a case of multiple perpetrator rape (*Regina v Msk Regina v Mak Regina v Mrk Regina v Mmk* [2004] NSWSC 319 (22 April 2004)), who said in a report of results of the Static-99 conducted on one of the perpetrators "Level of Risk: Although the Static-99 places MSK in a Low actuarial risk group, he is assessed as being at considerable personal risk of sexual or violent re-offending unless all relevant dynamic risk factors are addressed."

Dr Baron goes on to say in his report regarding one of the co-offenders:

Dynamic risk factors are those that relate to the offender's psychological status and history and to his recent life circumstances. They can be considered to be an index of current 'live risk', can change over time, and are therefore subject to intervention. Consideration of dynamic risk factors suggests that the actual risk level given above is an underestimate of MAK's risk. The following issues are of concern. The on-going denial of the offence in the face of the significant evidence to the contrary. The distorted family dynamics and the apparent dislocation of value systems that seems to underlie the offences. The sibling's lifestyle at the time of the attacks, including the possession of weapons, which appears to have been a significant factor in his offending behaviour. His apparently volatile intimate relationship, and lack of awareness of the dysfunctional aspects of the relationship. The nature of the offence itself, i.e. the fact that it was not perpetrated in isolation but in the company of both older and younger siblings and an affiliate. His ability to disregard the needs of others in the pursuit of his own, even where it means degradation and assault! His avowed admiration for his younger brother MMK, given MMK's involvement and use of aggression within the index offence. Further exploration is needed in this area in order to establish MAK's core beliefs surrounding the use of aggression and violence. His distorted ideas of what may be meaningful to the victims in terms of the damage done to them. His limited ability to reflect upon both the negative and positive aspects of his personal background and aggressive behaviour within his interpersonal relationships.

The excerpts of Dr Baron's psychological report present a great example of a balanced risk assessment. He demonstrates that, although he has conducted an actuarial risk assessment to anchor his prediction of risk, he must go on to consider the implication of dynamic risk factors and lack of protective factors in this case, given that they also affect the level of risk these offenders pose.

9.6 When harm is likely or imminent

When harm is likely or imminent, there is no clear set of guidelines or a manual that will outline a series of steps that must be followed in every particular context. This is because the situations in which harm might be imminent will vary depending on what the risk is. For instance, with an imminent or likely risk of suicide, psychologists are bound by their respective ethical codes to ensure no harm comes to their client. In addition, in this instance, there are generally crisis assessment teams that can be contacted for further advice.

In the instance of imminent or likely risk of a serious violent or sexual offense, the guidelines are less clear. One step toward a uniform set of guidelines in this area came from the tragic outcome of the case of Tatiana Tarasoff, who was stabbed to death by Prosenjit Podar, who had stalked her after she had romantically rejected him. Before her death, the perpetrator was engaged with a psychologist, to whom he had confided his plan to kill the victim. After the victim's death, and the perpetrator's controversial trial, the parents of the victim sought legal action against the psychologist for failing to warn the family of Podar's plan to kill Tarasoff. The court ruled in this landmark case of *Tarasoff v. Regents of the University of California (1974)* that the psychologist did have a duty to warn the family. The case was subsequently reheard in 1976, when a judge determined that the psychologist had a duty to protect the victim. This could be interpreted as the psychologist not necessarily needing to actually warn the victim, but contacting the police, for instance. Despite this, and that this is a US case, the case has set the aspirational principle for other clinicians. That is, if a psychologist becomes aware of a threat toward another that would put that individual's life in imminent risk, the psychologist should take all necessary steps to ensure the safety of that individual. However, is it always that straightforward? Do perpetrators always outlay a specific plan of who they plan to harm? Not always, which highlights why there are no specific guidelines on what to do when harm is likely or imminent.

The appropriate course of action can be further difficult to determine given that there may not be enough information about the endangered person to warn him or her. Perhaps the individual is just a danger to society in general, with no specific intended target. In this instance, concerned forensic clinicians can seek advice from supervisors and colleagues and contact local mental health services for advice. In addition, should the individual pose a more serious, but not imminent, risk to others, clinicians can seek what is known as a *Justice Examination Order* in Queensland, Australia. A Justice Examination Order allows for an authorized mental health service to conduct a nonurgent mental health assessment on an individual (other jurisdictions will have similar orders, and the reader should seek out the appropriate information for the jurisdiction(s) in which they operate). Of course, in the event of a more imminent risk, emergency services should be contacted.

Decisions regarding what to do when harm is likely or imminent depend on several factors, including the amount of information provided by the individual posing harm. In addition, these decisions will also be influenced by the overarching ethical guidelines and legislation. For instance, whether to report an imminent risk of child abuse or neglect will depend on the professional ethical guidelines of the clinician, and the state in which the clinician practices. Clinicians should always consult relevant guidelines and legislation, dependent on the type of risk situations presented to them. Otherwise, clinicians assessing risk should always consider their own safety, seek supervision where possible and appropriate, and communicate risk with other professionals, ethically and transparently.

Conclusion

Risk assessments are not without complications; however, risk assessment tools are not fundamentally inaccurate either (Rogers, 2000). Although there are limitations to all types of risk assessments, there are also advantages in each. The most accurate types of risk assessments are those that utilize actuarial risk assessment tools and/or SPJ, and present an objective evaluation of the individual being assessed.

This includes consideration of all types of factors (static, dynamic, and protective) that may increase and ameliorate the probability that they will re-offend, as well as consideration of the limitations of predicting future behavior. Balanced risk assessments can provide legal decision makers with insight into issues the individual offender faces, and provide guidance for future treatment and management. Risk assessment is not a perfect science; however, with continued research and development in the field, and transparent communication, risk assessment can ultimately be a worthwhile endeavor for the offenders themselves, potential victims, and society as a whole.

REVIEW QUESTIONS

1. Define risk assessment.

2. What is the difference between the three approaches of risk assessment?

3. What are the main problems with risk assessment?

4. What is the base-rate problem and how can it affect the prediction of risk?

5. Why are protective factors so important?

6. How should risk be communicated?

References

Andrews, D. A., Bonta, J., & Wormith, S. J. (2006). The recent past and near future of risk and/or need assessment. *Crime and Delinquency, 52*, 7–27.

Babchishin, K. M., Hanson, R. K., & Helmus, L. (2002). Communicating risk for sex offenders: risk ratios for Static-2002R. Retrieved from *Sex Offender Treatment, 7*, 1–12. http://www.sexual-offender-treatment.org/111.html.

Bartlett, T., & Ferrier, T. (July 22, 2011). *Underbelly image 'demonising' gold coast. The Sydney Morning Herald.*

Bryan, C. J., & Rudd, M. D. (2006). Advances in the assessment of suicide risk. *Journal of Clinical Psychology, 62*, 185–200. http://dx.doi.org/10.1002/jclp.20222.

Chadee, D. (2001). Fear of crime and the media: from perceptions to reality. *Criminal Justice Matters, 43*(1), 10–11.

Chapman, L. J., & Chapman, J. P. (1967). Genesis of popular but erroneous psychodiagnostic observations. *Journal of Abnormal Psychology, 72*, 193–204. http://dx.doi.org/10.1037/h0024670.

Chu, C. M., Thomas, S. D. M., Ogloff, J. R. P., & Daffern, M. (2013). The short- to medicum-term predictive accuracy of static and dynamic risk assessment measures in a secure forensic hospital. *Assessment, 20*, 230–241. http://dx.doi.org/10.1177/1073191111418298.

Coid, J., Yang, M., Ullrich, S., Xhang, T., Sizmur, S., Roberts, C., Farrington, D. P., & Rogers, R. D. (2009). Gender differences in structured risk assessment: comparing the accuracy of five instruments. *Journal of Consulting and Clinical Psychology, 77*, 337–348. http://dx.doi.org/10.1037/a0015155.

Cooper, B. S., Griesel, D., & Yuille, J. C. (2008). Clinical-forensic risk assessment: the past and current state of affairs. *Journal of Forensic Psychology Practice, 7*, 1–63. http://dx.doi.org/10.1300/J158v07n04_01.

Craig, L. A., Beech, A. R., & Browne, K. D. (2006). Cross validation of the risk matrix 2000 sexual and violent scales. *Journal of Interpersonal Violence, 21*(5), 1–22.

de Vries Robbé, M., de Vogel, V., & Stam, J. (2012). Protective factors for violence risk: the value for clinical practice. *Psychology, 3*, 1259–1263. http://dx.doi.org/10.4236/psych.2012.312A187.

Diamond, S. S., & Stalans, L. J. (1989). The myth of judicial leniency in sentencing. *Behavioral Sciences & the Law*, *7*, 73–89. http://dx.doi.org/10.1002/bsl.2370070106.

Douglas, K. S. (1996). *Assessing the risk of violence in psychiatric outpatients: The predictive validity of the HCR-20 risk assessment scheme* (Unpublished Master's thesis). Burnaby, British Columbia, Canada: Simon Fraser University.

Douglas, K. S., Ogloff, J. R., & Hart, S. D. (2003). Evaluation of a model of violence risk assessment among forensic psychiatric patients. *Psychiatric Services*, *54*, 1372–1379.

Falzer, P. R. (2013). Valuing structured professional judgment: predictive validity, decision-making, and the clinical-actuarial conflict. *Behavioral Sciences & the Law*, *31*, 40–54. http://dx.doi.org/10.1002/bsl.2043.

Glover, A. J., Nicholson, D. E., Hemmati, T., Bernfeld, G. A., & Quinsey, V. L. (2002). A comparison of predictors of general and violent recidivism among high-risk federal offenders. *Criminal Justice and Behavior*, *29*, 235–249. http://dx.doi.org/10.1177/0093854802029003001.

Grove, W. M., & Meehl, P. E. (1996). Comparative efficiency of informal (subjective, impressionistic) and formal (mechanical, algorithmic) prediction procedures: the clinical-statistical controversy. *Psychology, Public Policy, and Law*, *2*, 293–323.

Guy, L. S., Packer, I. K., & Warnken, W. (2012). Assessing risk of violence using structured professional guidelines. *Journal of Forensic Psychology Practice*, *12*, 270–283. http://dx.doi.org/10.1080/15228932.2012.674471.

Hanson, R. K. (1997). *The development of a brief actuarial risk scale for sexual offence recidivism*. User Report No. 1997–04. Ottawa, Ontario, Canada: Department of the Solicitor General of Canada.

Hanson, R. K., & Bussière, M. T. (1998). Predicting relapse: a meta-analysis of sexual offender recidivism studies. *Journal of Consulting and Clinical Psychology*, *66*, 348–362. http://dx.doi.org/10.1037/0022-006X.66.2.348.

Hanson, R., & Thornton, D. (2000). *Static-99: Improving actuarial risk assessments for sex offenders*. User Report 99–02. Ottawa: Department of the Solicitor General of Canada.

Hare, R. D. (2003). *Psychopathy checklist-revised technical manual* (2nd ed.). Toronto: MultihealthSystems, Inc.

Harris, G. T., Rice, M. E., & Camilleri, J. A. (2004). Applying a forensic actuarial assessment (the violence risk appraisal guide) to nonforensic patients. *Journal of Interpersonal Violence*, *19*, 1063–1074. http://dx.doi.org/10.1177/088626050428004.

Harris, G. T., Rice, M. E., & Cormier, C. A. (2002). Prospective replication of the violence risk appraisal guide in predicting violent recidivism among forensic patients. *Law and Human Behaviour*, *26*, 377–394. http://dx.doi.org/10.1023/A%3A1016347320889.

Harris, G. T., Rice, M. E., & Quinsey, V. L. (1993). Violent recidivism of mentally disordered offenders: the development of a statistical prediction instrument. *Criminal Justice and Behavior*, *20*(4), 315–335.

Heilbrun, K. (1997). Prediction versus management models relevant to risk assessment: the importance of legal decision-making context. *Law and Human Behavior*, *21*, 347–359.

Helmus, L., Hanson, R. K., Thornton, D., Babchishin, K. M., & Harris, A. J. R. (2012). Absolute recidivism rates predicted by Static-99R and Static-2002R sex offender risk assessment tools vary across samples: a meta-analysis. *Criminal Justice and Behavior*, *39*, 1148–1171. http://dx.doi.org/10.1177/0093854812443648.

Hood, R., Shute, S., Feilzer, M., & Wilcox, A. (2002). Sex offenders emerging from long-term imprisonment: a study of their long-term reconviction rates and of parole board members judgments of their risk. *British Journal of Criminology*, *42*, 371–394. http://dx.doi.org/10.1093/bjc/42.2.371.

Kahneman, D., & Tversky, A. (1982). The psychology of preferences. *Scientific American*, *246*, 160–173.

Kröner, C., Stadtland, C., Eidt, M., & Nedopil, N. (2007). The validity of the violence risk appraisal guide (VRAG) in predicting criminal recidivism. *Criminal Behaviour and Mental Health*, *17*, 89–100. http://dx.doi.org/10.1002/cbm.644.

Kruger, J., & Dunning, D. (1999). Unskilled and unaware of it: how difficulties in recognizing one's own competence lead to inflated self-assessments. *Journal of Personality and Social Psychology*, *77*(6), 1121–1134.

Leistico, A. M., Salekin, R. T., DeCoster, J., & Rogers, R. (2008). A meta-analysis relating the Hare measures of psychopathy to antisocial conduct. *Law and Human Behaviour*, *32*, 2–45. http://dx.doi.org/10.1007/s109 79-007-9096-6.

Monahan, J. (1981). *Predicting violent behavior: An assessment of clinical technique.*. London: Sage Publications.

Murphy, R., Schmeer, S., Vallée-Tourangeau, F., Mondragón, E., & Hilton, D. (2011). Making the illusory correlation effect appear and then disappear: the effects of increased learning. *The Quarterly journal of Experimental Psychology*, *1*(64), 24–40. http://dx.doi.org/10.1080/17470218.2010.493615.

Oxford dictionary and Thesaurus. (2004). Oxford: Oxford University Press.

Pedersen, L., Rasmussen, K., & Elsass, P. (2010). Risk assessment: the value of structured professional judgements. *International Journal of Forensic Mental Health*, *9*, 74–81. http://dx.doi.org/10.1080/14999013.2010.499556.

Prentky, R., & Burgess, A. (2000). *Forensic management of sexual offenders*. New York, NY: Kluwer Academic/Plenum.

Prentky, R. A., & Righthand, S. (2003). *Juvenile sex offender assessment protocol-II (J-SOAP-II) manual*. Bridgewater, MA: Justice Resource Institute. Available online https://www.ncjrs.gov/pdffiles1/ojjdp/202316.pdf.

Quinsey, V. L. (1998). *Violent offenders – appraising and managing risk*. Washington, DC: American Psychological Association.

Quinsey, V. L., Harris, G. T., Rice, M. E., & Cormier, C. A. (1998). *Violent offenders: appraising and managing risk*. Washington, DC: American Psychological Association.

Quinsey, V. L., Harris, G. T., Rice, M. E., & Cormier, C. A. (2006). *Violent offenders: appraising and managing risk* (2nd ed.). Washington, DC: American Psychological Association.

Rogers, R. (2000). The uncritical acceptance of risk assessment in forensic practice. *Law and Human Behavior*, *24*, 595–605. http://dx.doi.org/0147-7307/00/1000-0595$18.00/1.

Scurich, N., & John, R. S. (2012). Prescriptive approaches to communicating the risk of violence in actuarial risk assessment. *Psychology, Public Policy and Law*, *18*, 50–78. http://dx.doi.org/10.1037/a0024592.

Silver, E. (2006). Understanding the relationship between mental disorder and violence: the need for a criminological perspective. *Law and Human Behavior*, *30*, 659–674. http://dx.doi.org/10.1007/s10979-006-9018-z.

Steadman, H. J., Silver, E., Monahan, J., Appelbaum, P. S., Robbins, P. C., Mulvey, E. P., Grisso, T., Roth, L. H., & Banks, S. (2000). A classification tree approach to the development of actuarial violence risk assessment tools. *Law and Human Behavior*, *24*, 83–100.

Thornton, D., & Beech, A. R. (2002). Integrating statistical and psychological factors through the structured risk assessment model. In *Paper presented at the 21st annual research and treatment conference*. Montreal, Canada: Association of the Treatment of Sexual Abusers, October 2–5.

Towl, G., & Crighton, D. A. (1996). *The handbook of psychology for forensic practitioners.*. London: Routledge.

Tversky, A., & Kahneman, D. (1981). The framing of decisions and the psychology of choice. *Science*, *211*, 453–458.

US Department of Justice.2012). *FBI Uniform Crime Reports*. Washington, DC: Federal Bureau of Investigation.

Walsh, E., Buchanan, A., & Fahy, T. (2002). Violence and schizophrenia: examining the evidence. *The British Journal of Psychiatry*, *180*, 490–495. http://dx.doi.org/10.1192/bjp.180.6.490.

Ward, T., & Brown, M. (2004). The good lives model and conceptual issues in offender rehabilitation. *Psychology, Crime and Law*, *10*, 243–257. http://dx.doi.org/10.1080/10683160410001662744.

Webster, C. D., Douglas, K. S., Eaves, D., & Hart, S. D. (1997). *HCR-20: assessing the risk of violence. version 2*. Vancouver, Canada: Simon Fraser University and BC Forensic Psychiatric Services Commission.

Threat Assessment and Management

10

James S. Cawood, CPP

Factor One, Inc., San Leandro, CA, USA

Key Terms

Threat management
Risk assessment
Behavioral information
Passive management
Active management
First involvement
Information analysis
Intervention
Monitoring

INTRODUCTION

Threat assessment and management cover a great deal of ground, with a wide range of applied contexts and environments, and some shared aspects with other applied models. As discussed in this chapter, threat assessment and management will focus on the process of:

1. Determining the likelihood that an individual or individuals will commit an act of physical violence against an identified individual or group of individuals (targets or victims);
2. Within an identified period of time in the future (i.e., threat assessment);
3. The development and implementation of a practical threat management plan, based on the risk-level assessment; and
4. Disrupting the pathway to physical action, thereby increasing the safety of the individuals (i.e., threat management).

This process incorporates a wide range of knowledge and skills sometimes possessed by an individual, although it is more likely to be found in a multidisciplinary team of individuals dedicated to this process. The knowledge base incorporates understanding violent behavior in different contexts and environments, understanding empirically valid violence risk assessment methodologies, and understanding mental health, law enforcement, security, and legal interventions that can effectively influence behavior in a particular jurisdiction. The required skills include extensive knowledge and experience in information-gathering techniques (e.g., the location and review of various records for behavioral information, and effective interviewing of various parties) and applied behavioral analysis. Additional skills are critical thinking, an understanding of and experience in utilizing group dynamics to guide the development and implementation of planned interventions, focused report writing directed toward different audiences, and the delivery of expert testimony. Given other contributions in this book, we will focus primarily on skills that relate to the development of valid threat assessment and intervention plans, to provide the reader with a practical perspective and some concrete suggestions to enhance the overall process.

As context for this chapter, several items need to be raised. The first is that threat assessment practitioners are currently debating the similarities and differences between threat assessment and violence risk assessment (Meloy, Hart, & Hoffman, 2014). It is postulated by some that violence risk assessment is primarily the practice of assessing violence risk focused on an instigator, using static risk factors drawn from nomothetic data sets, and is used primarily in legal environments. Threat assessment is the practice of assessing the risk of harm to a victim, focused primarily on using dynamic risk factors relevant to the target or victim (idiographic focus) in community and nonlegal environments, such as organizations and other public settings. This discussion is a useful endeavor for continuing to refine the definition of terms and processes used to research this area. However, most practitioners understand that successful threat assessment and management require an extensive knowledge of both static and dynamic risk factors evidenced by both the instigator and the victim, individually and interactively (Skeem & Monahan, 2011). Furthermore, they understand that these are anchored to the biological, psychological, and sociological foundations of both parties, as well as the contextual and environmental factors present between and around them that influence human behavior, including violence (Arbach-Lucioni, Martinez-Garcia, & Andrés-Pueyo, 2012; Vitacco, Gonsalves, Tomony, Smith, & Lishner, 2012). This means that a threat assessment professional needs to have an advanced working knowledge of violence risk assessment principles and threat management skills to be successful in conducting a valid threat assessment, and subsequently implementing an appropriate threat management plan.

The second is that threat assessment and management have reached a point in maturity at which a substantial body of empirical knowledge has accumulated, and continues to accumulate on a yearly basis. Therefore, keeping up with the literature is essential for practitioners. This is reflected not only in the American Psychological Association (APA) launching the *Journal of Threat Assessment and Management*, beginning in 2014, but also in other empirical journals publishing articles on a monthly basis related to threat assessment and violence risk assessment research. These developments dovetail with the third and last point, which involves increasing expectations of how competent and defensible threat assessment and management activities are conducted by practitioners and organizations. Some evidence of this shift is reflected in the movement from guidelines (Association of Threat Assessment Professionals, 2006) to standards (ASIS International and the Society of Human Resource Management, 2011). This means that threat assessment and management practitioners cannot practice appropriately, or in a professional and defensible way, without adopting a level of practice that meets current standards and guidelines, incorporating appropriately researched, emerging empirical information and best practice. These developments make it a dynamic and exciting time to be a practitioner in this area of behavioral analysis and management.

10.1 **First involvement**

With these factors in mind, the process of threat assessment and intervention will serve as a framework for the material in this chapter.

The first event that needs to occur is for someone to be acting in a way that raises concerns about their behavior, and for this to come to the attention of a threat assessor. This can happen in an organizational setting (e.g., school, workplace, etc.) when policies exist to report predetermined types of behavior (e.g., intimidating behavior, threats, stalking, etc.), or when someone calls law enforcement to report concerning acts. In either case, this information has to be captured in some way, and either the person capturing it or someone to whom this information is transferred needs to make a determination about immediate next steps. As seen in Appendix A, the initial assessment determines whether the reported behavior is so concerning that action needs to be taken immediately to protect others. Often, this is not the case. However, if it is, this needs to be communicated with those at risk, with either security and/or law enforcement notified and directed to locations of concern either to prevent violence from initiating or to intervene and stop inappropriate behavior.

This determination of imminent risk is made using a combination of factors, including, but not limited to, the context of the behavior, the specificity of threats (i.e., more concerning are those that are specific by time, by target, and/or by methodology; see Cawood & Corcoran, 2009), knowledge that weapons are immediately available to the instigator, and whether the victim or victims are vulnerable (i.e., accessible) to attack. Once the immediate situation has been stabilized, further assessment takes place to understand the degree of ongoing risk posed, followed by the determination of a plan to reduce the level of assessed risk, implementation of the plan, and monitoring for new behavior. This leads to reassessment, plan evolution, new implementation, and further monitoring. This is all done with the intent that the threat, or risk of harm, continues to be reduced over time, until it reaches a level that can be tolerated by the individual, organization, or society as a whole. This level of acceptable reduction of risk depends on the nature of the threat or risk and the responsibilities of the threat assessor and management team.

If no imminent threat is assessed to be present or the immediate concern of harm has been addressed, more information is gathered, creating the foundation for a robust initial (threshold) assessment. This process requires the gathering and analyzing of information about the individuals of concern, from multiple and diverse sources (Heilbrun, 2009; Melton, Petrila, Poythress, & Slobogin, 2007), usually back to the time they were first adults, assuming the case does not involve a juvenile. The process of threat assessment is not behavioral profiling. Both use behavioral information, but the purpose of profiling is to narrow the investigative pool of individuals who might have done an act, so as to eventually identify an unknown perpetrator, while the purpose of threat assessment is to quantify the possible risk of future harm from a known subject to a protected person or population.

Therefore, if a case involves an unknown instigator, profiling might be the most appropriate first step to shrink the pool of possible subjects small enough to make the process of threat assessment viable (i.e., five subjects or less). However, no court-defensible process of threat assessment can be accomplished on an unknown subject, as the pool of available behavior would be too shallow (e.g., a limited series of anonymous text messages, e-mails, etc.) and too narrow (i.e., only the information available from one source or from one perspective, meaning what the subject chose to share in that limited pool of data) to allow for any valid threat assessment.

This does not mean that management actions cannot occur without a valid threat assessment. In an abundance of caution, individuals and organizations may choose to take preliminary steps to move vulnerable potential targets or strengthen security measures, or take other steps, on the basis of concerns over an unknown subject's actions. However, it does mean that the management actions are being taken "in the blind" and therefore should be viewed as "holding actions" rather than as well-grounded actions based on an appropriate process. Consequently, these immediate protective actions do not remove the obligation to continue with an investigation to identify the unknown person initiating the actions and then conduct a valid threat assessment; they just provide increased safety for the possible victims while those processes are being pursued.

10.2 **Sources of information**

Behavioral information is drawn from records and persons and encompasses both electronic and hard-copy records. Sources could include internet sites visited, social media use, blog posts, e-mails, text messages, browser history, photos, personnel records, civil and criminal court records, victim or witness statements, records of financial transactions, liens, judgments, surveillance video, Global Positioning System patterns, and so on. Behavioral information from people usually involves either interviews with individuals who have interacted with the subject of interest (e.g., victims, witnesses, coworkers, family members, service providers, neighbors, etc.) or direct interactions with the subject of interest through either interview or direct observation (i.e., surveillance). The important thing to remember for any behavioral information is that the sources and depth of the information available will vary based on the sources available to the assessor for that case (e.g., working for law enforcement, working for an organization, or working for an individual), as well as how long the case has been ongoing (i.e., the longer the case is worked, the more information is available).

Behavioral information is only of true value when enough information is obtained from any source to allow the assessor to understand the context of that which has been reported. As an example, assume a court record was located indicating that the subject had been accused of a battery of another person (who was not the current victim) at a licensed establishment. This information is interesting as it could indicate a willingness to use physical force. However, the true value of this information is learning the context in which the behavior occurred and the actual flow of events, then using that information to determine what, if anything, this information means in relation to the current assessment. Because this person committed battery on another patron at a bar one evening does not mean much regarding whether they will repeat the situation with another person at a place of business. However, this could change if the contextual and behavioral information from that battery can be linked or mapped to information concerning the pattern of conduct directed toward the potential job site victim. In other words, just because an instigator has a physical altercation with an individual in one context and environment does not necessarily mean they have an increased likelihood of the same behavior with a different person in another context or environment.

This is illustrated in situations where the vast majority of people who kill in combat situations in defense of their country pose no elevated risk to citizens when they return from war. Similarly, the vast majority of individuals who commit domestic violence in the privacy of their own homes do not attack people outside their residences (Dutton, 2006). Another important question to apply to the behavior is "Why didn't the person do more?" In many situations involving physical acting out, the instigator or perpetrator had an opportunity to hurt the person more severely or use a more harmful means or methods of attack (e.g., a choke versus slap, punch versus push, multiple hits versus once, etc.). So the question is "What are the cognitive, psychological, emotional, or contextual factors that stopped them from using more harmful means of attack or continuing the attack?" If factors that limited their conduct during the identified incidents, also known as "buffers," can be determined, we might be able to craft further interactions or interventions on similar vectors, increasing the effectiveness of the interventions. This is a result of these similar elements having already been shown to have an effect on their behavior in the past.

The final source of information that needs to be gathered for multifaceted threat assessment is what is going on around this behavior that could make management more unstable and therefore needs to be identified and managed as a part of the intervention process. This is particularly relevant in an organizational environment, although it can also apply in community cases. In an organizational context, this could include organizational members' fears of violence, fear-induced performance disruption or job avoidance, intentional provocation of the subject of concern by people in the organization or community, and counterproductive attempts by individuals to intervene, among others (Cawood & Corcoran, 2009). Just as the identification of buffers can be anchor points for the development and implementation of more effective interventions to prevent violence, the presence of destabilizing influences in the organization or community (environment) surrounding the current situation can prevent the intervention from being effective. Worse, these can increase the speed and severity of the escalation to violence. Therefore, the information gathering should include actively looking for information regarding destabilizing factors so they can be identified, and the intervention should address ways to manage these factors so they do not weaken the planned intervention or render it redundant.

Interviewing for behavioral information is one of the most important skills for practitioners to develop. There are numerous methodologies for interviewing individuals, but most interviewers develop their own style after being trained a number of different ways. The Enhanced Cognitive Interviewing (ECI) process is one of the most robust and empirically proven interview methodologies for these circumstances (Milne & Bull, 2014). ECI incorporates a seven-step interview process, based on hundreds of interview studies, many with high-emotion victims or witnesses or perpetrators facing high-risk outcomes. By semistructuring the interview, this seven-step process maximizes the ability of the interviewer to gain in-depth information while minimizing the risk that the interview process will alter the person's memory of actual events. These steps also help to minimize the use of problematic questions, including forced-choice questions, multiple questions, and leading questions, which can limit or bias the information provided or confuse the interviewee. Consequently, whether the ECI methodology is used (or other blended interviewing methodology versions), it is suggested that practitioners use a semistructured interview to enhance the quality and quantity of accurate information (Melton et al., 2007) that is obtained from victims, witnesses, and persons of concern in threat assessment.

10.3 Information analysis

Once all the information has been gathered, how can it be organized for maximum benefit? Many high-tech solutions have been developed for case management, including stand-alone electronic threat assessment management systems (e.g., www.Awareity.com), proprietary SQL databases, and simple flow chart programs (e.g., Microsoft Visio), among others. However, it is suggested that the reader start with a simple chronology as the initial step in organizing the information available, regardless of other methodologies or documentation that might be used. By placing information into a metachronology format (i.e., a timeline), a threat assessor can begin to see a number of important factors for analysis. These include, but are not limited to, whether the concerning behaviors have always been present or only recently surfaced; whether intimidating or threatening language has become more specific by time, target, or method; whether the level of physical violence has escalated, remained the same, or decreased over time until now; whether particular types of interactions lead to particular responses; whether contexts and environments seem to be associated with behaviors of concern; what has caused incidents to stop and whether those cognitions, emotions, or interactions seem to be the same or changing; how alcohol and other drugs have been involved, if at all; and whether the incidents increased since a significant physical trauma, such as a head injury or other traumatic brain injury, among others. The analysis of each of these factors, with an expanded view of information concerning contextual factors (i.e., victimology), potential buffers, and environmental factors, provides valuable information concerning the level of current violence risk the individual might pose, and to whom, where, and when.

As Chapter 9 covered risk assessments in detail, the reader is encouraged to refer to that chapter to understand how each of these factors can be weighed and appropriately balanced to reach a valid initial assessment of the level of potential violence risk.

Once an initial assessment of potential violence risk is ascertained for the person of concern, this information is coupled with specific information regarding contextual and environmental factors to drive the development of an intervention and monitoring plan. At the lowest level of assessed threat, options may be limited and it may only be deemed necessary to monitor future communications and interactions, looking for cues that would lead to future assessment and possibly more directed interventions. An example would be a person who is sending sporadic communications to another that are overly familiar or inappropriately personal, but are not displaying any intent to seek greater proximity or demanding any actions on the part of the recipient. If the proper background investigation has been done to determine what behaviors they have demonstrated toward others, with no ideation or behaviors of concern noted, it may be appropriate to log and review all future communications from the individual. Attention should be focused on noting changes in ideation and action orientation that would lead to more concerns for approach and/or confrontation. This would also mean that those receiving communications—or a proxy for them—would have to be trained for reception and initial analysis of the communications (e.g., logging them in, preserving the e-mails or texts for evidentiary use, and what language to look for that would signal changes in thinking or emotion that are of concern).

At higher levels of threat, it may be determined that someone should actively converse with the person of concern. This could be someone representing the focus of the behavior (i.e. victim), such as a threat assessment professional working either for them personally or for the organization the target is associated with (e.g., an employer, volunteer organization, religious organization, etc.), or a threat assessment professional working on behalf of the community, such as an appropriately trained social worker or police officer. It is rarely appropriate for the recipient (target or victim) to directly interact with the concerning person, even on a scripted and controlled basis, as this usually escalates behavior that comes to the attention of a threat assessment professional. At such an elevated level of concern, some would consider placing the person of concern under surveillance.

Although theoretically possible, in practice surveillance is seldom ideal in a threat management context (the reasons for this have been provided elsewhere; see Cawood & Corcoran, 2009). In brief, surveillance may increase the behavioral and legal risks in a particular case without any attendant benefits. On the one hand, protection of the threatened person is justified in guarding against inappropriate intrusions. On the other hand, surveillance is an intrusion into the threatener's world for the purpose of monitoring their behavior, even while they are engaged in nonthreatening behavior. The purpose often given for this intrusion is that knowing where the person of concern is provides comfort to the target.

However, providing this type of "comfort" significantly raises the risk that the person of concern discovers the surveillance, particularly when it endures for longer than a few days. This triggers a fear or anger reaction that causes escalation that may not have occurred otherwise. In other words, the person of concern may think, "You are hunting me, so now I have the right to self-defense and can rightfully escalate my hunt of you (the target)." Surveillance can also lead to the escalation of legal risk. People have the right to autonomy, unless they engage in behaviors that violate the laws of the community. Regarding civil litigation, if an individual invades another's privacy, harasses them, or takes actions that lead to an escalation in behavior that is later defended by claiming self-defense, the perpetrator of those intrusive acts (e.g., the person conducting surveillance) can be held responsible and may need to pay compensation, or be subject to sanctions for their conduct.

The watch phrase for threat assessment and management is "First, do no harm." In other words, the process of threat assessment and management should not escalate the risks to others, becoming the cause of the escalated inappropriate behavior or violence that the process is meant to deescalate or prevent. Therefore, the question that has to be asked is "What is the value of knowing what the person is doing during their day that will aid in protecting the victim and that justifies the risk of escalation?" The answer is, except in rare circumstances, nothing. If, on a higher risk case, the security perimeter is adequate around the potential victim or victims, then intervention can occur before the person of concern can get close to the victims. Intervention is justified, however, if the person of concern has sought out proximity to the victim, showing an intent to confront.

The most reasonable exceptions to surveillance for threat-related cases involve two very narrow sets of circumstances. One relates to a spot surveillance meant to locate the person of concern for immediate interaction with the threat assessment professional, such as a prelude to a "knock and talk" (i.e., an interaction initiated at the person of concern's location, such as their place of employment or residence), or the immediate service of a protective or restraining order. The other exception relates to gathering evidence for long-term, complex cases (e.g., domestic or international terrorism investigations, criminal cases involving drug trafficking or sales, criminal cases involving human trafficking, etc.) where gathering information on the individual's associates and associations, as well as their locations for different activities, is considered along with their inappropriate behavior. In short, surveillance is acceptable when an immediate need for the threat assessment or case management is required (i.e., spot surveillance) or when a larger, longer term need is being met by the use of surveillance.

It should be noted that this type of surveillance will normally be done by law enforcement personnel, who have the responsibility to conduct these types of investigations, the budget to sustain the effort, and, in many jurisdictions, limited, if not complete, immunity from legal claims arising out of their surveillance activities.

Finally, budgets for surveillance should not be overlooked as a concern for private sector threat management, as even a single-person surveillance effort can currently cost $25 to $55 U.S. per hour (meaning $600–$1320 per 24 hour period) in most industrialized countries in the world, plus expenses. Most surveillance efforts require a minimum of two persons to cover the front and back of any location if there is an opportunity to pursue the person as they move about the community, which raises the actual minimum to $1200 plus expenses for a 24 hour period.

This is an expense that also raises both behavioral and legal risk, while providing little or no information of value in return. This is especially true when these costs add to the cost of additional close-security personnel who are involved in protecting victims in higher risk cases where surveillance is considered.

10.4 **Intervention**

Tools for intervention and monitoring (i.e., management) of threat cases that are available include interviews with the subject of concern; voluntary mental health evaluations; voluntary mental health treatment options; administrative or disciplinary actions in organizational cases; verbal or written requests to cease or redirect certain communications; no-trespass orders; civil restraining orders; involuntary mental health evaluations, arrest; criminal case filing and criminal prosecution, with or without the additional issuance of an emergency protective order (EPO) or a protective order; and, finally, close monitoring of individuals on probation or parole. All interventions should be seen as individual tools in a threat manager's toolbox, applied at certain times and in a certain sequence, for individual cases.

The goal is to influence the person of concern to willingly choose, from their own perception of reality, first to not act violently and then, if possible, to end the inappropriate behavior toward the target. The key is willingly choosing. If the threat manager can use the process to determine how the threatener perceives the world and then craft direct or indirect interactions that get the threatener to stop their behavior because they perceive it is in their own best interest (as defined by them), this results in a higher level of permanent personal safety for the target of the behavior. Externally applied controls (e.g., restraining orders, involuntary commitments, or jail time) can be good short-term behavioral stabilizers in some higher risk situations that can provide time to conduct further assessment, develop longer term intervention plans, and implement those plans.

However, externally applied controls are not a long-term solution. If the person of concern does not come to their own conclusion that continuing their inappropriate behavior toward the target is not useful to them on a personal level, then the target always has to worry about what happens when the restraining order expires (if it works at all), the mental health hold is released, or the jail term is over. This is especially true if the person of concern perceives that the use of these external, disruptive tools was actually an "attack" on them by the target, which can add a tremendous amount of additional incentive to attack the target in "self-defense." These external controls, without internal changes in perception in the person of concern, only provide the opportunity for momentary stability of the situation, rather than permanent resolution.

This is not the only area of human behavior where internally derived shifts in perception have been found to provide long-term behavioral change. This is the theory behind Motivational Interviewing for the treatment of substance abuse and other mental health disorders. Miller and Rollnick (2002) developed the concept of Motivational Interviewing and conducted research showing its efficacy. They found the vast majority of heavy substance abusers were eventually able to curb their addictions and remain drug-free without the use of formal programs. The addicts just decided to quit for their own reasons (i.e., their own motivation). So the question for Miller and Rollnick was how to develop a style of interaction to get more people to discover their personal motivations for stopping the harmful behavior, and then helping them act on those motivations? Thus, Motivational Interviewing was created, was tested, and is now taught all over the world.

Like working with substance abusers, threat assessors use all the information they gather to develop an understanding of how the person of concern perceives the world, what is motivating them to act inappropriately, and what could motivate them to stop. If the threat assessor can determine those motivations, by transitioning into a threat manager role, it is then possible to take this information and craft interactions with the subject of concern that will influence (i.e., motivate) the subject to change course. A more permanent reduction in risk of harm is then provided. In those rare cases where no motivation can be found, maximum effort should be used to place that person in a controlled environment and contained for as long as possible. This is accomplished by using all influential and legal tools available to the threat manager, the target, and the organizations that may be involved.

This explanation of Motivational Interviewing is a natural segue to the use of the interview with the subject of concern as an intervention tool. Interviews with the subject are the most commonly used intervention tool in the threat manager's arsenal. Any interaction with the subject is possibly an important source of behavioral information and a positive intervention. This may also lead to negative interactions that motivate the person of concern to escalate their level of activity and pose a greater risk to their target. Consequently, great care needs to be taken in determining what behavioral and factual information the threat assessor or manager wants. This can include choosing the interview location, crafting the interview strategy, and preparing for the possible behaviors, deflections, diversions, and confrontations that could happen at the interview.

A detailed explanation of this process is beyond the scope of this chapter; however, other resources are available that provide more detail of these elements (Calhoun & Weston, 2013; Cawood & Corcoran, 2009; Meloy & Hoffman, 2014; Mohandie, 2000). Suffice to say, the best source of information to help determine the current level of risk posed by a person is, most often, the person of concern themselves regardless of whether they are operating in the range of what is considered "sane" behavior. Even people who are floridly psychotic (actively experiencing significant delusions and/or hallucinations) can provide important information that can help the threat assessor or manager determine what motivates them to act, providing they can communicate. Though this information may be clothed in illusions or allegory, the information that is gleaned from these interactions can allow the development of effective interventions.

An example of an effective intervention might be introducing specific information to a person experiencing paranoid ideation that eventually alters that person's paranoid delusions, motivating them to not approach certain targets or go to certain locations. This author has used this methodology on numerous occasions to reduce risk from actively psychotic people in the absence of the ability to have them engage in long-term, effective regimens of psychotropic and/or other therapeutic interventions.

Even though the best source of information about a person of concern is often the person themselves, that does not mean that in every instance the person of concern should be interviewed. It is important to keep in mind to first do no harm. In considering an interview with the person of concern, an assessment should be made whether the interview will provide more information than the potential elevated risk to the target in approaching this person in the first instance. In an employment context, that is rarely an issue. If a person's conduct in an employment setting has caused concerns about safety, it is often a requirement under employment law that the employee is offered an opportunity to provide an explanation of their behavior as a part of a fair disciplinary process. However, even with these circumstances, the organization may opt to take the increased legal risk of a wrongful termination lawsuit over the increased risk of physical harm that may occur with the direct interview of an employee under certain risk conditions.

There may also be other circumstances in which directly interviewing a person of concern would elevate risk higher than the current value of the information. Where a high-profile individual is being harassed or stalked by a person of concern, engaging with this person would confirm for them an actual connection with the individual (e.g., possibly deciding that because someone is responding to them on the high-profile person's behalf, the target or victim "must have noted my existence and wants a connection with me"), such as in romantic obsession or a case involving erotomania. In these circumstances, the contact might spur the person of concern to increase the frequency and intensity of attempts to get further attention from the target, or make actual contact. In many cases, this possible escalation in risk behavior supersedes the value of the information that can be gained by the direct contact.

If, after careful consideration, the interview is conducted, the topic areas of that interview will often include all the thematic areas required for the development of information needed to complete a valid violence risk assessment. These can include a thorough understanding of each incident of concern—as seen from the perspective of the person of concern; their perceived relationship with the target and others; their substance use; their ownership of or ability and intent to acquire weapons; their willingness to use weapons and when that would be reasonable to them; their medical history, mental health history, family history, history of trauma, and history of violence toward self and others; and their current ideation of violence toward self and others, among a host of factors. At the same time that the interviewer is exploring these areas, the interviewer is noting the changing emotional and cognitive states of the individual and mapping these changes to subject areas.

The interviewer should always be aware that as they are learning from the subject of the interview, the interviewee is also learning about the interviewer and what is of interest to them. In other words, the interview process is a two-way avenue of information gathering. Therefore, the interviewer should be mindful of how they ask questions and how to prevent communication of sensitive information about the victim, others, or possible intervention plans that could be detrimental to safety.

During this information-gathering process, there will also be an intervention component in which the interviewer, threat assessor, and manager begin to preview certain characteristics of possible future events, including interventions, to see how the person of concern reacts to those possibilities. At a high level of practice, the interviewer will be able to present these potential future possibilities in a way that matches the learning style (e.g., visual, auditory, tactile, etc.) of the subject of concern and also is congruent with the way that person perceives the world. A working hypothesis of the subject's worldview is developing and refined during the preceding parts of the interview. Presenting these possible future events in a manner that reduces resistance to their consideration maximizes the opportunity to understand whether possible future events are viable at this time with this person.

As discussed previously, attempting to force someone to do anything they are not motivated to do is largely a waste of time and resources. Not only does it fail to provide additional safety for the victims, but also it has potential to increase the risk to the targets. An example of using the interview as an intervention is when a person is employed by a company and has engaged in threatening behavior toward other coworkers. During this interview, a question might be "The company could chose to move forward in a number of directions from here; what is your sense of what should happen?" The person of concern might state "Well, the company could force me to take some time off, but I won't go to treatment—that hasn't worked for me!" The interviewer might follow with "Tell me about that," and then the person of concern may go on to explain that he had been asked to go to treatment as a condition for a prior job, he was not able to complete it, and employment was terminated. This information provides the assessor with a great deal of insight concerning cognition, personal insight, and reactivity levels, as well as potential reaction to that specific type of intervention. The interviewer might then ask, "So what do you think would be a next step here?" allowing the person to continue to outline what they believe would be appropriate for the company to do, and what they would be willing to do.

At some point in the interview, if the person of concern did not suggest ending the employment relationship, the assessor might state, "Some employers presented with a similar situation have considered other options, including ending the employment relationship; what is your sense of that?" A less concerning response would be "That would be completely unfair, but I have bills to pay and would go out and get another job and prove that they were wrong in letting me go," while a more concerning response would be "I would make it my life's mission to make them pay 10 times over for how they treated me!" As can be seen, the threat assessor or manager must have the ability to introduce topic areas in a general way, explore answers, and then use this information to help extend the assessment of risk. The information would also provide insight into how potential interventions might be viewed by the person of concern and whether particular interventions might increase or decrease the risk of harm to targets. This does not mean that the company would not still terminate the person's employment. Instead, they would do so with an understanding, depending on the assessment of answers given, of whether that would increase risk and require additional interventions. Alternatively, further action could be seen as unnecessary at the present time, and attendant monitoring sufficient for any additional behaviors of concern.

Regarding other possible interventions, the use of voluntary mental health evaluations and voluntary mental health treatment options requires the threat assessor or manager's belief that the level of risk allows for their use, that effective evaluation and treatment options are actually available to the person of concern, and that the person of concern will follow through to be evaluated and/or treated. This author has had a number of low-risk cases involving substance abuse-driven behavior or the initial onset of major mental illnesses (i.e., Axis I conditions) where these voluntary interventions have worked in decreasing the risk of inappropriate behavior. However, in higher risk cases, particularly those involving significant emotional levels, long-term major mental illness, or difficult personality disorders (e.g., borderline personality disorder, antisocial personality disorder, narcissistic personality disorder, paranoid personality disorder, and histrionic personality disorder), other interventions will most likely be necessary to create an environment where the person of concern decides they do not want to escalate the behavior because they don't want to experience negative outcomes.

Additional interventions, such as administrative or disciplinary actions in an organization (e.g., cases involving employees, students, customers, etc.) might include restriction of access to certain areas or facilities, suspension for a period of time, reduction in wages or rank, a suspension period for promotions, and reassignment to another unit or site. There might be a verbal or written request to cease or redirect certain communications to specific individuals or entities, with the understanding that further restrictions or actions could be considered for noncompliance. An escalation in intervention strength would be informing the individual that if they return to a particular location, their action would be considered trespassing and the appropriate authorities would be notified. In some locations, this "no-trespass" notification is required before a target (an individual victim or an organization) can seek civil restraining orders to create further sanctions for continuing intrusive, harassing, or threatening conduct.

In some cases, involuntary mental health evaluations can be a form of intervention; however, in most jurisdictions in industrialized nations, these require that the individual be seen as an immediate danger to self or others, or be gravely disabled (i.e., unable to take care of their immediate needs to survive), before they can be detained and placed on an involuntary hold. The questions for the threat assessor or manager before considering this possible intervention include "Does this person fit the criteria in this jurisdiction to be involuntarily committed?" "Who can involuntarily commit them?" "Will that entity be willing to do it, given these case facts, and do they have the resources to do it?" "Who will be evaluating them, and what criteria will they use to hold them?" "How long is the evaluation period?" and "How likely is it that they will be held for any significant period of time and get treatment that will change their perception of the situation so that the behavior will not continue upon their release?"

As with any intervention considered, no action should be taken before the appropriate questions have been asked and answered regarding the actual ability to both activate the intervention and sustain it for long enough to effect a positive change in the subject of concern's perception, so that actual change is made to reduce the risk of harm to the target. Threat assessor or managers should not set an intervention plan in motion without considering the reactions by the person of concern to the intervention and the ability to get a positive result. This should be thought of in terms of multiple outcomes, and intervention plans should include the ability to implement further interventions as needed, rapidly before the first intervention is even tried, in case the reaction to the first intervention requires additional support.

An example of this would be obtaining a restraining order. First, the question should be "If we got a restraining order, would the person honor it?" If not, except in extreme cases, it may be more effective not to seek one at all, and scrutiny should be given to considering an arrest for a crime and getting a protective order as a part of the criminal process, rather than diverting the intervention focus to an ineffective process. Assuming that the assessor believes that the person could be motivated to honor the restraining order, the next questions are: "Given the case facts, can we actually get a court to issue one?" "If so, who will be providing the affidavits?" "What type of personal information is required in this court process to be provided to the court and also to the person of concern?" And "Does the personal information required to be provided increase or decrease the risk to the targets?"

Assuming this does not increase the risk, if the order is violated, what would be the sanctions? Who will enforce the sanctions? How likely is it that these sanctions will be swiftly enforced, showing seriousness on the part of the court? Is personal service of the court case for the restraining order required? Is it or will it be possible to find and personally serve the person of concern? Who will come and testify in court? Does this testimony increase the risk to these persons? Would they be willing to testify? All of these questions would need to be considered and answered in a way that will positively move the case forward, before the order is sought.

Careful consideration must be exercised, keeping in mind that any ineffective interventions could empower the person of concern to perceive that they are beyond our ability to influence them. It is preferable to keep them uncertain regarding potential consequences. This leaves the door open to influencing their behavior later, rather than proving that attempted interventions are meaningless and therefore strengthen the perception that their conduct can continue toward the target or victim without consequence to them.

Certainly, similar considerations should be applied to any thoughts of using arrest, filing criminal charges, and criminal prosecution as an intervention. Is there a law or laws that have been broken? Who will arrest the person of concern and file a criminal case? Will the prosecutor's office be willing to prosecute it? If they did, what would be the likely result in that jurisdiction? How would the person of concern react to this process and the likely result? Could they get out on bail while the case is prosecuted? What would they do during that time? If they were convicted, will they be sufficiently monitored while on any probation or parole to reinforce their motivation to disengage from focus on the targets? How does this intervention affect the person of concern's ability to move on with a life that does not involve focusing on the targets? In other words, is the intervention making the person of concern virtually unemployable and more isolated from the community, so they will only be left with focusing on the target or victim as the source of their circumstances and feel they have more reasons to increase their inappropriate behavior? Will that increase or decrease the ongoing risk to the targets?

In summary, the use of interventions is case specific and guided by the risk assessment and the understanding of the person of concern's perception of the world, based on an analysis of their behavior. The use of interventions often starts with the interview of the person of concern, which is used as a means to develop a more complex understanding of their perception of the world, while also providing an opportunity to probe their response to possible future events. The application of interventions is made after careful consideration of both immediate and future responses to the intervention. Appropriate planning requires a consideration of whether the projected outcome of any intervention is worth initiating—always perceived through the lens of whether the intervention will make the target or victim safer or not.

10.5 Monitoring

This area of threat assessment, from the operational perspective, is the most important element of the process and the one that gets the least attention. When threat management fails, it is rarely from a failure in the initial assessment and initial intervention phase. Failures that occur often happen when the initial assessment and intervention have been successful, but the ongoing monitoring process that provides the ability to reassess risk and strengthen the intervention fails to identify new behaviors that signal a change in the risk level. A classic example of this is when, a few months after a person of concern becomes subject to a restraining order, new contacts are made by them in violation of the order. However, the target or victim does not report these to the appropriate parties, and the appropriate parties do not actively check in with the victim to maximize the possibility of hearing about new behaviors, so the situation deteriorates and the target is ultimately harmed. In retrospect, the new contact behaviors were significant, but nothing proactive could be done to prevent the violence because the monitoring system did not identify the new behaviors and, therefore, no reassessment and new intervention occurred.

The monitoring system, depending on the level of initially assessed risk, will be either passive or active, but most often will be active. Passive monitoring means that the threat assessor or manager or the Incident Management Team (IMT) has decided that the risk of violence is so low that they will wait for the target or potential victim, or people surrounding that individual, to initiate contact with them (i.e., the threat assessor remains passive) to inform them of new behavior of concern. Practically, this means that the risk has been assessed to be so low that the threat assessor or manager or the IMT hypothesizes that three to four new actions of interest, along an escalated trajectory toward violence, could be initiated by the person of concern toward the targets and the targets would still not be at immediate risk.

An example of this situation would be where a person is receiving an episodic series of "love notes" through the mail, all of which show an interest in the person, but no significant intensity of feelings or action orientation. The number of notes is concerning, given their persistence, but there is no attempt at physical approach and no indication of any being considered at the moment. This might justify a passive monitoring approach by educating the target on what language to look for in the new notes, as well as new noteworthy behaviors, and instructing them on thresholds for alerting the threat assessor should new items manifest. This level of monitoring might also be appropriate in an organizational setting in which a person is the target of one act of low-intensity harassment or verbal intimidation and the person of concern has been counseled. However, as can be understood from these examples, passive monitoring is only justified in very low-risk cases, because the margin of safety for the targets needs to be very large before it can be appropriately selected as an option.

The vast majority of cases, particularly in the initial months, require active monitoring. This means that the threat assessor or manager, or their designee, actively reaches out to those most likely to witness or be aware of new behaviors, and ask if new behaviors have occurred. The assessed risk level of the case moderates the number of contacts in a given time frame. In very high-risk cases, the level of contact could be several times per day, while in other cases (low, moderate, or high risk) it could be once a day, once every other day, once a week, several times a week, and so on. Selecting the frequency of contact and who to contact is practically determined. This occurs not just by the risk level but also by the resources available for monitoring, the level of interest or engagement in the assessment, and the management process demonstrated by the sources of new information. This is in addition to the risk tolerance of the threat assessor or manager and/or the IMT.

This means that in some cases, very active monitoring is indicated, such as a high-risk case of domestic violence. The victim is in denial about their level of physical risk of harm, thereby leading to significant concerns that they will not actively report new behaviors of concern. Resources to monitor could be low, so that even though an ideal active monitoring plan might involve contacts several times a day, it actually occurs only every other day. In this scenario, the threat assessor or manager and/or the IMT accepts the risk that something could go unreported, the target could be harmed, and they could be held responsible for allowing a preventable event to occur. The attendant potential for legal liability, loss of ongoing institutional support, and career impact could ensue.

It may surprise some readers that a target of high risk for violence would not report any new behavior the moment it occurs. Certainly, there are cases in which targets are capable of maintaining a steady degree of appropriate vigilance over a long period of time; however, that is the exception in the experience of this author. Most targets, over time, are fatigued by the process and want to move on with their lives, using cognitive distortion (i.e., minimization, rationalization, and denial) to lessen their sense of risk, even to their own detriment or that of those that they are responsible for, such as children. This is actually normal target or victim behavior and, therefore, the threat assessor or manager and/or the IMT needs to compensate for this natural tendency by using active monitoring, rather than relying on the target to report concerning changes in the behavior of the person of concern.

Though resources are seldom unlimited, available technology can greatly help the threat assessor or manager or their designee handle the monitoring process. Microsoft Outlook and other Personal Information Management (PIM) programs (e.g., ACT, Salesforce.com, etc.) can set up contact information for the sources of information on a case. These can also be used to create to-do lists and event notifications, remind the assessor to contact sources of new information, and even provide their contact information for immediate use. Other threat management programs, such as the TIPS platform at Awareity.com, have the additional capability to track the person responsible for contact and what information was learned from the contact, and distribute it to IMT members for review or for later documentation of what was done, by whom, and when. These tools are becoming increasingly sophisticated and can increase the capability of the threat assessor or manager, while not increasing the number of persons required to provide the same level of service.

Conclusion

Threat assessment and management require a wide range of knowledge related to the biological, psychological, sociological, contextual, and environmental factors associated with human violence, as well as practical skills related to critical thinking, information gathering, interviewing, behavioral information analysis, and the use of behavioral information to develop and implement intervention plans in a variety of environments using advanced communication skills to influence both individual perceptions and legal processes. Given the complexity of the field and the blending of both intellectual and practical factors, it is an exciting field to work in and provides the practitioner with an almost limitless opportunity to learn. It is hoped that this chapter both provided an overview of the practical applied process and raised points of interest that will encourage the reader to explore this field in more depth.

REVIEW QUESTIONS

1. Threat management and risk assessment are the same thing. True or false?

2. Multiple information sources are usually not needed for threat management. True or false?

3. If a person displays threatening behavior in one domain, it can be safely assumed they will be dangerous in other domains. True or false?

4. When threat management fails, it is usually during which stage?
   a. Monitoring
   b. Initial intake
   c. Examining sources of information
   d. Intervention
   e. It never fails

5. What is the difference between active and passive monitoring?

6. Why might it be prudent to undertake a risk assessment before a threat management?

References

Arbach-Lucioni, K., Martinez-Garcia, M., & Andrés-Pueyo, A. (September 2012). Risk factors for violent behavior in prison inmates: a cross-cultural contribution. *Criminal Justice and Behavior*, *39*(9), 1219–1239. http://dx.doi.org/10.1177/0093854812445875.

ASIS International and the Society of Human Resource Management. (2011). *Workplace violence prevention and intervention*. ASIS/SHRM WVPI.1-2011. Washington, DC: Author. ISBN: 978-1-934904-15-2.

Association of Threat Assessment Professionals. (2006). *Risk assessment guideline elements for violence*. California: Author.

Calhoun, F. S., & Weston, S. W. (2013). *Concepts and case studies in threat management*. Boca Raton, FL: CRC Press.

Cawood, J. S., & Corcoran, M. H. (2009). *Violence assessment and intervention: the practitioner's handbook* (2nd. ed.). Boca Raton, FL: CRC Press.

Dutton, D. G. (2006). *Rethinking domestic violence*. Vancouver, British Columbia: UBC Press.

Heilbrun, K. (2009). *Evaluation for risk of violence in adults*. New York: Oxford University Press.

Meloy, J. R., Hart, S. D., & Hoffman, J. (2014). Threat assessment and threat management. In J. R. Meloy, & J. Hoffman (Eds.), *International handbook of threat assessment* (pp. 3–17). New York: Oxford University Press.

Meloy, J. R., & Hoffman, J. (Eds.). (2014). *International handbook of threat assessment*. New York: Oxford University Press.

Melton, G. B., Petrila, J., Poythress, N. G., & Slobogin, C. (2007). *Psychological evaluations for the courts: a handbook for mental health professionals and lawyers*. New York: The Guilford Press.

Miller, W. R., & Rollnick, S. (2002). *Motivational interviewing: preparing people for change* (2nd ed.). New York: The Guilford Press.

Milne, R., & Bull, R. (2014). *Investigative interviewing: psychology and practice* (2nd ed.). New York: Wiley-Blackwell.

Mohandie, K. (2000). *School violence threat assessment: a practical guide for educators, law enforcement, and mental health professionals*. San Diego, CA: Specialized Training Services.

Skeem, J. L., & Monahan, J. (2011). Current directions in violence risk assessment. *Current Directions in Psychological Science*, *20*(1), 38–42. http://dx.doi.org/10.1177/0963721410397271.

Vitacco, M. J., Gonsalves, V., Tomony, J., Smith, B. E. R., & Lishner, D. A. (May 2012). Can standardized measures of risk predict inpatient violence? *Criminal Justice and Behavior*, *39*(5), 589–606. http://dx.doi.org/10.1177/0093854812436786.

Bibliography of suggested reading

Borum, R. (2009). Behavioral emergencies: an evidence-based resource for evaluating and managing risk of suicide, violence, and victimization [Children and adolescents at risk for violence]. In P. M. Kleespies (Ed.), *Children and adolescents at risk for violence* (pp. 147–163). Washington, DC: American Psychological Association.

Borum, R., & Verhaagen, D. (2006). *Assessing and managing violence risk in juveniles*. New York: The Guilford Press.

Calhoun, F. S. (1998). *Hunters and howlers: threats and violence against federal judicial officals in the United States 1789–1993*. Arlington, VA: United States Marshals Service. USMS Pub. No. 80.

Calhoun, F. S., & Weston, S. W. (2003). *Contemporary threat management: a practical guide for identifying, assessing and managing individuals of violent intent*. San Diego, CA: Specialized Training Services.

Calhoun, F. S., & Weston, S. W. (2009). *Threat assessment and management strategies: identifying the howlers and hunters*. Boca Raton, FL: CRC Press.

Campbell, J., Messing, J. T., Kub, J., Agnew, J., Fitzgerald, S., Fowler, B., LaFlair, L., Kub, J., Agnew, J., Fitzgerald, S., Bolyard, R., & Campbell, J. C. (January 2011). Workplace violence: prevalence and risk factors in the safe at work study. *Journal of Occupational and Environmental Medicine*, *53*(1), 82–89. http://dx.doi.org/10.1097/JOM.0b013e3182028d55.

Chu, C. M., Thomas, S. D., Ogloff, J. R., & Daffern, M. (August 2011). *The short-to medium-term predictive accuracy of static and dynamic risk assessment measures in a secure forensic hospital*. Assessment. Advanced online publication. http://dx.doi.org/10.1177/1073191111418298.

Conroy, M. A., & Murrie, D. C. (2007). *Forensic assessment of violence risk*. Hoboken, NJ: John Wiley & Sons, Inc.

Deisinger, G., Randazzo, M., O'Neill, D., & Savage, J. (2008). *The handbook for campus threat assessment & management teams*. Stoneham, MA: Applied Risk Management.

Fazel, S., Singh, J. P., Doll, H., & Grann, M. (July 2012). Use of risk assessment instruments to predict violence and antisocial behaviour in 73 samples involving 24827 people: systematic review and meta-analysis. *British Medical Journal*, *345*, 1–12. http://dx.doi.org/10.1136/bmj.e4692.

Fein, R. A., & Vossekuil, B. (1998). *Protective intelligence & threat assessment investigations: a guide for state and local law enforcement officials*. Washington, DC: U.S. Department of Justice, Office of Justice Programs, National Institute of Justice.

Fien, R., Vossekuil, B., Pollack, W., Borum, R., Modzeleski, W., & Reddy, M. (2002). *Threat assessment in schools: a guide to managing threatening situations and to creating safe school climates*. Washington, DC: U.S. Department of Education, Office of Elementary and Secondary Education, Safe and Drug-Free Schools Program and U.S. Secret Service, National Threat Assessment Center.

Guy, L. S., Packer, I. K., & Warnken, W. (2012). Assessing risk of violence using structured professional judgment guidelines. *Journal of Forensic Psychology Practice, 12,* 270–283. http://dx.doi.org/10.1080/15228932.2012.674471.

Jimerson, S., Nickerson, A., Mayer, M. J., & Furlong, M. J. (Eds.). (2011). *Handbook of school violence and school safety: international research and practice* (2nd ed.) Routledge.

Hanson, R. K., Helmus, L., & Bourgon, G. (2007). *The validity of risk assessments for intimate partner violence: a meta-analysis.* Public Safety Canada.

Hare, R. D. (1993). *Without conscience: the disturbing world of the psychopaths among us.* New York: The Guilford Press.

Hart, S. D., & Cooke, D. J. (2013). *Another look at the (im-) precision of individual risk estimates made using actuarial risk assessment instruments.* Behavioral Science and the Law. Advanced online publication. http://dx.doi.org/10.1002/bsl.2049.

Hart, S. D., & Logan, C. (2011). Formulation of violence risk using evidence-based assessments: the structured professional judgment approach. In P. Sturmey, & M. McMurran (Eds.), *Forensic Case Formulation* (pp. 83–105). Malden, MA: John Wiley & Sons, Ltd.

Hershcovis, M. S., Turner, N., Barling, J., Arnold, K. A., Duprè, K. E., Inness, M., LeBlanc, M. M., & Sivanathan, N. (2007). Predicting workplace aggression: a meta-analysis. *Journal of Applied Psychology, 92*(1), 228–238. http://dx.doi.org/10.1037/0021-9010.92.1.228.

Johnson, S. L., Burke, J. G., & Gielen, A. C. (April 1, 2012). Urban students' perceptions of the school environment's influence on school violence. *Children & Schools, 34*(2), 92–102.

Jones, N. J., Brown, S. L., & Zamble, E. (May 2010). Predicting criminal recidivism in adult male offenders: researcher versus parole officer assessment of dynamic risk. *Criminal Justice and Behavior, 37*(8), 860–882. http://dx.doi.org/10.1177/0093854810368924.

King, B. (2012). Psychological theories of violence. *Journal of Human Behavior in the Social Environment, 22*(5), 553–571. http://dx.doi.org/10.1080/10911359.2011.598742.

Kormanik, M. B. (2011). Workplace violence: Assessing organizational awareness and planning interventions. *Advances in Developing Human Resources, 13*(1), 114–127. Advanced online publication. http://dx.doi.org/10.1077/1523422311410658.

Kroner, D. G., Gray, A. L., & Goodrich, B. (December 1, 2011). Integrating risk context into risk assessments: the risk context scale. *Assessment, 20*(2), 135–149. http://dx.doi.org/10.1177/10731911114296613.

Lewis, K., Olver, M. E., & Wong, S. C. P. (May 2012). *The violence risk scale: predictive validity and linking changes in risk with violent recidivism in a sample of high-risk offenders with psychopathic traits.* Assessment. Advanced online publication. http://dx.doi.org/10.1177/1073191112441242.

McGowan, M. R., Horn, R. A., & Mellott, R. N. (2011). The predictive validity of the structured assessment of violence risk in youth in secondary educational settings. *Psychological Assessment, 23*(2), 478–486. http://dx.doi.org/10.1037/a0022304.

Meloy, J. R. (2000). *Violence risk and threat assessment: a practical guide for mental health and criminal justice professionals.* San Diego, CA: Specialized Training Services.

Meloy, J. R., Hart, S. D., & Hoffman, J. (2014). Threat assessment and threat management. In J. R. Meloy, & J. Hoffman (Eds.), *International handbook of threat assessment* (pp. 3–17). New York: Oxford University Press.

Meloy, J. R., Sheridan, L., & Hoffman, J. (2008). *Stalking, threatening, and attacking public figures: a psychological and behavioral analysis.* New York: Oxford University Press.

Messing, J. T., & Thaller, J. (2012). The average predictive validity of intimate partner violence risk assessment instruments. *Journal of Interpersonal Violence.* Advanced online publication. http://dx.doi.org/10.1177/0886260512468250.

Milczarek, M., & European Agency for Safety and Health at Work, & EU-OSHA (2010). *Workplace violence and harassment: a European picture.* http://dx.doi.org/10.2802/121198. https://osha.europa.eu/en/publications/reports/violence-harassment-TERO09010ENC.

Monahan, J., Steadman, H. J., Silver, E., & Et, A. (2001). *Rethinking risk assessment: the macarthur study of mental disorder and violence*. New York: Oxford University Press.

Otto, R. K., & Douglas, K. S. (2010). *Handbook of violence risk assessment*. New York: Routledge.

Pedersen, L., Rasmussen, K., & Elsass, P. (2010). Risk assessment: the value of structured professional judgments. *International Journal of Forensic Mental Health*, 9(2), 74–81. http://dx.doi.org/10.1080/14999013.2010.499556.

Lies, M. A., II (2008). *Preventing and managing workplace violence: legal and strategic guidelines. Section of state and local government law*. Chicago, IL: ABA Publishing.

Meloy, J. R. (1998). *The psychology of stalking: clinical and forensic perspectives*. San Diego, CA: Academic Press.

Rufino, K. A., Boccaccini, M. T., & Guy, L. S. (2011). Scoring subjectivity and item performance on measures used to assess violence risk: the PCL-R and HCR-20 as exemplars. *Assessment*, 18(4), 453–463. http://dx.doi.org/10.1177/1073191110378482.

Rugala, E. A. (2004). *Workplace violence: issues in response*. Washington, DC: Department of Justice, Federal Bureau of Investigation.

Schmidt, F. (May 2010). Detecting and correcting the lies that data tell. *Perspectives on Psychological Science*, 5(3), 233–242. http://dx.doi.org/10.1177/1745691610369339.

Scurich, N., Monahan, J., & John, R. S. (2012). Innumeracy and unpacking: bridging the nomothetic/idiographic divide in violence risk assessment. *Law and Human Behavior*, 36(6), 548–554. http://dx.doi.org/10.1037/h0093994.

Shaffer, D. K., Kelly, B., & Lieberman, J. D. (2011). An exemplar-based approach to risk assessment: validating the risk management systems instrument. *Criminal Justice and Behavior*, 22(2), 167–186.

Simon, R. I., & Tardiff, K. (2008). *Textbook of violence assessment and management*. Arlington, VA: American Psychiatric Publishing, Inc.

Singh, J. P. (February 2013). Predictive validity performance indicators in violence risk assessment: a methodological primer. *Behavioral Sciences and the Law*, 31, 8–22. http://dx.doi.org/10.1002/bsl.2052.

Singh, J. P., & Fazel, S. (September 2010). Forensic risk assessment: a metareview. *Criminal Justice and Behavior*, 37(9), 965–988. http://dx.doi.org/10.1177/0093854810374274.

Singh, J. P., Grann, M., & Fazel, S. (September 2013). Authorship bias in violence risk assessment? A systematic review and meta-analysis. *PLoS ONE*, 8(9). http://dx.doi.org/10.1037/journal.pone.0072424.

Davis, J. A. (2001). *Stalking crimes and victim protection: prevention, intervention, threat assessment, and case management*. Boca Raton, FL: CRC Press.

Turner, J. T., & Gelles, M. G. (2003). *Threat assessment: a risk management approach*. Binghamton, NY: The Haworth Press, Inc.

Virginia Tech Review Panel.(2007). *Mass shootings at Virginia Tech, April 16th, 2007: report of the Review Panel presented to Governor Kaine, Commonwealth of Virginia*. Richmond, VA: Virginia Tech Review Panel.

Vossekuil, B., Fein, R., Reddy, M., Borum, R., & Modzeleski, W. (2002). *The final report and findings of the safe schools initiative: implications for the prevention of school attacks in the United States*. Washington, DC: U.S. Department of Education, Office of Elementary and Secondary Education, Safe and Drug-Free Schools Program and U.S. Secret Service, National Threat Assessment Center.

Webster, C. D., & Hucker, S. J. (2007). *Violence risk assessment and management*. West Sussex, England: John Wiley & Sons, Ltd.

Yang, M., Wong, S. C. P., & Coid, J. (2010). The efficacy of violence prediction: a meta-analytic comparison of nine risk assessment tools. *Psychological Bulletin*, 136(5), 740–767. http://dx.doi.org/10.1037/a0020473.

Psychopathology and Criminal Behavior

Grant C.B. Sinnamon
James Cook University, Townsville, QLD, Australia

Key Terms

Psychopathology
Forensic psychopathology
Mental illness
Criminality
Offender
Victim
Personality disorder
Catathymic
Psychopath
Compulsion
Delusion
Obsession
Cognition
Emotion
Motivation
Behavior
Trauma
Epigenetic
Psychotic
Sadistic
Paraphilia
Social norm
Deviant
Abnormal
Erotomania
Stalking

INTRODUCTION

There is a common perception that mental illnesses and criminal behavior are ubiquitous. This is not necessarily the case, though if it were, many of us would sleep better at night knowing that crime and criminality was essentially an illness rather than a functional element of some far more complex social construction. It would indeed fit our increasing social need to pigeonhole differences and reduce things to a simple explanation (and therefore, presumably, a more easily "fixed" problem). The reality is far more complex and not readily espoused in a chapter or two of a single textbook. Certainly, by defini-tion, many thoughts (cognitions), emotional states (affect), and behaviors associated with psychopa-thology are deviant. However, deviance does not by extension mean criminal. Although it is true that all criminal acts are, by social tenet, deviant, not all deviant acts are criminal.

In spite of this, there are identified psychopathologies and cognitive, affective, and behavioral manifes-tations that, when present, collude with circumstances to make it more likely that an individual will engage in deviant behaviors. Furthermore, it is also true that many of these deviant acts may well turn out to also be criminal. It is equally true that these psychopathologies and psychobehavioral manifestations collude with a different set of circumstances to increase the risk of an individual becoming a victim of crime.

To go one further step, it is equally true that some of these characteristics, under the right circumstances, may collude to produce a corporate CEO, sports champion, or megastar actor or actress. This is perhaps, then, the first rule to keep in mind when proceeding through this chapter and applying its contents—it is not just the internal mechanisms but also the external supports and opportunities to express them that often determine the ultimate outcome. The corporate world is full of psychopathic personalities who have never committed a murder but have ruthlessly used their psychological makeup to allow them to pursue financial goals and objectives despite the potential impact on others. In fact, the study of what has become known as *corporate psychopathy* is a burgeoning field within the arena of organizational psychology.

In this chapter, we will explore the psychopathologies and various cognitive, affective, and behav-ioral characteristics that are associated with an increased risk of perpetrating a criminal act. The subject matter for this section of the chapter is sufficient to fill several textbooks, and therefore this section can, at best, only provide a cursory glance at a selection of psychopathologies associated with crime and an increased risk of criminal behaviors.

One disappointment to me owing to brevity is the need to exclude a section on the neuroscience of forensic psychopathology. The neurosciences—that is to say, the study of the structures, systems, and func-tions of the human central nervous system—are my great passion. Unfortunately, to do the topic justice would require far more than a few words in a single section of this chapter. This text is about the analysis of crime from a social science perspective, and as such I have elected to focus on the psychological aspects of the presented psychopathologies in order to make the information more accessible to the likely reader-ship, and to ensure it is in line with the remainder of the text. To any "neuro-nerds" out there, I apologize.

11.1 What is psychopathology?

When we talk of psychopathology what are we talking about? In its simplest terms, psychopathology refers to the understanding, study, or knowledge of any illness, disorder, dysfunction, or dysregulation of the mental processes that govern cognition, affect, and behavior. The word itself comes from three ancient Greek root words. *Psyche* means "mind" or "soul;" *pathos* means "suffering, feeling, emotion, calamity," or is literally translated as "what befalls one;" and *logia* means "to speak." Over time, suffixes derived from *logia* such as "logy"/"ology" have come to mean "study of," "science of," or "branch of knowledge [of]" (Harper, 2013). Therefore, for the purposes of this chapter, psychopathology will be defined as *the branch of knowledge and collective body of conditions relative to the suffering of the mind*.

In the clinical sense, for an individual to be diagnosed with a psychopathology they must meet a variety of guidelines both general and disorder specific. In general terms, the symptoms experienced must originate within the individual as opposed to being a situational reaction, and the symptoms must be outside of the individual's physical and/or mental control (American Psychiatric Association, 2013). It is often an area of disagreement between mental health professionals as to when a reaction moves from being a situationally driven response to a pathological condition. Psychopathology associated with traumatic events and bereavement are two examples where this can be a difficult distinction to make. On one hand, depression, anxiety, fear, cognitive distortion, behavioral dysfunction, and significantly disordered affect are all contextually appropriate responses to an acute experience of major trauma or grief. However, if an individual is still reacting to the event in this way several months postevent, then it may be that the responses have become pathological. If, at this time, the individual is unable to control their physical and mental reactions, then this would qualify them as meeting the general criteria for psychopathology symptoms.

Disorder-specific criteria are dependent upon the characteristic symptomology of each individual disorder (for more detail about disorder-specific criteria, please see American Psychiatric Association, 2013; World Health Organisation, 1992). For example, in the case of major depressive disorder (MDD or unipolar depression), an individual must meet several diagnostic criteria (Table 11.1) that relate to the known key characteristics and symptoms of the disorder such as increased negative affect, anhedonia, lassitude, changes to sleep and dietary practices, psychomotor changes, cognitive impairment, feelings of guilt and worthlessness, and suicidal ideation.

When assessing an individual for psychopathology, there are four factors or themes that form the lens through which the "abnormality" of symptoms should be interrogated. Also known as *the 4Ds of Abnormality*, these are deviance, distress, dysfunction, and danger (Table 11.2; Comer, 2005). Assessing any possible psychopathology using these themes allows the assessor to make evaluations of deviance and abnormality based on the potential for negative impact on the individual, their family and peers, and society in general. It is ultimately the potential for harm or serious dysfunction that dictates whether a cluster of symptoms will be labeled as pathological.

Increasingly, mental processes, emotions, and resulting behaviors that are different or inconvenient have been ensnared in the web of psychopathology. It appears that the world, as it shrinks through the advances in technology, is also striving for homogeneity, and any ways of thinking, experiencing, and interacting with the world that do not readily "fit" are being pathologized and labeled as deviant. Indeed, many social commentators and social science academics are contributing to a contrary discourse lamenting the loss of individuation, social choice, and the celebration of alternative experience (Blažič, & Vojinovič, 2002; Brown, 2012; Butler, 2013; Lelandais, 2013). This phenomenon is reflected in the evolution of psychopathology where many personal characteristics and behaviors long experienced by individuals have now become recognized pathologies. Some of these, such as those characterizing the attention deficit disorders, have progressed in this fashion amid considerable controversy. This is perhaps one of the differentiating factors of psychiatry over many, if not most, other areas of medicine. That is the reality that diagnosing psychopathology is often an arbitrary process in which individual opinion and experience, as much as any "hard" evidence, plays a large role in the diagnostic outcome proffered by the attending mental health professional.

Table 11.1 General List of Diagnostic Criteria for MDD Based on the Diagnostic Criteria Established in the DSM-5 and ICD-10

General Diagnostic Criteria for Major Depressive Disorder (MDD)

A. SYMPTOMS MUST:
1. Cause impaired function: The symptoms cause significant distress and/or functional impairment in social, occupational, educational, or some other important area
2. Not be a direct result of substance use or a general medical condition

B. MUST HAVE: Depressed mood (negative affect) and/or a loss of interest or pleasure in daily activities (anhedonia) for at least the previous two weeks
1. These changes must represent a definite change from the person's usual mood/level of interest (i.e., a change from baseline)

C. MUST HAVE: Specific symptoms, at least five of these nine, present nearly every day for at least the previous two weeks:
1. Significant negative affect most of the day, nearly every day, as indicated by either subjective report (e.g. feels sad or empty) or observation made by others (e.g. appears tearful). In children, this may include a hypomanic state.
2. Anhedonia (decreased interest or pleasure in most activities)
3. Significant weight change (~5%) and/or change in appetite
4. Change in sleeping patterns: Insomnia or hypersomnia
5. Change in levels of activity: Psychomotor agitation or retardation
6. Lassitude (fatigue or loss of energy, physical and/or mental weakness)
7. Feelings of guilt and/or worthlessness: Feelings are excessive and/or inappropriate and may be delusional (possibly indicating accompanying psychosis)
8. Poor concentration and/or indecisiveness: Not thinking clearly and unable to concentrate or make decisions either at all or without great difficulty
9. Suicide ideation: Actively thinking about death, feeling that they would be better off dead, and/or of killing themselves. May or may not have a suicide plan and the resources to complete the act.

American Psychiatric Association (2013) and World Health Organisation (1992).

Therefore, I offer a word of caution for all who venture into the world of psychopathological diagnosis, labeling, stereotyping, and profiling under any circumstances—be they forensic, academic, medical, or policy related. The advice I offer is this:

The differential, evanescent, and heterogeneous nature of many psychopathologies means that collectivist and generalistic processes should be avoided at every opportunity. Instead, one should always maintain the awareness that there is potential for considerable individual difference in every situation. Therefore, in the case of crime analysis, there is little to be gained from thinking along the lines of "This person is schizophrenic" or "This person is an alcoholic." Rather, it will serve you much more effectively to think in terms of specific factors that can be individuated. For example, "This person is paranoid" or "This person does not empathize." Individuating symptoms of psychopathology or deviant cognition, affect, and behavior (or behavior management) will allow you to create far more accurate representations, profiles, and predictive models than any singular psychopathology label will.

This idea is not unique, and I cannot claim to possess some wonderful insight of which others proceed unawares. In fact, within spheres of psychological and neuropsychological activity, psychopathology is increasingly being investigated in the context of specific symptoms, characteristics, states, and

Table 11.2 The Four Key Themes of Abnormality Used in Consideration of Psychopathological Symptomology (Also Known as the 4Ds of Abnormality).

Factor	Description
Deviance	Refers to the extent to which an individual's specific cognitions, affect. and behaviors are considered to be outside of what is "acceptable" and/or "normal" within their common social environment.
	When making this consideration, it is important to recognize that there are many subcultural groups whose thoughts, beliefs, emotional responses, and behaviors may bear little resemblance to the wider cultural norms of a society. Therefore, the measure of deviance must be considered in the appropriate context of the individual and not simply by a predefined mainstream cultural normative set.
Distress	Refers to the extent to which the presenting symptoms cause emotional distress to the individual experiencing them.
Dysfunction	Refers to the extent to which the presenting symptoms impair the individual from engaging in their "normal" daily activities such as work, study, self-care, or engaging in general activities of daily living like washing, preparing a meal, or dressing appropriately. Additionally, dysfunction is increasingly used to refer to systemic neurophysiological deficits that are known (or thought) to be complicit in the presenting psychological impairments.
	It is important to be aware that not all dysfunctional behavior has a pathological origin. Some individuals may choose to engage in dysfunctional behaviors such as self-harm, hunger strikes, or sadomasochistic sexual endeavors (although certain sexually deviant behaviors may qualify as pathological).
Danger	Refers to the inherent risk of harm either to the individual or others, usually by the individual, as a result of presenting symptoms.
	Examples of high-risk symptomology include aggressive paranoia, delusions or hallucinations involving persecution or directives to harm self or others, and suicide ideation.

Comer (2005).

traits. It is now widely acknowledged that singular or smaller subsets of symptoms may present unique problems for sufferers. Furthermore, this realization extends to the acceptance that certain symptoms may have a below-diagnostic presentation (subthreshold psychopathology), while still producing significant cognitive, affective, and behavioral challenges. Ultimately, this has led to many psychopathologies being moved from the traditional model of diagnostic absolutism in which an individual either has the condition or they do not, based upon a diagnostic threshold of symptoms and symptom severity. Today, more psychopathologies are being moved into spectrum models in which individuals are placed upon a continuum according to symptom presentation and may be moved up or down the spectrum as symptoms subside or increase.

11.2 What is forensic psychopathology?

While psychopathology in general takes in the purview of all areas of mental dysfunction, forensic psychopathology is a specialized area in which the focus rests on those psychological maladies and their symptoms and consequences that, for one reason or another, intersect with the legal system.

Simply put, forensic psychopathology is interested in clinical psychopathologies with criminal implications. This includes any psychological or neuropsychological condition that impacts the perpetrators of criminal acts as well as their victims.

Forensic psychopathology plays an increasing role within the legal system and holds a precarious place within the overall scheme of the criminal and legal systems. One of the greatest challenges for psychopathology in a forensic environment is its often contrary view of a situation when compared to the legal system in general. The legal system is grounded in precedent with historical rulings and outcomes providing the guidance for future decisions, strategies, arguments, and judicial considerations. Forensic psychopathology, however, is grounded in the scientific approach, which is a science- and emergent-knowledge-based process. Historical "knowledge" is continually being superseded, reconsidered, and discarded in the face of the emergent evidence. This creates unique challenges for both fields when they find themselves confined to the same space and the same set of circumstances requiring attention.

This divide is becoming increasingly evident in many present-day settings. Many Western countries are seeing the sociopolitically driven renewal of punitive, revenge-based, sociolegal policies and practices in which deviance and crime are being treated harshly in an ever-increasing manner. Funding for preventative and rehabilitative projects is on the decline, while prisons and their populations are becoming larger. Psychosocially derived rationales for crime and criminality are being rejected in favor of a fundamental attribution error-laden rationale that places blame and responsibility squarely on the individual and their personal "weaknesses" or personality flaws. This is in contradiction to the emergent science showing increasing evidence for a confluence of environmentally shaped, biopsychosocially derived factors strongly associated with both criminality and victim risk profiles. The immediate future relationship between forensic neuroscience and associated pathologies, and the political, legal, and criminal justice systems, seems destined for a prolonged feud.

Ironically, within this atmosphere, forensic psychopathology as a science is making significant inroads into understanding the circumstances that surround the relationship between crime and the processes of the brain and mind. Neuroscience has uncovered many links between cognitions, emotions, and behaviors associated with increased criminality and the structures, systems, and functional outputs of the brain. Despite the controversy, the construct of forensic psychopathology has become ingrained into the Western psyche as is evidenced by the plethora of popular media offerings relating to the subject. From Thomas Harris's Hannibal Lector (immortalized by the great Sir Anthony Hopkins) to contemporary shows such as *Dexter* and stranger-than-fiction reality such as "The Rotenburg Cannibal," Armin Meiwes, who advertised and found on the internet a willing victim that he could kill and eat, Western culture has become enamored with the field of forensic psychopathology. In this chapter, we will explore forensic psychopathology in two ways. Firstly, we will look at some of the mainstream diagnosed psychopathologies that are associated with crime and criminal behaviors. Secondly, and of far more use, we will explore the constituent parts of these psychopathologies that result in an increased risk for criminal association—that is, the various cognitions and emotions that initiate motivations as well as the resultant behaviors that manifest in the attempt to achieve resolution. It is these constituent parts that are of greatest value in crime analysis as they provide a direct insight into the mind and motivation of the individual, thereby giving an opportunity to better understand a given crime or crime scene and, perhaps even more importantly, to predict future criminal acts.

11.3 The nature versus nurture fallacy: the interaction between programing, predisposition, and experience

Traditional Psych 101 classes often commence with the tutor dividing the class of first-year students into two groups and instructing them to take opposing sides in an impromptu debate as to whether nature or nurture was the more important factor in development. Today, this has largely been discarded as we have come to understand that the argument between nature and nurture is a false divide. Instead, we have come to appreciate that the relationship is one of symbiosis. Nature and nurture operate within a bidirectional relationship in which nature provides the blueprints and nurture provides the experiential resources necessary for development. Therefore, to understand why a crime occurred, how to predict similar events in the future, and ultimately how to determine preventative measures that will mitigate the risk of future events, it is important to include in any analysis the antecedent factors that lead to the circumstances such that the crime was committed. A full and effective analysis must therefore include an accounting of the potential impact that personal developmental and continuing environmental factors may have had on the course of events.

Our contemporary social mentality is one that demands immediate, simple, and absolute answers to questions of causality, risk of recidivism, and problem resolution. In a psychopathological sense, this often means we look to simple, homogeneous, internal causes for abnormal behavior and to simply "Take a pill" or "Lock them up and throw away the key" resolutions. Indeed, it is increasingly the case that crime is attributed to people being innately bad and lacking in social morality or self-control. The underlying message is that some people are just born "bad." This mentality is indicative of the populist belief in Western society that genetics is the primary mechanism through which we, as people, are created. This is nonsense, and any individual who is serious about using scientific methodology to analyze crime and criminal phenomena must be wary of this kind of simplistic, reductionist thinking.

In a very real sense, our environment dictates the way that our brains function. The function of neurons—the basic operational cells of the central nervous system—is a result of the type and amount of proteins that are present on the cell at any time. Neurons are generally covered in millions of proteins that are present and prepared to be utilized by the cell to produce a host of endogenous chemistry that will engage the cells in a specific function as required. The amount and type of proteins that are present in the cell are dependent upon two things: (1) the amount of nutrients and oxygen that are available to the cell, and (2) the amount, nature, and intensity of stimulation the neuron receives from the environment.

The amount of nutrient and oxygen is extremely important to brain cells in the same way that they are to the rest of the body. Appropriate environment and nutrition are essential to brain function as up to 20–25% of the energy used by your whole body is used by the brain alone. The nutrient and oxygen availability is largely determined by environmental factors relating to the characteristics of your immediate surroundings, diet, general health, and exercise and fitness levels.

Cellular stimulation is directly related to the environment and the way in which an individual interacts with their surroundings: Are neurons and neuronal pathways being enriched as a result of the stimulation they are receiving? Are neurons being overstimulated and therefore taxed beyond their physical capacity to cope (performance threshold)? Alternatively, is the amount of stimulation the neurons are receiving below the minimum activation threshold, thereby inhibiting the development and performance of the cell? Is the brain being stressed and therefore the specific patterns of stimulation are triggering proteins that produce threat response chemistry rather than other, more long-term beneficial chemistry required for "normal" performance of emotions, learning, memory, and other important functions? These factors play a major role in brain development, function, and ongoing health. Any prolonged over- or understimulation of brain cells can result in serious neural pathway damage and even cell death. Moreover, chronic stress-induced stimulation of neurons can result in a host of damaging results, such as a proliferation of unproductive and ultimately deleterious cell receptors, high levels of production and secretion of neurotoxic chemistry, and ultimately cell death and atrophy of brain tissue. Psychopathology is often characterized by many of these deficit features, and increasingly, evidence is illustrating the strong correlation between psychological well-being and the structure, systems, and function of the brain (Beblo, Sinnamon, & Baune, 2011).

It is for these reasons that chronic stress and trauma in early life, while the brain is engaged in rapid organization and hyperplastic growth, have been shown to be very damaging to brain structure, physiology, and function (Pechtel & Pizzagalli, 2011). As a kind of double whammy, the environments that are most often associated with negative patterns of stimulation and nutrition are also associated with poor relational attachment and significant differential learning (Sutherland, 1947), in which individuals are exposed to a variety of deviant behavioral and social norms that are presented to them as "normal" and acceptable. Differential environments accompanied by suboptimal neural development and/or function collude to influence personality, behavioral norms, and cognitive orientations that make it more likely for an individual to engage in activities outside of what is accepted by wider society. This invariably includes criminal activity.

11.4 Genetics

Genetics is a complex and poorly understood area regarding the extent to which it is directly responsible for behaviors and higher-order cognitive and physiological processes. While the debate rages over how much influence genetics has, it is generally thought that, at best, our genes may have about a 30–40% influence on our development, with our physiological function with the environment playing a far greater role. For factors such as personality, cognitive performance, behavioral choices, and morality, this influence is thought to be considerably lower still. Therefore, it is highly unlikely that genetics can be held responsible for criminality and one's risk for criminally associated psychopathology. The greater likelihood is that our genetics provide the basic physiological blueprints for our anatomical development, regeneration, and function. Our myriad environmental factors then provide the orchestration that results in the way these psychological and physiological factors mature and are ultimately oriented.

11.4.1 Epigenetics

While genetics in and of themselves are not likely to play a singularly major role in shaping specific personality and behavior, the role of epigenetics appears to be far more likely to be involved in the process of shaping our psychological orientations and characteristics of mind. Invariably, this includes those factors that relate to crime, criminality, and criminally oriented psychopathology.

Epigenetics literally means "above, outside, or around" genetics and refers to the way in which genetic expression or activity is manipulated by heritable factors that are neither caused by or cause the alteration of the DNA sequence (Holliday, 2006). In this way, epigenetics refers to nongenetic influences on gene activity that can either last through cell divisions for the entire lifetime of an organism or can survive procreative transfer and be inherited across multiple generations without alteration to the underlying DNA sequence of the specific gene (Bienertova-Vasku, Necesanek, Novak, Vinklarek, & Zlamal, 2013; Schlichting & Wund, 2014; Zaidi et al., 2014). The term was initially coined to refer to the linking of genetics and developmental biology (Waddington, 1939), and in that early context was broadly defined as "the sum of all those mechanisms necessary for the unfolding of the genetic program for development" (see Holliday, 2006, p. 76). In the contemporary context, epigenetics refers not only to the mechanics of development but also to the mechanics of more generalized physiological and psychophysiological functions.

In this contemporary context, the field of epigenetics is a relatively new scientific arena and is fast becoming a very important field within the clinical neurosciences. Specifically, in the area of forensic neuroscience, epigenetics is being identified as a factor in a host of neuropsychiatric pathologies that are associated with criminality risk such as borderline personality disorder (Dammann et al., 2011; Franklin et al., 2010; Teschler et al., 2013), psychotic and manic disorders (Connor & Akbarian, 2008; Guidotti et al., 2011; Read, Bentall, & Fosse, 2009; Wockner et al., 2014), and addictive disorders (Maze & Nestler, 2011; Wong, Mill, & Fernandes, 2011).

While much research has been conducted and continues to be undertaken in epigenetics and its mechanisms, the inheritance processes of epigenetics is much less understood than those of the more traditional genetic inheritance processes. In traditional genetics, inheritance is based upon cellular lineage and heritability through a natural process of cellular cloning (Holliday, 2006). Therefore, if changes occur to a cell, it is expected that future clones of the cell would also have the same changes or genotype. In contrast, epigenetics is more likely to occur in cellular groups rather than individual cellular lineages. For example, a group of cells with the same receptor type may be impacted epigenetically by an alteration to the chemistry involved in the activation of the cell via that specific receptor. This functional alteration then results in changes to a specific phenotypical expression relating to the associated mechanisms. In the case of potential forensic psychopathology, this change may, for example, relate to any number of neurotransmitters or other neurochemistry such as dopamine, serotonin, or corticotrophin-releasing factor, which, in turn, may lead to high-risk psychopathology symptoms such as paranoia, aggression, hallucinations, or increased susceptibility to addiction.

In an epigenetic process, these changes create significant functional variation in the phenotype of the individual while producing no alterations to the underlying genotype. In some cases, the epigenetic changes may reverse, while in others they may remain for the duration of the individual's life and even be inherited by any children produced. While considerable work is being undertaken to improve our understanding of epigenetic mechanisms such as histones, DNA acetylation and methylation, and chromatin structure or configuration, the specifics of these processes and their importance to the mechanisms of epigenetics remain largely unelucidated.

11.4.2 Early-life stress and childhood trauma

Perhaps one of the most significant findings in the field of epigenetics thus far is that one of the major influences on epigenetics is stress and trauma, and in particular stress and trauma in early life (Ehlert, 2013; Malan-Muller, Seedat, & Hemmings, 2014; McGowan, 2013; Murgatroyd, Wu, Bockmuhl, & Spengler, 2010). Research into the impact of early-life adversity such as trauma, neglect, abuse, and chronic stress (collectively termed early-life stress, or ELS) has shown that the impact of these experiences on the developing individual is significant, pervasive, and highly deleterious (Clark, Caldwell, Power, & Stansfeld, 2010; Green et al., 2010; Strine et al., 2012). ELS refers to any event a child experiences in early life (generally in their first six years or the "critical period" of development, but can extend to include up to the onset of puberty) that exposes the child to psychological or physical stimuli that exceed their ability to successfully cope (Gunnar & Quevedo, 2007). Among the most common stressors are those that are associated with variations to the themes of abuse, neglect, and trauma. The stressor may be a single major event, a chronic stressor, or a succession of traumatic events.

Whether or not exposure to such an event or string of events leads to a deleterious outcome is dependent upon a myriad of other factors, including environment, psychosocial supports, and individual characteristics, rather than the details of the event itself. For some, a single exposure to a stressor may be sufficient to produce deleterious outcomes to the machinations of the brain and/or mind. For others, even multiple exposures may be mitigated as a result of individual resilience and/or social supports offered by carers. Ultimately, the spectrum of the consequences of ELS extends from the development of high resilience and resourceful adaptation to severe levels of biopsychosocial maladaptation with a significant prevalence of psychological and physiological pathologies (often comorbidly) in children, youth, and adults (Ehlert, 2013).

ELS experiences increased criminality risk (perpetrator and victim) through two primary mechanisms: (1) biologically, via maladaptive neurodevelopmental processes; and (2) socially, via differential reinforcement (Sutherland, 1947), poor attachment, and lack of empathy. Biologically, risk is increased due to the negative impact that ELS has on neurobiological development. The intricacies of this impact are complex, however. Any adverse impact on neurological systems during early life will have significant repercussions in the capacity for the brain to develop and mature. The result can be a host of psychopathologies and neuropsychopathologies that manifest due to neural pathway deficits that prevent the brain from operating as an integrated unit. For example, the brainstem and midbrain areas, responsible for autonomic regulation and sensory reception, may be hyperresponsive and unable to be downregulated by the insightful and rational processes of the higher-order regions of the frontal cortex. This results in hypervigilance and survival-mode reflexive behaviors of the fight–flight fear response. Impulsive, defense-oriented reflexive behaviors include extreme (and abrupt) violence such as reactionary homicide and impulsive defense behaviors that can quickly get out of hand, such as the escalation of a domestic argument to a hostage situation. Other neurological deficits associated with ELS have been shown to be associated with the development of personality disorders (particularly, but not limited to, antisocial and borderline personality disorder), psychopathy, recidivistic violence, victim-precipitated crimes, lack of empathy, lack of remorse or guilt, impulsive behaviors, and attention deficit disorders.

Socially, ELS is associated with a higher risk of psychopathology that results in criminal activity due to the way in which a child may be exposed to poor relationships with their associated lack of empathic development, and criminally deviant behaviors that may be reinforced as "normal" or acceptable behaviors. Children who have experienced significant ELS are often exposed to these stressors as a result of abuse, trauma, and neglect at the hands of parents, primary caregivers, or adult relatives in a caregiver role. One consequence of this betrayal by primary adults in the child's life is the inability of the child to be able to establish secure attachments. Socially, then, the child does not go on to develop the ability to establish positive and mutually protective and rewarding relationships with others.

Moreover, poor attachment results in a lack of social connection to other individuals and to the community at large. This leads to a lack of empathy for others, a lack of social conscience, and a disregard for social norms, rules, and conventional morals and ethics. The lack of empathy, in turn, means that the child (and later the adult) does not feel a moral connection to the society in which they live, nor are they restrained by feelings (or potential feelings) of guilt or remorse as a result of wrong actions or the harm their actions may befall on others. These personality characteristics are decidedly pathological and result in an extremely high risk of criminal behavior. Insecure and reactive attachments are linked to myriad psychopathologies, including personality disorders, psychotic and manic disorders, addiction, paraphilia, and a host of other conditions associated with antisocial, aggressive, self-destructive, and deceptive or manipulative characteristics.

Differential reinforcement as a gateway to criminal deviance was first posited by Sutherland (1947) in his Differential Association Theory. Put simply, Sutherland believed that a person's behaviors can become deviant or "criminally delinquent" when they are presented with more positively reinforced behavioral choices that are favorable to violations of law than they are with choices that support upholding it. According to Sutherland (1947), criminal behaviors are differentially reinforced as acceptable by exposure to social circumstances in which antisocial conduct is espoused as a positive social message. In this environment, a child (or, later, a primed or impressionable adult) will see evidence that criminal behavior is not only socially acceptable but also rewarded within the social strata in which they are embedded. These behaviors may be rewarded either directly, such as a child being socially praised and physically rewarded for stealing, bullying, or engaging in sexual activity, or they may be rewarded indirectly, such as a child receiving gifts as a result of their parents receiving a financial windfall from illegal activities (e.g. selling stolen goods or drugs, or even cheating on a tax return).

The risk of ELS resulting in criminal psychopathology is enhanced if a child remains within a chronically stressful environment and continues to be exposed to differential reinforcement of criminal and/or antisocial beliefs, attitudes, and behaviors. In this environment, a child is exposed to the multiplying factors of the burden of deficit neurodevelopment, poor attachment and social connection, and differential reinforcement of criminal and/or delinquent behaviors. For example, a child who is neglected and cannot rely on an adult to provide for their basic needs may learn that theft, deception, and self-preservation are behaviors that result in reward. Combined with a neuropsychopathology that results in diminished cognitive capacity in domains such as social perception and executive function, this sort of differentially reinforced learnt behavior is more likely to result in criminal activity.

While criminally oriented psychopathology that has its roots in ELS is often amplified by the social conditions in which the ELS occurred, if a child experiences stress, abuse, neglect, or trauma in an environment that can mitigate the impact of these experiences via strong attachments, security, and stability—even if this is offered after the event through relocation of the child—then the child has a much lower risk of engaging in, or becoming a victim of, crime. While both the risk of developing psychopathology and the risk of engaging in criminal behaviors are reduced, even if they still develop psychopathology, the risk of criminal deviance will be significantly mitigated by the positive social structures and behavioral "norms" to which the child has been exposed and enculturated.

While ELS represents a major challenge due to the vulnerability of the developing brain, stress and trauma also play significant roles in adult mental health. Research shows that an overwhelming number of adults in prison also suffer from a variety of psychopathologies, many of which have directly contributed to the behaviors that resulted in their incarceration (Goldenson, Geffner, Foster, & Clipson, 2007; Henrichs & Bogaerts, 2012).

11.5 Criminal psychopathology: mental states and characteristic pathology associated with the perpetration of crime

Criminal psychopathology is an attractive term for law enforcement as it implies that there may be some matrix of conditions and behaviors that is available to assist in identifying perpetrators of crime and predicting those who may be more likely to commit criminal acts in the future. To some extent, we can identify pathologies or personal characteristics and ascribe to those affected various degrees of inherent risk of perpetrating certain behaviors. Indeed, because of this capacity, we have the fields of forensic psychology and criminal profiling. However, as stated, this is not as simple as it sounds.

There is significant individual difference involved in any train of psychological and social events that lead up to a person acting in one way or another. Historical experience (learning, memory, and reflexive conditioning), cognitive orientations, emotional factors, motivations, and ultimately physical capacity to engage in a particular action all collude to create an outcome. The influence of specific psychopathology or pathological symptoms runs through several of these factors and is therefore more complex as a predictor than a lineal relation such as physical capacity or motivation. Instead, it is the specific factors that can be broken down from within the psychopathology that are more beneficial to any analysis of a crime.

While there are many psychopathologies that can result in criminality, in this next section the focus will be on a relatively small cross-section of those with a stronger association to behaviors relating to person-to-person criminal activity (e.g. sexual and/or violent physical assault, sadistic homicide, and stalking). These pathologies will be discussed in the context of the psychopathology, as well as the specific behavioral, cognitive, emotional, and motivational characteristics that may be associated (delusion, obsession, compulsion, paranoia and persecution, helplessness, rumination, fear, cognitive irrationality, irritability, risk taking, aggression, etc.). There is no doubt that many other pathologies and characteristics could be included if this chapter were simply a forensic psychopathology text with no other function and if there was several hundreds more pages to utilize. The defining factor that has helped compile the contents for the next section is whether or not the psychopathology or characteristic to be included is of ultimate value in the realm of applied crime analysis.

11.6 Typical psychopathologies, mental states, and psychopathological characteristics associated with perpetrators of crime

11.6.1 Catathymic process

For any text dedicated to forensic mental health, catathymic pathology has a very strong legitimate claim for the first position in the list of conditions under consideration. A catathymic process or crisis is an unexpected, impulsive, and explosive violent outburst that is not readily understandable from the situation in which it occurred but rather is only able to be understood in the context of individual unconscious (or outside of conscious awareness) motivations (Maier, 1912; Schlesinger, 2004). The term comes from the Greek roots *kata* and *thymos*, which are perhaps best defined together as "in accordance with emotion/s." In this way, then, a catathymic behavioral response occurs when an individual is overwhelmed by emotion to the point that psychological rationality and control are compromised, resulting in a primitive reflexive outburst of aggression. When Maier forwarded the idea of catathymia (*Katathyme*) in 1912, he posited that the catathymic process was a psychological response brought about by the strong and unyielding emotional attachment to underlying schemas (Maier, 1912). These schemas do not necessarily represent their subject matter with any accuracy or even reality. Indeed, Maier believed that some psychotic characteristics could be catathymic in nature (e.g. paranoid delusions) if they were derived from underlying psychological phenomena such as maladaptive cognitions to trauma. In contrast, a drug-induced psychosis would not be catathymic because it was brought about by altering brain function rather than from some process of psychogenesis (Schlesinger, 2007).

The original idea of catathymia forwarded by Maier (1912), Wertham (1937), and other contemporaries was that the condition constituted its own clinical identity. Within the parameters of current diagnostic frameworks, catathymia, the catathymic process, or a catathymic crisis is not directly translatable to a single pathology. While catathymia was recognized in earlier versions of the Diagnostic and Statistics Manual of Mental Disorders (DSM), it was not included in the DSM-IV (American Psychiatric Association, 2000) and there remains no diagnostic entity called "catathymia" within the DSM-5 (American Psychiatric Association, 2013) or ICD-10 (World Health Organisation, 1992). Instead, catathymia is a situation brought about via circumstances found in many psychopathologies. Therefore, in its modern incarnation, catathymia is more likely to be considered as a syndrome, symptom, or consequence within the confines of other mental disorders such as posttraumatic stress disorder, psychotic disorders (particularly those with delusional characteristics), mood disorders (and mood disorders with psychotic features), and several of the personality disorders.

In contemporary circles, catathymia is used to describe any explosive or impulsively violent behavior that appears to have its roots in heightened emotionality charged by irrational thought processes or beliefs. The manner in which catathymia is represented within the criminal sphere can be quite varied. For example, violent serial crime, familial homicide, stalking-related violence, and sudden (often) inexplicable interpersonal violence perpetrated against a stranger or casual acquaintance are often laced with catathymic crisis.

Catathymic crisis has been described in many ways. For our purposes here, I will describe the catathymic process within three subcategories. The first I have called *delusional catathymic process* (DCP), and it is, in essence, a general description of the original catathymic crisis espoused by early proponents such as Maier (1912) and Wertham (1937); and the second is the dual categorization suggested by Revitch and Schlesinger (1978) in which catathymic crisis may occur within either a *chronic catathymic crisis* (CCC) or an *acute catathymic crisis* (ACC).

11.6.2 Delusional catathymic process

DCP is closely aligned with both Maier's original ideas of catathymia and those espoused by Frederic Wertham in the decade prior to the outbreak of the Second World War (Wertham, 1937). Wertham's catathymic process involved the process in which underlying thoughts and ideas that are highly charged with emotional content transform the emergent stream of cognitions in a way that (usually) results in irrational wishes, desires, fears, and ambivalent striving (Wertham, 1937). Through this process, the (prospective) offender develops the delusional notion that they must engage in a violent act against another person or against themselves in order to achieve or resolve the situation to its conclusion. This idea manifests in the mind as a definite plan with an accompanying strong motivation to see it through; however, the plan may be resisted at some level within the individual's mind and therefore meet with behavioral hesitation or delay. According to Wertham, "The violent act usually has some symbolic significance over and above its obvious meaning…[and the cognitive processes of the individual] may have an almost delusional character in its rigidity and inaccessibility to logical reasoning" (Wertham, 1937, p. 974).

Wertham proposed an environmental etiology for the pathogenesis of the psychological conditions required for a catathymic crisis to manifest. Wertham asserted that an extreme emotional turmoil is manifest as a result of an unresolvable inner conflict brought about by a traumatic experience. This experience and consequent inner conflict and its associated hyperemotionality are blamed on external influences, and the individual's thought processes become increasingly self-absorbed, disturbing, and irrational until they suddenly come to a decision (generally embedded within a delusion) that some violent act is the only solution (Wertham, 1937). This decision becomes increasingly obsessional with related thought processes becoming increasingly irrational, reinforcing the delusional thoughts and validating the obsession. The person becomes convinced that this violent act is the only solution to their internal conflict. Although the individual may resist the urge to act, eventually they will commit (or attempt to commit) the offense. Wertham asserts that committing the act produces a (usually) temporary catharsis during which the offender will feel relief from the inner conflict and emotional turmoil, and during which they may experience a sense of normality. The individual may even achieve some form of insight into their condition. For some, psychological recovery may ensue, while in others the relief is temporary and they are eventually "recalled" to violence to relieve the inner tension again.

This form of catathymic process is often seen in serial crime, and Wertham believed that the associated violence was not limited to violence against another person but could also extend to lighting fires, or arson; self-harm, such as self-castration or other genital mutilation; self-blinding, and other forms of significant nongenital self-mutilation; and suicide (which, in this context, may be construed as a form of intrinsically engaged homicide).

Wertham suggests that there are five stages within this form of catathymic process (Wertham, 1978):

1. The initial disordered thinking
2. Development of a plan and increased emotional turmoil
3. Heightened emotional turmoil resulting in the committal of the violent act
4. A sense of superficial calm and normality
5. (Re)establishment of an inner calm

In the event of a temporary resolution only, then a sixth stage may be posited in which the individual comes full circle and is once again accosted by the same psychological and emotional turmoil that they could not resolve previously. This is then followed by the re-experiencing of the five stages above.

11.6.3 Chronic catathymic crisis

Revitch and Schlesinger (1978) do not identify catathymia within a construct of an independent clinical entity, but rather as a psychological phenomenon that can take either a chronic or acute form. From a behavioral standpoint, both CCC and ACC are therefore elements of the motivational processes within the psychodynamics of broader psychopathology.

The construct of CCC is essentially the same as that forwarded within DCP (above) and espoused by Maier (1912), Wertham (1937), and their contemporaries. Structurally, the main variation between DCP and CCC is that Revitch and Schlesinger have reduced Wertham's five stages down to just three (Revitch & Schlesinger, 1989):

1. The incubation period
2. The committal of violence
3. The achievement of relief

The incubation period may last from a few days to years, and during this time, the (future) offender becomes progressively more emotionally disturbed, irrational (and possibly delusional) in their thoughts, and obsessive about their (future) victim. It is during this phase that violence becomes the answer to their problems of inner turmoil. If the (future) victim has a relationship with the (future) offender, then it is often the case that as the inner turmoil and intense emotional instability increase and start to push toward the surface, then disturbed, irrational, and obsessional thought processes will be noticed by the victim and they will withdraw from the offender. This in turn often exacerbates the irrationality, tension, anger, and conflict within the offender, hastening the progression toward the inevitable violence. In the instance of stalking, this relationship may not be real but rather may be a delusion on the part of the offender. The psychological characteristics of this period include depression, and what Schlesinger (2007) refers to as schizophrenic-like thinking. Suicidal and homicidal thoughts are prevalent and, if the offender does not receive relief from the violent act, suicide is often a further outcome.

CCC can manifest over a prolonged period in one of three ways. As in the example above, CCC may occur in the confines of a long-term relationship in which the future perpetrator and future victim are known to one another, often intimately. Various incarnations of familial homicide and familial murder-suicide are prime examples of this form of CCC. CCC may also occur within the confines of a one-way obsessional-delusional interaction in which the future perpetrator forms an irrational attachment to another person (most often, it is a male offender and a female victim except in cases where the delusional thinking is erotomanic) and delusional thought processes conspire to evolve a belief that violence either against the individual or against another on behalf of the individual will resolve the perpetrator's internal conflict. Myriad stalking cases can be seen in this light, perhaps one of the more famous being that of John Hinckley Jr.'s obsession with the actress Jodie Foster and his attempted assassination of then-President of the United States of America Ronald Reagan. Finally, CCC may occur as a result of obsessional or delusional thinking by the future offender that, at least in the early stages of the process, is not targeted to a specific individual. Instead, the future offender is fixated on the act of violence itself rather than on a specific "victim." Alternatively, the future offender may be fixated on both the act and a victim "type," which then assists them to eventually progress to the selection of a specific victim. Some serial killers could be seen within this manifestation of CCC.

11.6.4 Acute catathymic crisis

In contrast to the progressive build-up to violence that occurs in CCC, ACC takes place as a sudden outburst that is triggered by an interaction with an individual who is either a casual acquaintance (e.g. the girl behind the counter at the local coffee shop) or a stranger whom the perpetrator has just met. In the case of ACC, the offender is triggered into a violent reaction as a result of some action or perceived action by the stranger. An example of this form of catathymic crisis is when a "spontaneous" violent act is perpetrated against a stranger in an unlikely situation such as when a chance sexual encounter results in homicide. A stereotypical situation could be where a young male with underlying sexual issues (e.g. latent or conflicted homosexual feelings, or early-life abuse or attachment factors that, while not contributing to an obvious psychological abnormality, are creating significant emotional conflict) is confronted by a woman (or man) with a sexually assertive personality. The (about to be) victim may behave or make a comment that triggers the underlying aggressive response (think of it as a maladaptive fight–flight defense mechanism) with the end result being homicide.

According to Schlesinger, "the acute catathymic process taps deeper levels of emotional tension and is triggered by an overwhelming emotion attached to an underlying conflict" (Schlesinger, 2007, p. 11). Although the offender may sometimes be able to provide an explanation for their behavior and a detailed account of what transpired, it is often the case that they have no or only partial recollection of the violent act itself.

According to Schlesinger, there are several similarities between chronic and acute catathymic events that result in homicide. These similarities include (this is a descriptive list based on Schlesinger's original list; for Schlesinger's specific list of similarities, see Schlesinger, 2004, p. 138):

1. With a few exceptions, cases primarily involve men killing women.
   a. The main instance where females become offenders against a male victim is in cases of domestic violence between intimate partners, although less than 3% of all men who are murdered are killed by their female intimate partner (Schlesinger, 2004).
   b. Male-to-male catathymic homicide is most likely to occur in the context of acute catathymic crisis.
   c. Female-to-female catathymic homicide is extremely rare, with most cases occurring between incarcerated women.
2. The victim is generally symbolic or representative of the deeper seated conflicts. The victim is not the "real" issue.
3. The victim's behavior (or perceived or imagined behavior) triggers the highly volatile existing emotion-laden internal conflicts.
4. Internal conflicts are generally egocentric in that they are focused on thoughts and emotions around real or imagined personal inadequacies. These are often relational in nature (often stemming from early-life-generated attachment issues) and in adult circumstances tend to extend to the sexual arena.
5. The homicidal act works to redact the internal turmoil and emotional tension.
6. In the immediate posthomicide period, the offender will generally experience a time of psychological balance or homeostasis. This generally includes feelings of relief and a stabilizing (or flattening) of affect.
7. The offender does not generally try to avoid capture for very long. They will often give themselves up to police or tell a friend or trusted authority such as a priest.
8. Mental health professionals are generally very poor at recognizing the risk that is inherent in the offender's conflicted emotionality during the prehomicide incubation period.
9. Investigation and crime analysis may miss the sexually derived offender motivations due to them often being relatively obscure. The act may not present in any way like a sex crime, and the scene may be more indicative of a more general violent crime. Indeed, the sexual motivations may not become apparent until the analysis can include direct contact with the (cooperative) offender.

Similarly, Schlesinger has identified a number of differentiating characteristics between acute and chronic catathymic crisis. Table 11.3 presents the differentiating characteristics of the two catathymic forms.

11.7 Compulsive homicide

Compulsive homicide differs from catathymic homicide in that the event is not an attempt to resolve some internal cognitive-emotional conflict but is a manifestation of a long-standing, deep-seated compulsion to kill. The offender is not attempting to use homicide to relieve some inner turmoil, but rather is using homicide to satisfy a long-standing (often) sexually sadistic fantasy.

Table 11.3 Differentiating Characteristics of Acute and Chronic Catathymic Processes

Characteristic	Acute	Chronic
Activation of process	Triggered by a sudden overwhelming emotion attached to underlying sexual conflicts of symbolic significance	Triggered by a buildup of tension, a feeling of frustration, helplessness, and inadequacy sometimes extending into the sexual area
Relationship to the victim	Usually a stranger	Usually a close relation such as an intimate or former intimate partner
Victim symbolization	Often a displaced matricide	Rarely a displaced matriarch but the victim may have symbolic significance.
Incubation period	Several seconds	One day to one year or more; may involve stalking
Level of planning	Unplanned	Planned, frequently in the form of an obsessive rumination
Method of attack	Sudden, violent; often overkill	Violent but not sudden
Crime scene	Very disorganized, reflecting a lack of planning	Less disorganized
Sexual activity	Occasional sexual activity just before the attack; impotency common	Sexual activity rare at time of homicide
Postmortem behavior	Sometimes necrophilia and occasionally dismemberment	Rarely necrophilia or dismemberment
Feeling following the attack	Usually flattening of emotions	Usually a feeling of relief
Memory of the event	Usually poor	Usually preserved

Schlesinger, (2004), p. 162.

Compulsive homicide can be either *unplanned*, in which the offender murders in a crime of opportunity, or it can be *planned*, in which the offender takes significant time to plan, prepare, and execute the murder.

11.7.1 Unplanned compulsive homicide

Like ACC, unplanned compulsive homicide (UCH) is a behavioral outcome resulting from disordered internal thought processes and motivations. However, unlike ACC offenders whose emotionality is charged by internal conflict (the combination of which results in the motivational trigger for violent behavior), offenders in cases of UCH have no such internal "conflict" driving emotion and motivation. Rather, UCH offenders have an internal desire to kill, often associated with a fusion of violence and sexual arousal, which they may use within the confines of fantasy for a number of years before finally acting on their desires. As the name implies, unplanned compulsive homicide is generally a crime of opportunity, and therefore, the ultimate enactment of their homicidal desires is often due to a confluence of opportunity and some situation that reduces self-control sufficient to allow the impulsive explosion of violence to ensue. This may involve environmental factors such as being alone with a vulnerable victim candidate while intoxicated (thereby reducing their behavioral inhibitions), or it may involve opportunity concurrent with endogenous factors such as comorbid psychopathology like depression, hypomania, or mania.

11.7.2 Planned compulsive homicide

Just like CCC, planned compulsive homicide (PCH) is the culmination of an extended period of obsessive rumination involving the killing of another person. This is, however, where the similarities stop. PCH involves an intense desire or compulsion to kill in order to emulate a sexually sadistic fantasy construction. The killing of another person is not motivated by a belief that it is a means of resolving a problem (internal conflict), but rather the act of killing, usually in some ritualistic manner, is the motivation in itself. In this way, PCH is a (usually) sexually motivated compulsive homicidal act perpetrated because of a deep inner desire to kill and not because the murder will achieve any purpose (whether logical, irrational, or delusional). Offenders who commit PCH may be classically psychopathic and often show severe traits of personality disorders, particularly narcissistic and antisocial. Rare females who commit PCH may also present with borderline, paranoid, or histrionic personality traits (Table 11.4).

11.8 Sadistic aggression

Sadistic behavior in the psychological sense and sadistic behavior in the criminal sense are two quite different entities. Psychopathologically, sadistic aggression or sadistic behavior is defined clearly in the DSM (American Psychiatric Association, 2013) and the ICD-10 (World Health Organisation, 1992) as a paraphilia or atypical sexually arousing deviance. In this way, sadism has been, psychologically speaking, placed firmly into the realm of the sexual; it is not about the enjoyment of inflicting pain on others, but rather it is about the sexual pleasure derived from doing so. In this situation, a diagnosis or label requires only that a person receives sexual gratification from engaging in behavior that inflicts physical or psychological pain on another person—whether or not the other person is consenting.

In a forensic sense, however, this is not the case. Sadistic aggression, in the legal sense, refers to the process of inflicting physical pain on another person, or by restricting the freedom of another person, for one's own pleasure (not necessarily sexual) in the absence of consent. In general terms, courts are not interested in what two *consenting adults* do in the privacy of their own home (there are exceptions if the activities progress toward the endangerment of life, a minor is involved, or the "consent" given by the *receiving* party is suspect due to their having a reduced cognitive capacity to do so either organically or as a result of ingestion of a substance). Therefore, criminally speaking, sadistic aggression occurs when, for their own pleasure, one person restrains the liberty of, or inflicts physical pain upon, another individual who is an *unwilling* partner or on a partner for whom there are reasonable doubts as to their legal and/or functional capacity to provide an *informed consent*, regardless of whether or not the offender receives or is motivated by the potential to receive sexual gratification. The crossover of psychopathological sadistic aggressor and criminally sadistic aggressor then exists somewhere within a very wide and significantly gray line. As forensic psychopathology must, by definition, include the existent quality of a criminal act, forensically psychopathological sadistic aggression must therefore contain the qualities ascribed it by law more rigidly than those ascribed to sadistic aggression under mental health diagnostic criteria. The essential element of criminal pathological sadistic aggression, then, is that an offender receive pleasure (which may or may not be sexual in nature) from inflicting physical pain or restraint (of

Table 11.4 Characteristics of Behavior and Motivations Associated with Acute Catathymic Crisis, Chronic Catathymic Crisis, Unplanned Compulsive Homicide, and Planned Compulsive Homicide

Characteristic	Impulsive Homicidal Behavior		Planned Homicidal Behavior	
	ACC	UCH	CCC	PCH
Desires and motivations	No clear conscious desires, but motivated by intense feelings of inadequacy. May be derived from early-life attachment issues and often extend to become sexually oriented.	Strong (compulsive) desire to kill present, usually for an extended period of years. Motivated by sexual fantasy involving sadistic violence.	No clear conscious desires, but motivated by intense and progressive emotion-fueled internal conflicts resulting in increased feelings of helplessness and inadequacy. May be derived from early-life attachment issues and often extend to become sexually oriented. Has increasingly obsessive thoughts about killing that are victim specific. Believes that killing the proposed victim is a solution to their problem (internal conflicts and emotional turmoil).	Strong (compulsive) desire to kill that becomes progressively stronger, usually for an extended period of years. Motivated by sexual fantasy involving elaborate (but not victim-specific) sadistic violence.
Relationship with victim	Victim is most often a stranger or very casual acquaintance who inadvertently taps the underlying cognitive-emotional conflict, triggering a catathymic response.	Often known to the offender. The victim does not tap any internal conflict but rather becomes a victim because they have a vulnerability that provides the opportunity for the offender to murder.	Victim is most often either an intimate partner or a (relatively) recent former intimate partner, or a stranger with whom the offender has developed an obsession. The offender links the victim to their internal cognitive-emotional conflict through increasingly obsessional, irrational, and (often) delusional thought processes. The offender progresses to the belief that murdering the victim is the solution to their problem.	Victim is most often a stranger, but may occasionally be an acquaintance. It is extremely rare for the victim to be in, or have formerly been in, an intimate relationship with the offender for any extended period of time. However, given the highly sexualized construction of the fantasy and offense, some element of sexuality may be present in the manner in which the offender and victim become acquainted. For example, the offender may lure the victim through a sexual liaison (or the promise or possibility of one).

| Behavioral or visible indications of risk | No significantly obvious indicators. The offender is often introverted or withdrawn, although they may sometimes make threats that they will be violent or "explode." These are usually ignored by those who witness them. | Does not make overt threats; however, there are often a number of indicators in the background. These include sadistic sexual fantasies with a compulsion to act on them, outbursts of anger, disordered mood (e.g. depression and anxiety), environmental factors that increase personal stress levels (relationship breakdown, job loss, and financial problems), pathological dishonesty, manipulative behavior, animal cruelty, high levels of voyeurism and possibly fetishism and sexually motivated burglary, previous domestic violence or (if male) other physical acts against women, a general attitude of misogyny, compulsive or ritualistic behaviors, and early-life adversity such as childhood abuse and childhood sexual abuse (often at the hands of a maternal figure). | Generally, the offender makes no threats or shows obvious external indicators of intent; however, the offender may confide in a friend or respected authority figure such as a priest. Disclosures are often ignored or underestimated by those who are confided in. | Offender will rarely, if ever, make threats or provide an overt warning of intent; however, as with UCH, there are myriad behavioral indicators in the background (see the UCH "Behavioral or visible indicators of risk" column in this table). Cruelty to animals may extend to using animals as "practice" for the "real event." Indicators may therefore include a succession of pets that meet with untimely "accidents" or "illnesses" and a string of missing pets within an extended neighborhood area near the offender (although killers who fit into this group are often very good at planning and therefore may take precautions to ensure they abduct or obtain animals in locations appropriately distant from their own neighborhood). |

Continued

Table 11.4 Characteristics of Behavior and Motivations Associated with Acute Catathymic Crisis, Chronic Catathymic Crisis, Unplanned Compulsive Homicide, and Planned Compulsive Homicide *Continued*

Characteristic	Impulsive Homicidal Behavior		Planned Homicidal Behavior	
	ACC	UCH	CCC	PCH
Sexual behavior at time or scene of the crime	Necrophilia and/or body mutilation is relatively common—offender may dismember the victim postmortem.	Offender may sexually assault the victim during the attack or postmortem, either by personally penetrating the victim or by inserting various objects into the victim.	Overtly sexual behaviors while committing the offense are rare.	Sexual behaviors are a hallmark of this form of homicide and are therefore commonly associated with the offense. Behaviors may include those similar to UCH; however, they are likely to include more ritualistic or symbolic characteristics. Remember that this is usually a planned-in-detail fulfillment of a long-term sexually sadistic fantasy.
Postcrime behavior	Does not generally attempt to escape from the police for very long. Will often turn themselves in to police or confide in a friend or trusted authority figure who will assist them to do so.	Makes a concerted effort to escape the police; however, the unplanned and disorganized nature of the offender and the crime usually mean they are unsuccessful and are apprehended relatively quickly.	Generally makes no attempt to escape the police. Will often attempt suicide in concert with the crime (murder-suicide is often a distinguishing feature of intimate spouse homicide events), or in the immediate aftermath of the offense (offender will be found dead in their home or other favorite location, or will suicide while in prison).	Offender plans well and works hard to avoid personal detection. Careful planning often results in successful evasion of authorities, resulting in repeat offending and multiple victims. Offender is often caught through serendipity rather than meticulous police work.

Note: Schlesinger presents two tables that show the differentiating characteristics of ACC and UCH and of CCC and PCH, respectively; Schlesinger's tables provide a good alternative source of information in this area (to see Schlesinger's tables, please refer to Schlesinger (2004), pp. 162 and 294).

freedom) on a nonconsenting individual (now a victim). The secondary component that must also be satisfied is the notion of the behavior being pathological in nature. That is, the condition must adhere to guidelines for classification as a mental disorder. That is to say that, in general terms, the symptoms being experienced must originate within the individual and not be a situational reaction, and they must be outside of the individual's physical or mental control (American Psychiatric Association, 2013).

Therefore, the first hurdle that sadistic aggression must meet to be considered a pathological entity is the desire to victimize another person by inflicting physical pain, and the pleasure derived from engaging in such an act must be generated internally (a craving of sorts) and not be a reaction to a situational stimulus. For example, wanting to punch someone for crashing into the back of your car and then feeling really good about doing it afterward, does not qualify you as a sadistic aggressor (no matter what the onlookers may yell at you). Secondly, the internally generated desire must be motivating to the point of *compulsion* such that the desire is *uncontrollable*. Again, really wanting to punch the other driver, but instead taking a deep breath and driving away after exchanging details, does not constitute an uncontrollable compulsive desire. The desire must be so strong as to become obsessive and almost painful to withhold from enacting.

The second general hurdle for our version of sadistic aggression is the question of how it fares against the *4Ds of Abnormality*: *deviance*, *distress*, *dysfunction*, and *danger* (Table 11.2; see Comer, 2005). To make a determination that our definition of sadistic aggression can meet the general criteria for a psychopathology and therefore qualify as a forensic psychopathology, it must be evaluated for its extant levels of *deviance*, *distress*, *dysfunction*, and *danger*.

Certainly, the notion of receiving pleasure from inflicting physical pain on another is antagonistic to the prevailing broad cultural social norms of the vast majority of the known world and is therefore deviant. There is some argument that activities such as sadism, while certainly deviant to the wider cultural norms, are in fact an accepted normalcy for the subculture in which it thrives and therefore should not be summarily categorized as it is. There is certainly some sympathy for this position in the context of sadism as a sexual paraphilia within the confines of general psychopathology. If indeed we were exploring sadism in the pure context of psychopathology, then we could distinguish between sadism perpetrated upon a consensual partner (a masochist) versus that perpetrated on an unwilling partner who, by the defining elements of sadistic behavior, can be readily characterized as a victim. However, here in the context of forensic psychopathology, we are limited to considering the deviance inherent in the act of receiving pleasure from inflicting pain on a victim and not on a willing confederate. In that context, there is little doubt as to the significantly deviant nature of the behavior and its underlying psychological machinations.

The second of the *4Ds* is *distress*. In most cases, the extent to which the offender is distressed by their sadistic desires and behaviors appears limited. In this element of the test, then, sadistic aggression fails the psychopathology test. However, the extent to which the offender's behavior produces distress in their victims and the broader community is significant, and this distress is also an element of the evaluation process. We can therefore surmise that forensically defined sadistic aggression can and does certainly produce significant distress to those impacted by its behavioral manifestation. It is an aspect of a significant number of forensic psychopathologies in which offenders are less distressed by the behavioral characteristics of their pathology than are their victims and others around them. Certainly, psychopathy, antisocial personality disorder, and planned compulsive homicide, to name a few, are all examples of psychopathologies with the potential for significant forensic implications in which the individuals with the pathology appear far less distressed about their characteristics than do those with whom they interact.

The third of the *4Ds* is *dysfunction* and refers to the extent to which the syndrome results in dysfunction in one or more important areas of the offender's life. Sadistic aggression has the potential to create significant dysfunction, particularly within the social sphere of an offender's life. In the context of sexually sadistic aggression, the sadistic behavior often becomes an overwhelming element of sexual gratification for the sufferer. This need for sadistic engagement for sexual fulfillment represents a high degree of sexual dysfunction in the context of our forensic definition as any sadistic behavior must be nonconsensual. In the more general forensic definition of sadistic aggression, the offender may engage in other compulsive acts of sadistic aggression to obtain pleasurable feedback such as assault and battery. In this context, an individual may be a regular antagonist in local pubs and bars, or they may actively prowl the streets looking for unwitting (and unwilling) victims. This again represents a high degree of social dysfunction and, as such, illustrates that sadistic aggression certainly has the potential to produce extensive dysfunction.

The final element of the *4Ds* is *danger* and refers to the extent to which the condition represents an active danger to the sufferer-offender and/or others. Given the very nature of the act of sadistic aggression in the forensically defined context, there is a clear danger of significant harm to others as a result of the offender's behaviors. Therefore, forensically defined sadistic aggression meets the "danger to self or others" test of psychopathology. Given the assessment performed in this chapter, it is clear that sadistic aggression, as it is defined by criminal law, has the capacity to also meet the criteria to establish it as a pathological entity.

Note: This process not only has been an exercise to assess sadistic aggression but also has provided us an opportunity to undertake a practical evaluation of a criminally defined act in respect to its pathological identity. This form of assessment can be undertaken on any number of specific behaviors, symptoms, characteristics, or criminally defined behavioral conditions. While not always of use in and of itself, unless for some academic perspective, it is useful as a means of linking criminal elements and psychopathological characteristics when analyzing crimes, criminals, and predictive models. This is particularly true when there is no directly corresponding pathological entity under the constraints of either the DSM or ICD.

The psychopathological form of sadistic aggression does not correlate with psychopathy but is more akin to a social deviance. It has been argued that sadistic aggression as a paraphilia in which engagement is consensual should not be considered a mental illness at all (Laws & O'Donohue, 2008). However, the forensically defined sadistic aggression has several correlates with the psychopathic personality. The diagnostic criteria for *antisocial personality disorder* (ASPD) in the *DSM-5* (American Psychiatric Association, 2013) are shown in Table 11.5 with primary correlates of criminal sadistic aggression shaded. Other elements of ASPD may correlate depending on individual circumstances.

Crime statistics show that the majority of sadistic crimes are not overtly sexual in nature, and so perhaps there is an argument for the establishment of two distinct groups of criminally sadistic aggression—sexually sadistic crime and nonsexually sadistic crime. The common thread is that both groups use sadistic aggression that is not consensual. From the criteria set out in Table 11.5, it is clear that individuals who engage in criminal sadistic aggression have several elements of psychopathy. Therefore, any crime analysis could possibly view a criminally sadistic act—whether sexual or nonsexual—as a behavioral manifestation of psychopathy.

11.9 Addiction and criminal behavior

Addictions take many forms. Studies have shown a wide pattern of criminal behavior among individuals who suffer from various addictions, although criminal behavior is most associated with addiction associated with drugs, alcohol, and gambling. Most commonly, crime is associated with efforts to obtain more of the item to which the offender is addicted. While the nature of the crime can vary, they

Table 11.5 General List of Diagnostic Criteria for ASPD Based on the Diagnostic Criteria Established in the DSM-5

General Diagnostic Criteria for Antisocial Personality Disorder

The essential features of a personality disorder are impairments in personality (self and interpersonal) functioning and the presence of pathological personality traits. To diagnose antisocial personality disorder, the following criteria must be met:

A. Significant impairments in personality functioning manifest by:
   1. Impairments in self-functioning (a or b):
      a. Identity: Ego-centrism; self-esteem derived from personal gain, power, or pleasure.
      b. Self-direction: Goal setting based on personal gratification; absence of prosocial internal standards associated with failure to conform to lawful or culturally normative ethical behavior.
      and
   2. Impairments in interpersonal functioning (a or b):
      a. Empathy: Lack of concern for feelings, needs, or suffering of others; lack of remorse after hurting or mistreating another.
      b. Intimacy: Incapacity for mutually intimate relationships, as exploitation is a primary means of relating to others, including by deceit and coercion; use of dominance or intimidation to control others.

B. Pathological personality traits in the following domains:
   1. Antagonism, characterized by:
      a. Manipulativeness: Frequent use of subterfuge to influence or control others; use of seduction, charm, glibness, or ingratiation to achieve one's ends.
      b. Deceitfulness: Dishonesty and fraudulence; misrepresentation of self; embellishment or fabrication when relating events.
      c. Callousness: Lack of concern for feelings or problems of others; lack of guilt or remorse about the negative or harmful effects of one's actions on others; aggression; sadism.
      d. Hostility: Persistent or frequent angry feelings; anger or irritability in response to minor slights and insults; mean, nasty, or vengeful behavior.
   2. Disinhibition, characterized by:
      a. Irresponsibility: Disregard for—and failure to honor—financial and other obligations or commitments; lack of respect for—and lack of follow-through on—agreements and promises.
      b. Impulsivity: Acting on the spur of the moment in response to immediate stimuli; acting on a momentary basis without a plan or consideration of outcomes; difficulty establishing and following plans.
      c. Risk taking: Engagement in dangerous, risky, and potentially self-damaging activities, unnecessarily and without regard for consequences; boredom proneness and thoughtless initiation of activities to counter boredom; lack of concern for one's limitations and denial of the reality of personal danger.

C. The impairments in personality functioning and the individual's personality trait expression are relatively stable across time and consistent across situations.

D. The impairments in personality functioning and the individual's personality trait expression are not better understood as normative for the individual's developmental stage or sociocultural environment.

E. The impairments in personality functioning and the individual's personality trait expression are not solely due to the direct physiological effects of a substance (e.g. a drug of abuse, or a medication) or a general medical condition (e.g. severe head trauma).

F. The individual is at least age 18 years of age.

American Psychiatric Association (2013).

commonly include fraud and theft-related activities such as check forgery, embezzlement, theft, larceny, armed robbery, and loan fraud as well as a variety of offenses related to illegal prostitution, drug production and distribution, and a range of cons and hustles. In short, anything that can provide quick cash flow or direct product to feed their habit.

In the course of crime perpetrated in order to obtain more "product," violence is generally perpetrated as a secondary factor. Violence may occur because an addict is not thinking clearly, is desperate, and may be suffering severe withdrawal effects causing them significant physical and mental distress. This combination and its accompanying irrationality are a recipe for disaster, and if the addict has a weapon then there is a high risk of it being used either accidentally (e.g. accidental discharge of a firearm) or, in the case of desperation, delusion, or paranoia, as an uncontrolled and impulsive threat response. A particularly relevant exception to this is in domestic violence and assault, where alcohol abuse and dependence, as well as other drugs such as amphetamines, are identified as being a factor in the vast majority of cases in almost every location where statistics have been obtained. From a psychopathological standpoint, many addictions are recognized as independent clinical entities, including alcohol and other substances, gambling, and sex addiction (American Psychiatric Association, 2013). An exploration of the origins of an addiction within the realm of psychopathology can quickly turn into a case of "What came first, the chicken or the egg?" This is because addiction and addictive traits are commonly comorbid with other psychopathologies, and the initial directionality of the relationship is not always clear. Alcohol and substance addiction may arise due to a process of self-medication by an individual already suffering from any number of disorders, including mood disorders (depression, bipolar, and hypomania), anxiety disorders (generalized anxiety disorder, panic disorder, and social or other phobia), posttraumatic stress disorder, schizophrenia or other psychotic conditions, and a raft of personality disorders. Alternatively, the high use of alcohol and other drugs as well as the neurological alterations associated with addictions such as gambling, sex, and even video game addiction are associated with the onset of a range of other psychopathologies, including mood and anxiety disorders, eating disorders, cognitive impairment, learning and behavioral disorders, and psychotic disorders.

The issue, then, is whether addiction presents a criminal risk by itself, or whether it is the confluence of addiction and other comorbid psychopathology that presents the risk. While it is certain that the pressures of "need" that are fueled in addiction present strong criminal motivations in themselves if an addict lacks the legitimate means to fund their addiction, it is also the case that certain psychopathological characteristics will most certainly exacerbate the addiction risk factors to enhance the likelihood of criminality. Many of the psychological comorbidities associated with addiction have characteristics that present them with (under the right circumstances) a propensity toward criminally oriented behavior. The addition of alcohol or substance addiction (for example) may exacerbate this propensity through both a removal of inhibition and an increased inclination toward higher risk behavior, as well as increasing the need for financial resources in order to fund the addiction itself. Psychopathology with characteristics including impulsivity, obsession, compulsion, cognitive impairment, social-moral impairment, lack of empathy, narcissism, and other antisocial elements may collude with increased disinhibition and "need" to result in a significant risk for criminal behavior. Similar pathological symptoms have been linked to other addictions, including those that do not involve the ingestion of exogenous chemicals such as gambling, sex, and video game addictions. Finally, in both addiction-related and many of the other connected comorbid psychopathologies, a number of associations have been made with a host of neurostructural, neurochemical, and neurofunctional correlates such as dopamine dysregulation, with the increased psychological symptoms and behavioral risks.

Addiction as a psychopathology, then, has several possible mechanisms by which it may influence criminal behavior. It may have a direct influence through higher cortical factors such as disinhibition, social-moral judgment, and executive control impairments, and via other subcortical neurophysiology alterations impacting behavior, emotion, and motivational mechanics. Addictions may impact criminal behaviors through their mediating role on other comorbid psychopathologies by amplifying associated characteristics. Finally, addictive psychopathology may increase criminality through the need of the addict to acquire resources to help feed their addiction. It is unusual for any of these factors to exist in isolation, and therefore, addictive psychopathology tends to have a number of factors influencing the risk of criminal behavior that combine to exacerbate the risk and further increase the likelihood of offending.

11.10 Disordered thinking

Disordered thinking is a key characteristic of forensic psychopathology as individuals often engage in violent, threatening, or offensive behavior as a result of obsession, delusion, paranoia, hallucination, or fantasy. These forms of distorted cognitive processing are characteristic of several psychopathologies associated with increased forensic risk, including the personality disorders, anxiety, depression and bipolar disorder, conduct disorder, and schizophrenia. Any study that undertakes a prevalence analysis or "head count" of the pathology types that collectively make up any major group of criminals will undoubtedly find these psychopathologies common within the members of the participation group. There are strong cross-over elements between these forms of cognitive disorder with many obsessional thinkers also displaying delusions, paranoia, or fantasy. Similarly, were we to examine delusional thinkers, it would be highly likely that a large percentage of them would also display a variety of combinations of the other listed disordered-thinking processes. Therefore, any lineal and individuated examination of these thought processes will invariably contain much in the way of duplication. With this challenge in mind, the following section explores elements of these key disordered thought processes that are associated with criminal behavior.

11.10.1 Obsession

The notion of obsession within a criminal framework has probably become most associated with *Obsessional Following* or *Stalking*. That is, the obsessional and unwanted following of another person. Miller (2012, p. 495) defined as stalking, is "an intentional pattern of repeated intrusive and intimidating behaviors toward a specific person that causes the target to feel harassed, threatened, and fearful, or that a reasonable person would regard as being so." The act of obsessional following is age-old, having been written about by many historical figures from the time of the ancient Greeks, through to the romantic literature of the eighteenth, nineteenth, and twentieth centuries. Indeed, what we have come to consider the criminal act of stalking has previously been regarded in the context of obsessive romantic advances. Legislation making obsessional following or stalking a criminal act has only become commonplace in the last 25 years. For example, the first state in the United States to criminalize stalking was California in 1990; however, it did take just a further six years for the remaining 49 states to pass similar antistalking laws (Davis & Chipman, 1997). While stalking has now been recognized as a forensic psychopathology, the ability to "pigeonhole" the act is more difficult. Stalking and its obsessional components are quite unique to each individual offender. This is because it is a confluence of the individual psychodynamics of thoughts, emotions, and motivations that ultimately go on to create, drive, and maintain the stalker's behavioral manifestations. Essentially, the macromotivation is to control the victim; however, the micromotivations are driven by disordered thoughts and associated emotions that result from individual histories and mental health derivations. What is known is that, if left unchecked and unrequited, the obsessional thoughts will lead to progressive behaviors that all too often end in violence and tragedy.

The extent of the obsessional thoughts generally determines the strength (salience) of the emotional responses that are generated by the target of the obsession. In turn, it is the salience of the emotional responses that drives and directs motivation (motivational valence), which again, in turn, drives the valence (direction) or character of the behavioral outcome. The emotional response also feeds back into the obsessive thought process, providing additional impetus to stimulate a strengthening and progression of the obsessive thoughts. Figure 11.1 shows the basic character of this process.

Figure 11.1 shows the direction and causal relationship between internal psychodynamics and behavior.

Stalking behavior can be analyzed for its potential threat to the target of obsession. From a psychopathological perspective, the degree of obsession (disordered thinking) can be linked to the severity of the associated mental disorder and the nature and extent of the potential injury to the target of the obsession (Davis & Chipman, 1997). In this way, then, stalking can be best viewed within the context of a spectrum or continuum in which the degree of obsession can be used as a yardstick to assess the potential risk of behaviors escalating to violence.

While the popular view is that continued rejection in the face of clinical *erotomania* (also known as *de Clerambault's syndrome*)—the irrational belief that an essentially unattainable stranger is passionately in love with you—is the most common cause of violence in stalking cases, the reality is that obsessional following of a stranger leading to violence is relatively rare. Instead, it is far more common that stalking that results in violence or homicide is a case of intimate partner (or ex-partner) violence (IPV; Miller, 2012; Norris, Huss, & Palarea, 2011). Stalking is most common when a relationship has soured, usually in the face of some form of domestic violence. This form of stalking is known as *simple obsessional stalking*, is the most common form of stalking right across the socioeconomic and cultural spectra, and accounts for as much as half of all stalking cases. It is often difficult to identify unless the victim asks for help as the stalker may not present any outward symptoms of abnormality and can usually maintain an otherwise "normal" existence. A third form of stalking takes the middle ground between *erotomania* and

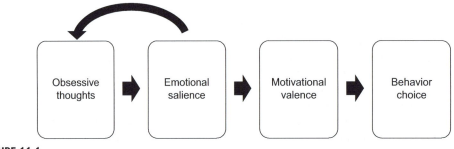

FIGURE 11.1

Important foundational role of thoughts in obsessional following (stalking).

Simple Obsessional Stalking. Known as *love obsessional stalking*, this form of obsession takes the form of an offender engaging in an ongoing campaign to make the target of their obsession aware of them. The two parties are most often known to one another. The two may work together, they may interact through a service relationship (e.g. local shop staff and a customer), they may know each other through a local club or service organization, or they may have a mutual friend through whom they are acquainted. The stalker's obsessive thoughts are often accompanied by creative, though irrational, plans to get the target to notice them. These plans are often quite bizarre and, as such, are not able to be understood by the victim. When attempts to be noticed are unsuccessful, the stalker will escalate their efforts, and behaviors will become increasingly irrational and bizarre until the obsessive thoughts and associated emotions move into compulsions. At this time, the risk of violence by the stalker against the target of their obsession (victim) becomes extreme. These basic types of obsessive following can be further broken down and dissected; however, the characteristics of these three remain steadfastly consistent throughout.

Obsessional thinking is a major characteristic of many forms of forensic psychopathology. It is a factor within a variety of diagnosable psychopathologies, including many of the personality disorders, depression and bipolar (manic) disorders, conduct disorder, and schizophrenia. Moreover, it is often an identifiable pathological feature of criminal activity even in the absence of diagnosable psychopathology. Stalking is perhaps a good example of where this is often the case. Clinical research has provided extensive detail on psychopathologies related to stalking (Davis & Chipman, 1997; McEwan & Strand, 2013; Thompson, Dennison, & Stewart, 2013). However, the research has also shown that stalkers come from a variety of backgrounds and psychological profiles, many of which do not readily include full-syndrome psychopathology. In this instance, there is evidence that the individual characteristics of stalkers still include pathological characteristics such as obsessional thinking, maladaptive social capacity, and codependent or controlling personality traits.

Obsession and relational access are two key characteristics that can assist in the analysis of stalking and risk of behaviors escalating to violence and homicide, with empirical data showing the greater the obsession and degree of relational accessibility by the stalker to the target, the greater the risk of physical violence. In contrast, the lower the level of relational access the stalker has to their obsession target and the greater the delusional characteristics of the accompanying psychopathology—although delusional thinking generally carries an increased risk of aggression (Appelbaum, Robbins, & Monahan, 2000)—the lower the risk of physical violence by the stalker against their target. The only exception to this rule appears to be when stalking behavior is associated with delusional jealousy in IPV (Silva, Derecho, Leong, and Ferrari, 2000) and in erotomania (Carabellese et al., 2013; McGuire & Wraith, 2000).

11.10.2 Delusion and paranoia

Delusional thinking has long been considered to be associated with an increased risk of violent behavior, and the last two decades have seen an increase in the empirical evidence supporting this assertion (Appelbaum et al., 2000; Carabellese, Rocca, Candelli, & Catanesi, 2014; Spencer & Tie, 2013; Van Dongen, Buck, & Van Marle, 2012). When we talk of delusional thinking, what are we referring to? Delusional thinking is characterized by the presence of reoccurring and persistent delusions that are nonbizarre in nature. A delusion is an irrational belief that is held with strong conviction, even in the face of contradictory evidence. Delusional thinking must involve nonbizarre delusional thoughts, which means that the delusions must be plausible—they must be of a kind that could possibly be true. One example is a belief that you are being spied on by government agents. This form of paranoia is a good and common example of nonbizarre delusional thinking. This differs from more severe psychosis in which delusions held are often bizarre in nature—that is to say, they involve beliefs that would be impossible to be true, such as a belief that you could fly.

People who experience delusional thinking often organize these disordered thoughts into a logical (although improbable and untrue) and consistent schema or view of the world. This means that delusional thinking can often be challenging to identify early enough to mitigate the risk of criminal aggression against another person. This is because the person suffering from the delusions will often appear quite "normal" even when they are in the middle of a delusional episode. The obvious signs may include the individual making odd behavioral choices, or they may reflect their delusional beliefs in their comments. For example, if an individual had delusions about being spied on by government agents, they might always use cash when making purchases or they may refuse to use a telephone or computer for fear of being compromised in some way. They may make comments such as "I always use cash because it makes it harder for them to keep track of me." These individuals will often pass unnoticed for an extended period until their delusional thoughts either progress to include a paranoia that they believe is physically threatening to them and so they start to react in more unusual (including potentially aggressive) ways to "protect" themselves, or they either seek help themselves or are identified by an astute family member or friend. Hallucinations are not a common characteristic of delusional thinking, and they may only occur infrequently or may not occur at all.

11.11 Delusional subtypes with forensic association

11.11.1 Jealousy

Jealousy delusions are the form of delusional thinking most commonly associated with violence—usually IPV. When delusional disorder is featured in court cases involving instances of person-to-person violence, it is most often associated with this subtype. In the forensic pathology of this form, the offender usually becomes convinced that their intimate partner is having an affair. The delusion is fed by very circumstantial evidence—such as their partner not answering the phone when they call—and they will still hold to the delusion even in the face of evidence to the contrary. The jealous delusional individual will become focused on gathering "proof" and often become increasingly controlling of their partners' movements and contact with others, often to the point of trying to confine them to the home.

11.11.2 Paranoid or persecutory

While the jealousy subtype is the most common forensic form of delusional thinking, the most common form of delusional thought disorders overall is paranoid delusions or persecution-type delusions. In these conditions, individuals are under the delusion that others are conspiring to harm them in some way. This form of the disorder is also a high risk for escalation to aggression and violence if the individual identifies a specific person or persons as the conspirators. If the delusion becomes fixated on a target, then the individual has the potential to progress to "self-defense" against those they perceive as wanting to harm them. Self-defense in this instance often includes a preemptive strike against the perceived aggressors or a heightened threat sense that may lead to them misinterpreting some interaction between the delusional individual and their apparent aggressors as a threat (to their life) and overreacting with extreme violence.

11.11.3 Erotomania

Erotomania has been described briefly under obsessional thinking in this chapter; however, erotomania also includes significant elements of delusional thinking in which the sufferer is under the delusion that some individual—often, a stranger or an important person or celebrity—is in love with them. This form of delusion is more common in females and often results in stalking. While rare, this form of delusion does result in both nonfatal violence and homicide against the love interest or against others who may be perceived as potential rivals or who the stalker believes may have some other irrationally derived role to play in bringing the stalker and their target together.

11.12 Delusion as a feature of other forensic psychopathology

Delusions often occur as a feature of other psychopathologies. Psychotic disorders such as schizoaffective disorder and schizophrenia often have delusional thinking as a prominent characteristic, as can some personality disorders and mood disorders (both unipolar and bipolar depression) with psychotic features. Moreover, delusions often appear comorbid with neurological conditions such as brain tumor, stroke, acquired brain injury, and some neurodegenerative disorders such as dementia and Alzheimer's (Crail-Melendez, Atriano-Mendieta, Carrillo-Meza, & Ramirez-Bermudez, 2013; Echávarri et al., 2013; Pareés et al., 2013; Zugman et al., 2013). Alzheimer's and dementia conditions are often associated with increased levels of aggression, and therefore any expression of atypical aggression—even physical assault against loved ones—is often an indicator of neurodegenerative disease in an aging or otherwise high-risk (for neurodegenerative disorder) offender. Finally, delusions are a common side effect of a number of drugs—both illicit and prescription (Aggarwal, Banerjee, Singh, Mattoo, & Basu, 2012; Cameron Ritchie, Block, & Lee Nevin, 2013; Signorelli, Battaglia, Costanzo, & Cannavò, 2013). In a forensic setting, drug-induced or medically derived delusional thinking should always be considered and either confirmed or ruled out before moving to other psychopathological explanations.

Ultimately, because disordered thinking, such as delusions, is a high-prevalence feature in so many psychopathologies, there is little to be gained from focusing on the diagnosis of the full-syndrome condition; rather, it is the delusion, obsession, paranoia, or the like that is the characteristic important to the analysis of the criminal act or the risk for perpetration of a criminal act by a potential offender.

11.13 Psychotic disorders

It is a popular belief that individuals with psychotic disorders are at a much higher risk to commit violent person-to-person crimes than other groups within the population. However, much of the research that was initially conducted in this area was highly biased due to the nature of the sampling groups. Further, the media often dedicate much more time and column space to crime stories in which the perpetrator suffers from a mental illness. This form of demonizing mental health sufferers has led to the general consensus that it is the marauding psychotics who are our biggest threat to life and limb. The reality is somewhat more complex. Generally speaking, the greatest risk to our life and limbs comes from those we already know. In the general population, there are by far more people assaulted and murdered by their intimate partners or ex-intimate partners than any other group. For example, in the United States, the single greatest cause of death of pregnant women is *uxoricide*, or murder of a wife.

To add to the complexity, we now "stir" into the mix violent crime committed by individuals during a psychotic episode. While the overall number of violent crimes committed by people with a psychotic disorder as a percentage of all violent crime and homicide is very small (about 5%; Fazel & Grann, 2006), as a group, individuals suffering from psychotic illness or who have other mental or physical disorders in which psychotic features are present—either through drug inducement or other factors—have a much higher risk for committing a violent person-to-person crime than those of the general population (Appelbaum et al., 2000; Nestor, 2002; Rezansoff, Moniruzzaman, Gress, & Somers, 2013; Tiihonen, Isohanni, Räsänen, Koiranen, & Moring, 1997; Whittington et al., 2013). Therefore, although in the big picture, psychotic individuals do not commit a significant proportion of the violent crimes committed, they do commit more than their "fair share" when compared to the general population and other subgroups. Criminal risk in psychotic disorders is linked to specific symptomology such as delusions, hallucinations, and fantasy thinking, and therefore treatment adherence is a vital element in mitigating risk in this group. Treatment adherence is a major problem in this mental health group, particularly among younger sufferers, which means that psychopathology with psychotic features is likely to continue to feature regularly in crime and criminal analysis.

The issue of psychotic disorders and crime is even more complex than the issue of risk. Criminal conviction and sentencing are designed as both a punishment for committing the act and a deterrent in order to prevent reoffense or others from committing a similar offense. Therefore, most Western criminal legislation defines criminality as requiring some form of intent to commit or negligence that results in a criminal outcome, and both the knowledge that an act is criminal and the personal capacity to refrain from committing the act. If an individual has no criminal intent, serious negligence, knowledge that their behavior is criminal, or personal capacity to stop themselves from performing the act, then enforcing punishment or enacting deterrents is of no fundamental benefit (other than as revenge). The restricting of the free movement of an individual as a protective mechanism for the benefit of others is a different issue to that of punishment and deterrent. It is this pool of muddy water in which offenses committed by individuals with serious psychotic disorders or who are experiencing psychotic episodes often reside.

Criminal justice is a culturally constructed process that differs widely across nation-states and cultural groups. In some locations, mental illness is viewed with a high degree of compassion, and psychological factors are taken into consideration by police and prosecutors even before a case is brought before the judiciary. In others, mental illness is effectively ignored and even individuals with clear cases of severe mental impairment are treated with the full force of a punitive legal system, including in many cases the death penalty. For much of the developed world, the judicial reality sits somewhere in between these two extremes. Western judicial systems often show a sympathy bias toward certain psychological impairments while having no sympathy for others. For example, an individual who suffers a psychotic episode and commits an act may have a more lenient sentence as a result of the mitigating circumstances of their illness. Conversely, an alcoholic or drug addict who takes control of a motor vehicle and kills another person as a result of being intoxicated is unlikely to be shown any leniency.

Returning specifically to psychotic disorders; as previously stated, the criminal risk that is inherent in psychosis is related to certain symptomology. These include factors such as reduced impulse control, aggression, and severe affective reactivity underpinned by severe disordered thinking and hallucinations. The elements of psychotic disordered thinking are similar to those already discussed in this chapter's section on disordered thinking; however, severe psychosis often results in more severe formations of these characteristics.

Disordered thinking in severe psychosis often extends from nonbizarre to bizarre delusions in which the delusions are of a kind that are implausible and fantastic in nature. These may be further reinforced by hallucinations that provide validation for the delusions. For example, delusions of severe persecution or grandiosity may be reinforced by hallucinations in which the sufferer is "receiving" secret messages encrypted in the radio or television signal, or they may be experiencing auditory hallucinations in which they are "receiving" direct instructions from God or aliens. In these cases, the increasing obsessional thoughts associated with the delusions and the hallucinations drive stronger emotional responses (which feed back to strengthen the disordered thinking), and in turn increase the valence of the motivation to act on the disordered thoughts and the phantom instructions "received" via hallucination. Eventually, this progresses the internal psychological maelstrom until the desire to act (motivation) becomes a compulsion and the behavioral outcome ensues. Alternatively, sufferers may fight hard to resist the disordered thoughts and developing compulsions. In either event, the neurological deficits associated with these conditions prevent any long-term concerted capacity to resist. These neurological deficits are complex and include increased activation of subcortical activity in areas such as the *brain stem* and *reticular activating system* accompanied by impaired higher-order neural activity in the areas of *limbic* and *frontal lobe* function, resulting in retarded impulse control; increased reflexive responses to stimuli—particularly threatening or perceived-to-be threatening stimuli; increased aggression responses; and reduced capacity for empathy, social affiliation, information processing, and rationality.

11.14 **Personality disorders**

Personality disorders (PDs) are generally viewed as multifactorial syndromes resulting from a confluence of biopsychosocial mechanics, including hereditary factors (both genetic and epigenetic), early-life experience (stress, trauma, abuse, neglect, illness, and attachment and relational interaction with primary adult caregivers), environmental characteristics, individual anatomical and physiological factors, and innate character traits. The generic symptomology of PDs include (Petherick & Sinnamon, 2013, p. 421):

> *Emotional dysregulation, inability to maintain positive relationships, social isolation, anger outbursts, suspicion and lack of trust, inability to delay gratification, poor impulse control, and often there is a history of alcohol and/or substance abuse. The thoughts and behaviors of those with a personality disorders are characteristically considered odd, eccentric, melodramatic, overly emotional, anxious, and/or fearful. Many signs and symptoms of specific personality disorders "bleed" into one another, and it is often difficult to proffer an accurate diagnosis that could not be differentially provided by another clinician.*

Everybody has individual "quirks" of personality. Some are endearing, while others are downright annoying; however, neither endearment nor annoyance qualifies a personality "quirk" as pathological. So, at what point does a quirky personality become a pathologically disordered personality? Previously, PDs were coded as Axis II disorders (American Psychiatric Association, 2000), which means that they were listed along with conditions that represented deep-rooted problems that were difficult to treat or modify. Within the Axis II coding, PDs were broken into 10 separate disorders of personality, with these being classified into three categories or clusters of characteristics and symptoms—clusters A, B, and C. Cluster A included PDs in which sufferers were considered to be odd or eccentric—paranoid, schizotypal, and schizoid PDs were listed in this cluster. Cluster B—antisocial, borderline, histrionic, and narcissistic PDs—was grouped to include the individuals who are seen as being dramatic, emotional, or erratic. The final cluster, cluster C, included those PDs where individuals were viewed as essentially anxious or fearful—these are avoidant, dependent, and obsessive-compulsive PDs. The new edition of the Diagnostic and Statistics Manual of Mental Disorders (DSM-5; American Psychiatric Association, 2013) has done away with the Axis classification system, and so PDs are no longer coded in terms of being an Axis II disorder.

The criminal risk from PD stems largely from the dysfunction that is inherent in cognition and affect regulation, and the impact that these have on motivation (desires, compulsions, and perceived needs and threats) and consequent behavior. For example, paranoid PD is characterized by thoughts of paranoia and persecution that drive behavior that is a combination of self-preservation and "get them before they get you"; antisocial PD is essentially characterized by the lack of empathy and social awareness, and a focus on self-interest that drive behaviors intended to obtain self-serving outcomes both physical and emotional with little or no regard to the welfare of others; narcissistic PD, by the presence of inflated self-worth that drives behaviors designed to preserve self-esteem and keep the focus of attention fixed on self; and borderline PD, by inconsistent, self-defeating, and defensive cognitions that drive codependent, approval-seeking, and self-preserving behaviors in a circular process involving relational attraction followed by often violent and destructive relational breakdown. Extreme affective reactivity, often with underlying comorbid mood disorder, is a hallmark common to all PDs.

PDs are of particular relevance to criminality as motivation and personality are inextricably linked, and it is ultimately the characteristics of motivation that determine behavior. Motivation provides the valence or direction in which effort will be placed as well as the determinants that establish what the nature of any course of action will be—friendly, violent, aggressive, defensive, fight, or flight, and so on. Personality characteristics underpin the way in which we perceive our surroundings and interpret our experiences. Motivation is the means by which we create impetus to pursue those items we deem necessary to protect, build, or maintain elements of our personality that require attention. A disordered personality then will utilize distorted, dysregulated, and deviant internal mechanisms, such as (disordered) thoughts and hyperregulated and dysregulated emotionality, to build and direct motivation in order to behaviorally seek out that which is desired in an attempt to repair the disenfranchised elements of self that are the hallmark of the particular PD. In any crime analysis, it is important to understand the strong concordance between the emotional dimensions of motivation, personality, and behavior.

According to Petherick and Sinnamon (2013), there are five common elements that are associated with personality-disordered perpetrators of crime:

1. Poor self-esteem: either very low or high but fragile
2. Emotional negativity bias and lack of capacity to modulate emotional expression
3. High-risk, antisocial, self-serving, or differentially reinforced behavioral tendencies
4. An unwillingness or incapacity to accurately assess and mitigate risk
5. An unwillingness or incapacity to modulate behavior responses

There is a strong positive correlation between PDs and violent recidivism. Violent serial offenders often suffer PDs that are characterized by an inability or unwillingness to assess risk and adhere to established and elementary social norms. Instead, offenders with PDs pursue self-serving and/or self-preservation motives with little reflection on morality, social acceptance, or the impact on those around them. In some subcategories of PD such as antisocial PD, the socially deviant behaviors are essentially possible because internal mechanisms, desires, and personal behavior are not subject to the same empathic modulation as they are in the wider population (Raposo, Vicens, Clithero, Dobbins, & Huettel, 2011; Seitz, Nickel, & Azari, 2006).

The link between PD, motivation, and criminal behavior is described further in this section and is derived from the previously published work of Petherick and Sinnamon (2013) on the motivational role of PDs within victim and perpetrator typology.

11.14.1 Reassurance-driven PD and criminality

Some PD offenders may be *reassurance-driven* and are therefore motivated by a deep-seated need for approval and acceptance from others. Four PDs fit within this form of criminal motivation: *borderline*, *histrionic*, *avoidant*, and *dependent*.

Borderline and histrionic PDs are characterized by thoughts and emotions that are dramatic and hyperemotional. These two PDs are at risk for criminal behaviors due to very low self-esteem, high levels of impulsivity, a strong need for attention and approval, and excessive emotionality and extremely volatile and reactionary behavioral responses. Borderline and histrionic personalities have very poor predictive capacity and are therefore unable to interpret their surroundings and assess the risks that may be present. In the case of borderline and histrionic PD, the propensity for impulsiveness, volatility, reactive aggression, and lack of capacity to mitigate the hyperemotionality and negative cognitions that are associated with an intense fear of rejection all combine to increase the risk of aggressive outbursts and violent reactions toward others when they feel emotionally or physically threatened. The resulting crimes are generally person-to-person violence, destruction of property, and arson. These individuals are also unlikely to learn from their experiences and therefore have a very high rate of recidivism.

Avoidant and dependent PDs experience disordered thinking that is extremely anxious, in turn resulting in strong negative emotions that are characterized by high levels of fear. Offenders who fall into these PD types often find themselves in codependent relationships. They will submit themselves to the other person and remain even in the face of extreme demands, adversity, and abuse. With deep-seated feelings of inadequacy, these individuals often need to obtain self-esteem by proxy. That is, they need to gain approval from others in order to feel good about themselves. These feelings of inadequacy and the need for acceptance from others, combined with other characteristics such as timidity, shyness, submissiveness, and high tolerance for abuse, mean that these individuals often place themselves in situations where they are at high risk for personal violence. Reassurance-driven criminal behaviors perpetrated by offenders with avoidant and dependent PDs are usually motivated out of a desire to retain or obtain the acceptance and/or approval of others. Commonly, these two personality types will become involved in criminal behavior such as theft, prostitution, participation in a gang rape or assault, or the manufacture or distribution of drugs or other illegal goods in order to gain entry to or acceptance by an individual or group (e.g. a codependent relationship or gang membership). As violence is not a natural behavior for this group, for violence against people or property to occur, the individual would need to believe that the target of their approval warrants such an act.

11.14.2 Assertive-driven PD and criminality

PD offenders who are *assertive driven* are similarly motivated by the need to obtain or restore self-esteem through other people (just like the reassurance-driven PD types). The two groups, however, differ in the emotional content that drives their motivation. Rather than fear driving their motivation and subsequent behavior, the assertive-driven PDs are primarily motivated by anger. PDs with this orientation tend to have an underlying belief that they are superior to others, therefore deserve more than they already have, and, furthermore, are owed something by others (or society).

This group is angry because they believe that they, in some way that may be clear or unclear to them, are not getting everything they deserve. Paradoxically, this group tends to have very low self-esteem or a high self-esteem that is very fragile. Motivation therefore comes from the desire to restore their self-esteem through obtaining accolades from others. Alternatively, they will attempt to restore self-esteem by emotionally dominating others. They do this by deliberately setting out to make other people feel inferior and in turn make themselves feel more superior. This type of personality is hyper-emotional, melodramatic, and extremely intense and is characteristic of narcissistic PD.

Narcissistic PD offenders express these flamboyancies in a highly aggressive and dominating "look at me" manner that is ultimately compensatory in nature. The attitude and behaviorally expressed belief that they are superior are not underpinned by substance but rather are constantly undermined by their low or fragile self-esteem. In this way, their behaviors are motivated by the need to repair the incongruence. They do this through a variety of mechanics, including fantasies about success, power, and their attractiveness in the eyes of others. The result is a blurred reality in which exaggerations of personal achievements, success, and talents are made, resulting in an increased need for external praise and recognition to "prop up" their assertions. The self-absorption means that narcissists usually make these ongoing high-maintenance demands without the ability to recognize or acknowledge the feelings of those around them.

The selfish, aggressive, and "I deserve more" personal agenda-driven attitudes of the narcissistic personality often result in criminal behaviors aimed at obtaining what is believed to be rightfully theirs (e.g. financial gain through fraud, theft, or embezzlement in an attempt to obtain material items that they perceive will make them more respected by others). Often, the criminal behavior will be purposefully directed at specific individuals in an attempt to take what is theirs as both a means of gaining material possessions and an act of domination over a rival (whether or not that person is aware of an existing rivalry or even aware of the offender in any personal sense). This could be an employer, sibling, teammate, neighbor, or other person who the offender has identified as a rival. Furthermore, the aggressive and volatile elements of their personality can lead to violence if their fragile self-esteem is threatened (e.g. someone belittles them or embarrasses them in some way, particularly if there is an audience). Any perceived threat to their need for dominance can result in retaliatory aggression.

11.14.3 Anger retaliatory-driven PD and criminality

PD offenders driven by retaliatory anger are motivated as a result of some exaggeration and distorted interpretation of the circumstances surrounding an event or events—whether real or perceived. Two groups of personality-disordered offenders fall into this group. The first includes offenders with paranoid PD. Paranoid PD offenders act because of a belief that they have been wronged in some way. Very low self-esteem coupled with paranoid disordered thoughts result in a self-preservational process of externalizing strong angry emotions onto another. Disordered thinking provides a plausible, though likely untrue, rationale that allows the offender to lay blame elsewhere (e.g. the government, a spouse, a sibling, an employer, or a coworker). Events are interpreted with a paranoid negativity bias, and therefore the offender begins to see conspiracy in the actions of those around them, further increasing the level of anger. The lack of capacity to modulate anger responses and inhibit impulsive urges ultimately leads to acts of "retaliation," which can include extreme violence and homicide.

The second group of retaliatory anger offenders is likely to fall within the sphere of either borderline or narcissistic PD. Just like offenders with paranoid PD, these offenders also have very poor capacity to modulate hyperemotional anger responses, assess risk, or manage impulsive behavior. These individuals are usually found to be in existing volatile relationships that, when combined with their unstable mood, fear of rejection, and impulsiveness, provide a high-risk environment for violence. These personality types become offenders when they feel threatened and react with a "shoot first" mentality in retaliation for the wrong that they believe is going to be done to them. In the event that a wrong is actually done to them, they are likely to retaliate impulsively and with extreme violence without regard to risk.

Interestingly, paranoid and narcissistic offenders have characteristically externalized blame onto others and then set about to seek retaliation accordingly; however, borderline offenders are more likely to have internalized blame creating an intense self-loathing. The borderline offender seeks to harm the "other" in retaliation because they blame the other person for "making" them feel that way.

11.14.4 Pervasive anger-driven PD and criminality

Pervasive anger is a primary motivating feature of paranoid, antisocial, borderline, histrionic, and narcissistic offenders. The key that drives this pervasive hyperresponsive anger is that the disordered personality appears to develop with a schema that generalizes the negativity bias inherent in all PDs, in an aggressive manner. This means that offenders with these PDs are able to find fault everywhere and an "other" is used as a convenient displacement for the internal hyperemotionality that the sufferer cannot modulate or rationalize using internal mechanisms. Any externalized factor can become the lightning rod for the sufferer's anger. These personality types often become offenders when they initiate violent exchanges as a means of exerting control and reestablishing self-esteem. Alternatively, they may damage property if it is a contextually appropriate displacement of their anger.

11.14.5 Excitation-driven PD and criminality

A complex and challenging offender type, excitation-driven offenders are most likely to have either antisocial or dependent personality characteristics. The antisocial personality often fits this motivational type as the lack of empathy and social affiliation toward others allows this offender the emotional space to be able to place themselves in situations where the pain or suffering of others is irrelevant compared to the pleasure obtained for themselves. Sadistic aggression behaviors (described earlier in the chapter), whether consensual or nonconsensual, may therefore appeal to them.

The strong desire for approval from others is the motivating factor that often results in the dependent personality placing themselves in relationships of unequal power in which they are subservient to others. A sadistic partner may entice a dependent personality into increasingly sadistic "play," and the dependent personality may acquiesce due to fear of being rejected and a desire to be accepted. Increasingly violent "play" may move into abuse if the dependent personality cannot (by fear or physical incapacity) extricate themselves from the escalation. The greatest risk occurs when the dependent personality encounters the antisocial personality. In this case, the victim–perpetrator relationship is epitomized.

While more likely to be a victim than an offender in this form of criminal motivation, dependent personalities may become offenders if they finally "crack" in the face of ongoing abuse and retaliate reflexively. There have been numerous cases in which placid, dependent personality types react aggressively in the face of chronic provocation. This can happen because, although they are largely fear driven, they also have impaired impulse control and when provoked even the mildest mannered of individual or animal may resort to a reflexive fight response and engage in highly aggressive survival behavior. Dependent PD offenders may also be prime candidates for violent crime within an acute catathymic crisis.

In addition, schizoid or schizotypal PD types can be at high risk for excitation-driven offending. These personality types have characteristically low levels of emotionality and reward stimulation. In these individuals, altered perceptions and distorted reality may combine with limited emotional expression to motivate a desire to experience greater affect. This can lead to self-harming behaviors as well as thrill-seeking encounters with others. A lack of capacity to assess and mitigate risk makes these personality subtypes potentially at high risk to inflict significant harm on others (as offenders) as well as become the unwitting victim of others.

11.14.6 Materially driven PD and criminality

All PDs are at risk for offending through this form of criminal motivation. Low or fragile self-esteem, distorted perceptions, paranoia, fear, a need for instant gratification, poor impulse control, needs arising from alcohol and/or substance abuse, anger and feelings of missing out, feelings that others are preventing you from obtaining objects or outcomes, fear of rejection, and need for approval can all bring about this motivation. An offender may use criminal means to obtain material gain, whether through violence, theft, or an illicit commercial activity (e.g. drug dealing, gambling, prostitution, or fraud).

11.14.7 Self-preservation-driven PD and criminality

As with materially driven PD offenders, all PDs may be driven by self-preservation motives to commit crime. Self-preservation is a fundamental survival instinct and, in cases where an individual is impaired in their capacity for social reasoning and risk assessment, the natural desire for self-preservation can result in a variety of scenarios in which behavior may become criminal in nature. The desire for self-preservation can lead to the commission of a variety of criminal behaviors aimed at restoring some kind of injustice (real or perceived) or protecting the well-being of self or others (again, whether real or perceived). These forms of motivations are particularly high risk for those who perceive the world through the thick lens of paranoia, through distorted and delusional schizotypal perceptions, through the "cover my tracks" manipulations of the antisocial personality, or through the myopic view of narcissism whereby elevated beliefs about self-worth and the right to have, and expectations of praise, admiration, and material reward, may promote aggressive pursuit of these goals.

Conclusion

This chapter has presented a variety of elements that fall within the scope of forensic psychopathology. No single chapter can ever fully explore what is a rich, diverse, and often controversial collective of subject matter; however, we have journeyed through a range of topics that are directly applicable to the task of applied crime analysis. After all, that is the ultimate goal of this text.

Ultimately, psychopathology is closely associated with criminal behavior. Mental illness is on the rise, and therefore we can assume that crime associated with mental disorders will also become more prevalent. It is therefore incumbent on all of us who work within the mental health, criminal justice, and law enforcement arenas to better understand the relationship between brain, mind, and behavior. It is only through improvements in our knowledge and the careful and considered application of evidence-based principles that we can ever hope to make inroads to improving the welfare of the mentally ill, reduce the risk of them becoming offenders, and ultimately improve our capacity to protect society from what should be preventable criminal behavior.

To conclude the chapter, it is perhaps appropriate to reiterate to the reader that the core principle in applying forensic psychopathology to crime analysis is to examine the psychopathological characteristics of a criminal, crime scene, or predictive algorithm from the perspective of the pathological components and not from the lofty heights of clinical diagnostic labels. That an offender is suffering from depression tells you nothing of value for your analysis. However, knowing that your offender was experiencing severe feelings of hopelessness, guilt, and suicidal ideation may provide some explanatory material for what has transpired. In the same way, knowing that your potential offender is likely to be paranoid, delusional, hyperemotional (primarily in an angry and hypomanic state), and lacking in impulse control may assist you to establish a profile of likely offenders and/or a suspect list, develop predictive models of crime risk, or assess risk in a bail or parole hearing.

Chapter Summary

- Psychopathology refers to our understanding, study, or knowledge of any illness, disorder, dysfunction, or dysregulation of the mental processes that govern an individual's cognition, affect, and behavior. A simple definition of *psychopathology is the branch of knowledge and collective body of conditions relative to the suffering of the mind*. To be diagnosed with a psychopathology, an individual must possess a variety of general and disorder-specific criteria. In general terms, the symptoms being experienced must originate within the individual as opposed to them being a situational reaction, and the symptoms must be outside of the individual's physical and/or mental control. Disorder-specific criteria are dependent upon the characteristic symptomology of each individual disorder and are set out in diagnostic manuals such as the Diagnostic and Statistical Manual of Mental Disorders (DSM-5) and the International Statistical Classification of Diseases and Related Health Problems (ICD-10). Forensic psychopathology is a specialized area in which the focus rests on those psychological maladies and their symptoms and consequences that, for one reason or another, intersect with the legal system. Simply put, *forensic psychopathology is interested in clinical psychopathologies with criminal implications*. This includes any psychological or neuropsychological condition that impacts the perpetrators of criminal acts as well as their victims.
- *Nature and nurture operate within a bidirectional relationship in which nature provides the blueprints and nurture provides the experiential resources necessary for development*. Therefore, to understand why a crime occurred, how to predict similar events in the future, and ultimately how to determine preventative measures that will mitigate the risk of future events, it is important to include in any analysis the precedent factors that lead to the circumstances being such that the crime was committed. A full and effective analysis must therefore include an accounting of the potential impact that personal developmental and continuing environmental factors may have had on the course of events.

- Genetics is a complex and poorly understood area when it comes to the extent to which it is directly responsible for behaviors and higher-order cognitive and physiological processes. While the debate rages over how much influence genetics has, it is generally thought that at best, our genes may have about a 30–40% influence on our development and our physiological function with the environment playing a far greater role. Therefore, it is highly unlikely that genetics can be held responsible for criminality and one's risk for criminally associated psychopathology. The greater likelihood is that our *genetics provide the basic physiological blueprints for our anatomical development, regeneration, and function.* Our myriad *environmental factors then provide the orchestration that results in the way these psychological and physiological factors mature and are ultimately oriented.* Epigenetics literally means "above, outside or around" genetics and refers to the way in which genetic expression or activity is manipulated by heritable factors that are neither caused by or cause the alteration of the DNA sequence. In this way, *epigenetics refers to nongenetic influences on gene activity that either can last through cell divisions for the entire lifetime of an organism or can survive procreative transfer and be inherited across multiple generations without alteration to the underlying DNA sequence of the specific gene.*

- Research into the impact of early-life adversity such as trauma, neglect, abuse, and chronic stress (collectively termed early-life stress, or ELS) has shown that *the impact of these experiences on the developing individual is significant, pervasive, and highly deleterious. ELS sufferers experience increased criminality risk (perpetrator and victim) through two primary mechanisms: (1) biologically, via maladaptive neurodevelopmental processes; and (2) socially, via differential reinforcement, poor attachment, and lack of empathy.* While criminally oriented psychopathology that has its roots in ELS is often amplified by the social conditions in which the ELS occurred, if a child experiences stress, abuse, neglect, or trauma in an environment that can be mitigated via strong attachments, security, and stability—even if this is offered after the event through relocation of the child—then the child has a much lower risk of engaging in, or becoming a victim of, crime. They may still suffer from quite severe psychopathology; however, the risk of criminal deviance is mitigated by the social structures and behavioral "norms" to which the child is exposed and enculturated.

- *A catathymic process or crisis is an unexpected, impulsive, and explosive violent outburst that is not readily understandable from the situation in which it occurred but rather is only able to be understood in the context of individual unconscious (or outside of conscious awareness) motivations.* A catathymic behavioral response occurs when an individual is overwhelmed by emotion to the point that psychological rationality and control are compromised, resulting in a primitive reflexive outburst of aggression. In its modern incarnation, catathymia is more likely to be considered as a syndrome, symptom, or consequence within the confines of other mental disorders such as posttraumatic stress disorder, psychotic disorders (particularly those with delusional characteristics), or several of the personality disorders. In contemporary circles, catathymia is used to describe any explosive or impulsively violent behavior that appears to have its roots in heightened emotionality charged by irrational thought processes or beliefs. The manner in which catathymia is represented within the criminal sphere can be quite varied. For example, violent serial crime, familial homicide, stalking-related violence, and sudden (often) inexplicable interpersonal violence perpetrated against a stranger or casual acquaintance is often laced with catathymic crisis. Catathymic crisis has been described in many ways. For our purposes, we have described it using three categories: delusional catathymic process (DCP), chronic catathymic crisis (CCC), and acute catathymic crisis (ACC).

- *Compulsive homicide differs from catathymic homicide in that the event is not an attempt to resolve some internal cognitive-emotional conflict but is a manifestation of a long-standing, deep-seated compulsion to kill.* The offender is not attempting to use homicide to relieve some inner turmoil but, rather, is using homicide to satisfy a long-running (often) sexually sadistic fantasy. Compulsive homicide can be either unplanned, in which the offender murders in a crime of opportunity, or planned, in which the offender takes significant time to plan, prepare, and execute the murder.

- Sadistic behavior in the psychological sense and sadistic behavior in the criminal sense are two quite different entities. Psychopathologically, sadistic aggression or sadistic behavior is defined as a paraphilia or atypical sexually arousing deviance. In this way, sadism has been, psychologically speaking, placed firmly into the realm of the sexual; it is not about the enjoyment of inflicting pain on others, but rather it is about the sexual pleasure derived from doing so. In this situation, *a diagnosis or label requires only that a person receives sexual gratification from engaging in behavior that inflicts physical or psychological pain on another person—whether or not the other person is consenting.* In a forensic sense, however, this is not the case. *Sadistic aggression, in the legal sense, refers to the process of inflicting physical pain on another person, or restricting the freedom of another person, for one's own pleasure (not necessarily sexual) in the absence of consent.* Therefore, criminally speaking, sadistic aggression occurs when one person restrains the liberty of, or inflicts physical pain upon, another individual who is an unwilling partner or on a partner for whom there are reasonable doubts as to their legal and/or functional capacity to provide an informed consent, whether or not the offender receives or is motivated by the potential to receive sexual gratification.

- Addictions take many forms. Studies have shown a wide pattern of criminal behavior among individuals who suffer from various addictions, although criminal behavior is most associated with addiction associated with drugs, alcohol, and gambling. Most commonly, crime is associated with efforts to obtain more of the item to which the offender is addicted. Addiction as a psychopathology may have a direct influence on criminal behavior through higher cortical factors such as disinhibition, social-moral judgment, and executive control impairments, and via other subcortical neurophysiology alterations impacting behavior, emotion, and motivational mechanics. Addictions may impact criminal behaviors through their mediating role on other comorbid psychopathologies by amplifying associated characteristics. Finally, addictive psychopathology may increase criminality through the need of the addict to acquire resources to help feed their addiction. It is unusual for any of these factors to exist in isolation, and therefore, addictive psychopathology tends to have a number of factors influencing the risk of criminal behavior that combine to exacerbate the risk and further increase the likelihood of offending.

- Disordered thinking is a key characteristic of forensic psychopathology as individuals often engage in violent, threatening, or offensive behavior as a result of obsession, delusion, paranoia, hallucination, or fantasy. These forms of distorted cognitive processing are characteristic of several psychopathologies associated with increased forensic risk, including the personality disorders, depression and bipolar disorder, conduct disorder, and schizophrenia. The notion of obsessional thinking within a criminal framework has probably become most associated with obsessional following or stalking, which is the obsessional and unwanted following of another person. *Stalking is an intentional pattern of repeated intrusive and intimidating behaviors toward a specific person that causes the target to feel harassed, threatened, and fearful, or that a reasonable person would regard as being so.* Stalking is most common when a relationship has soured, usually in the face of some form of domestic violence. This form of stalking is known as *simple obsessional stalking*, is the most common form of stalking right across the socioeconomic and cultural spectra, and accounts for as much as half of all stalking cases. *Love obsessional stalking* takes the form of an offender engaging in an ongoing campaign to make the target of their obsession aware of them. The two parties are most often known to one another. The popular view is that continued rejection in the face of clinical *erotomania* (also known as *de Clerambault's syndrome*)—the irrational belief that an essentially unattainable stranger is passionately in love with you—is the most common cause of violence in stalking cases; however, the reality is that obsessional following of a stranger leading to violence is relatively rare. Instead, it is far more common that stalking that results in violence or homicide is a case of intimate partner (or ex-partner) violence (IPV). Obsessional thinking is a major characteristic of many forms of forensic psychopathology. It is a factor within a variety of diagnosable psychopathologies, including many of the personality disorders, depression and bipolar (manic) disorders, conduct disorder, and schizophrenia. Moreover, it is often an identifiable pathological feature of criminal activity, even in the absence of diagnosable psychopathology.

- Delusional thinking has long been considered to be associated with an increased risk of violent behavior, and the last two decades have seen an increase in the empirical evidence supporting this assertion. *A delusion is an irrational belief that is held with strong conviction even in the face of contradictory evidence. Delusional thinking must involve nonbizarre delusional thoughts,* which means that the delusions must be plausible—they must of a kind that could possibly be true.
- The criminal risk that is inherent in psychosis is related to certain symptomology, including factors such as reduced impulse control, aggression, and severe affective reactivity underpinned by severe disordered thinking and hallucinations. *Disordered thinking in severe psychosis often extends from nonbizarre to bizarre delusions in which the delusions are of a kind that are implausible and fantastic in nature.* These may be further reinforced by hallucinations that provide validation for the delusions.
- The generic symptomology of PDs include: *emotional dysregulation, inability to maintain positive relationships, social isolation, anger outbursts, suspicion and lack of trust, inability to delay gratification, poor impulse control, and often a history of alcohol and/or substance abuse. The thoughts and behaviors of those with personality disorders are characteristically considered odd, eccentric, melodramatic, overly emotional, anxious, and/or fearful. Many signs and symptoms of specific personality disorders "bleed" into one another, and it is often difficult to proffer an accurate diagnosis that could not be differentially provided by another clinician.* The criminal risk from PD stems largely from the disorder that is inherent in cognition and affect regulation, and the impact that these have on motivation (desires, compulsions, and perceived needs and threats) and consequent behavior. Extreme affective reactivity, often with underlying comorbid mood disorder, is a hallmark common to all PDs. PDs are of particular relevance to criminality as motivation and personality are inextricably linked, and it is ultimately the characteristics of motivation that determine behavior. There is a strong positive correlation between PDs and violent recidivism. Violent serial offenders often suffer PDs that are characterized by an inability or unwillingness to assess risk and adhere to established and elementary social norms.

REVIEW QUESTIONS

1. What is the difference between psychopathology and forensic psychopathology?

2. What are the developmental factors that increase risk for criminal deviance?

3. Describe catathymic homicide and its three subtypes as described in the chapter.

4. What are the defining elements of criminally sadistic aggression, and how does this compare to general psychopathological sadistic aggression?

5. Name the five common elements that Petherick and Sinnamon suggest are associated with personality-disordered perpetrators of crime.

6. Personality disorders are often closely associated with certain motivations that increase the risk of engaging in criminal behavior. Describe three of the seven motivations associated with personality disorder and criminal behavior risk, and name the personality disorders that are most associated with each of the three you describe.

References

Aggarwal, M., Banerjee, A., Singh, S. M., Mattoo, S. K., & Basu, D. (2012). Substance-induced psychotic disorders: 13-Year data from a de-addiction centre and their clinical implications. *Asian Journal of Psychiatry, 5*(3), 220–224.

American Psychiatric Association. (2000). *Diagnostic and statistical manual of mental disorders (text revised)* (4th ed.). Arlington, VA: American Psychiatric Publishing.

American Psychiatric Association. (2013). *Diagnostic and statistical manual of mental disorders* (5th ed.). Arlington, VA: American Psychiatric Publishing.

Appelbaum, P. S., Robbins, P. C., & Monahan, J. (2000). Violence and delusions: data from the MacArthur violence risk assessment study. *American Journal of Psychiatry, 157*(4), 566–572.

Beblo, T., Sinnamon, G., & Baune, B. T. (2011). Specifying the neuropsychology of affective disorders: clinical, demographic and neurobiological factors. *Neuropsychology Review, 21*(4), 337–359. http://dx.doi.org/10.100 7/s11065-011-9171-0.

Bienertova-Vasku, J., Necesanek, I., Novak, J., Vinklarek, J., & Zlamal, F. (2013). "Stress entropic load" as a transgenerational epigenetic response trigger. *Medical Hypotheses*. http://dx.doi.org/10.1016/j.mehy.2013.12.008.

Blažič, M., & Vojinovič, B. (2002). .Globalisation and the media. *Informatologia, 35*(3), 187–192.

Brown, M. (2012). Suppressing ideological diversity: John Reed and the threat of injustice. International Journal of Diversity in Organisations. *Communities and Nations, 11*(4), 99–104.

Butler, J. C. (2013). Authoritarianism and fear responses to pictures: the role of social differences. *International Journal of Psychology, 48*(1), 18–24.

Cameron Ritchie, E., Block, J., & Lee Nevin, R. (2013). Psychiatric side effects of mefloquine: applications to forensic psychiatry. *Journal of the American Academy of Psychiatry and the Law, 41*(2), 224–235.

Carabellese, F., La Tegola, D., Alfarano, E., Tamma, M., Candelli, C., & Catanesi, R. (2013). Stalking by females. Medicine. *Science and the Law, 53*(3), 123–131.

Carabellese, F., Rocca, G., Candelli, C., & Catanesi, R. (2014). Mental illness, violence and delusional misidentifications: the role of Capgras' syndrome in matricide. *Journal of Forensic and Legal Medicine, 21*, 9–13.

Clark, C., Caldwell, T., Power, C., & Stansfeld, S. A. (2010). Does the influence of childhood adversity on psycho-pathology persist across the lifecourse? A 45-year prospective epidemiologic study. *Annals of epidemiology*, *20*(5), 385–394. S1047-2797(10)00031-1 [pii] http://dx.doi.org/10.1016/j.annepidem.2010.02.008.

Comer, R. (2005). *Fundamentals of abnormal psychology* (4th ed.). New York: Worth.

Connor, C. M., & Akbarian, S. (2008). DNA methylation changes in schizophrenia and bipolar disorder. *Epigenetics*, *3*(2), 55–58.

Crail-Melendez, D., Atriano-Mendieta, C., Carrillo-Meza, R., & Ramirez-Bermudez, J. (2013). Schizophrenia-like psychosis associated with right lacunar thalamic infarct. *Neurocase*, *19*(1), 22–26.

Dammann, G., Teschler, S., Haag, T., Altmuller, F., Tuczek, F., & Dammann, R. H. (2011). Increased DNA methylation of neuropsychiatric genes occurs in borderline personality disorder. *Epigenetics*, *6*(12), 1454–1462. http://dx.doi.org/10.4161/epi.6.12.18363.

Davis, J. A., & Chipman, M. A. (1997). Stalkers and other obsessional types: a review and forensic psychological typology of those who stalk. *Journal of Clinical Forensic Medicine*, *4*(4), 166–172.

Echávarri, C., Burgmans, S., Uylings, H., Cuesta, M. J., Peralta, V., Kamphorst, W., Rozemuller, A. J., & Verhey, F. R. (2013). Neuropsychiatric symptoms in Alzheimer's disease and vascular dementia. *Journal of Alzheimer's Disease*, *33*(3), 715–721.

Ehlert, U. (2013). *Enduring psychobiological effects of childhood adversity.Psychoneuroendocrinology*, *38*(9), 1850–1857. http://dx.doi.org/10.1016/j.psyneuen.2013.06.007.

Fazel, S., & Grann, M. (2006). The population impact of severe mental illness on violent crime. *American Jounal of Psychiatry*, *163*(8), 1397–1403. http://dx.doi.org/10.1176/appi.ajp.163.8.1397.

Franklin, T. B., Russig, H., Weiss, I. C., Graff, J., Linder, N., Michalon, A. (, et al. (2010). Epigenetic transmission of the impact of early stress across generations. *Biological Psychiatry*, *68*(5), 408–415. http://dx.doi.org/10.1016/j.biopsych.2010.05.036.

Goldenson, J., Geffner, R., Foster, S. L., & Clipson, C. R. (2007). Female domestic violence offenders: their attachment security, trauma symptoms, and personality organization. *Violence and Victims*, *22*(5), 532–545.

Green, J. G., McLaughlin, K. A., Berglund, P. A., Gruber, M. J., Sampson, N. A., Zaslavsky, A. M., & Kessler, R. C. (2010). Childhood adversities and adult psychiatric disorders in the national comorbidity survey replication I: associations with first onset of DSM-IV disorders. *Archives of General Psychiatry*, *67*(2), 113–123. 67/2/113 [pii] http://dx.doi.org/10.1001/archgenpsychiatry. 2009.186.

Guidotti, A., Auta, J., Chen, Y., Davis, J. M., Dong, E., Gavin, D. P., Tueting, P., Grayson, D. R., Matrisciano, F., Pinna, G., Satta, R., Sharma, R. P., Tremolizzo, L., & Tueting, P. (2011). Epigenetic GABAergic targets in schizophrenia and bipolar disorder. *Neuropharmacology*, *60*(7–8), 1007–1016. http://dx.doi.org/10.1016/j.neuropharm.2010.10.021.

Gunnar, M., & Quevedo, K. (2007). The neurobiology of stress and development. *Annual Review of Psychology*, *58*, 145–173. http://dx.doi.org/10.1146/annurev.psych.58.110405.085605.

Harper, D. (2013). *Online Etymology Dictionary. Electronic Edition: Douglas Harper*. Available from http://www.etymonline.com/.

Henrichs, J., & Bogaerts, S. (2012). Correlates of posttraumatic stress disorder in forensic psychiatric outpatients in the Netherlands. *Journal of Traumatic Stress*, *25*(3), 315–322. http://dx.doi.org/10.1002/jts.21706.

Holliday, R. (2006). Epigenetics: a historical overview. *Epigenetics*, *1*(2), 76–80.

Laws, D. R., & O'Donohue, W. T. (2008). Introduction. In D. R. Laws, & W. T. O'Donohue (Eds.), *Sexual deviance: theory, assessment, and treatment* (2nd ed.) (pp. 1–20). New York: Guilford Press.

Lelandais, G. E. (2013). Citizenship, minorities and the struggle for a right to the city in Istanbul. *Citizenship Studies*, *17*(6–7), 817–836.

Maier, H. W. (1912). Katathyme Wahnbildung und Paranoia [Catathymic delusions and paranoia]. *Zeitschrift fur die gesamte Neurologie und Psychiatrie*, *5*, 545.

Malan-Muller, S., Seedat, S., & Hemmings, S. M. (2014). Understanding posttraumatic stress disorder: insights from the methylome. *Genes Brain Behavior*, *13*(1), 52–68. http://dx.doi.org/10.1111/gbb.12102.

Maze, I., & Nestler, E. J. (2011). The epigenetic landscape of addiction. *Annals of the New York Academy of Sciences, 1216,* 99–113. http://dx.doi.org/10.1111/j.1749-6632.2010.05893.x.

McEwan, T. E., & Strand, S. (2013). The role of psychopathology in stalking by adult strangers and acquaintances. *Australian and New Zealand Journal of Psychiatry, 47*(6), 546–555.

McGowan, P. O. (2013). Epigenomic mechanisms of early adversity and HPA dysfunction: considerations for PTSD research. *Frontiers in Psychiatry, 4*(110). http://dx.doi.org/10.3389/fpsyt.2013.00110.

McGuire, B., & Wraith, A. (2000). Legal and psychological aspects of stalking: a review. *Journal of Forensic Psychiatry, 11*(2), 316–327.

Miller, L. (2012). Stalking: patterns, motives, and intervention strategies. *Aggression and Violent Behaviour, 17,* 495–506.

Murgatroyd, C., Wu, Y., Bockmuhl, Y., & Spengler, D. (2010). Genes learn from stress: how infantile trauma programs us for depression. *Epigenetics, 5*(3), 194–199.

Nestor, P. G. (2002). Mental disorder and violence: personality dimensions and clinical features. *American Journal of Psychiatry, 159*(12), 1973–1978.

Norris, S. M., Huss, M. T., & Palarea, R. E. (2011). A pattern of violence: analyzing the relationship between intimate partner violence and stalking. *Violence and Victims, 26*(1), 103–115.

Pareés, I., Saifee, T. A., Kojovic, M., Kassavetis, P., Rubio-Agusti, I., & Sadnicka, A. (2013). Functional (psychogenic) symptoms in Parkinson's disease. *Movement Disorders, 28*(12), 1622–1627.

Pechtel, P., & Pizzagalli, D. A. (2011). Effects of early life stress on cognitive and affective function: an integrated review of human literature. *Psychopharmacology (Berl), 214*(1), 55–70. http://dx.doi.org/10.1007/s00213-010-2009-2.

Petherick, W., & Sinnamon, G. C. B. (2013). Victim and Offender Motivations. In W. Petherick (Ed.), *Serial crime: theoretical and practical issues in behavioral profiling* (3rd ed.). Boston: Anderon.

Raposo, A., Vicens, L., Clithero, J. A., Dobbins, I. G., & Huettel, S. A. (2011). Contributions of frontopolar cortex to judgments about self, others and relations. *Social Cognitive and Affective Neuroscience, 6*(3), 260–269. http://dx.doi.org/10.1093/scan/nsq033.

Read, J., Bentall, R. P., & Fosse, R. (2009). Time to abandon the bio-bio-bio model of psychosis: exploring the epigenetic and psychological mechanisms by which adverse life events lead to psychotic symptoms. *Epidemiologia e psichiatria sociale, 18*(4), 299–310.

Revitch, E., & Schlesinger, L. B. (1978). Murder: evaluation, classification, and prediction. In I. L. Kutash, S. B. Kutash, & L. B. Schlesinger (Eds.), *Violence: Perspectives on murder and aggression* (pp. 138–164). San Francisco: Jossey-Bass.

Revitch, E., & Schlesinger, L. B. (1989). *Sex murder and sex aggression.* Springfield, Il: Charles C. Thomas.

Rezansoff, S. N., Moniruzzaman, A., Gress, C., & Somers, J. M. (2013). Psychiatric diagnoses and multiyear criminal recidivism in a canadian provincial offender population. *Psychology, Public Policy, and Law, 19*(4), 443–453.

Schlesinger, L. B. (2004). *Sexual murder: catathymic and compulsive homicides.* Boca Raton, Fl: CRC Press.

Schlesinger, L. B. (2007). The Catathymic Process: Psychopathology and psychodynamics of extreme interpersonal violence. In L. B. Schlesinger (Ed.), *Explorations in Criminal Psychopathology: clinical syndromes with forensic implications* (2nd ed.). Springfield, Il: Charles C Thomas Publisher Ltd.

Schlichting, C. D., & Wund, M. A. (2014). *Phenotypic plasticity and epigenetic marking: an assessment of evidence for genetic accommodation.* Evolution http://dx.doi.org/10.1111/evo.12348.

Seitz, R. J., Nickel, J., & Azari, N. P. (2006). Functional modularity of the medial prefrontal cortex: involvement in human empathy. *Neuropsychology, 20*(6), 743–751. http://dx.doi.org/10.1037/0894-4105.20.6.743.

Signorelli, M. S., Battaglia, E., Costanzo, M. C., & Cannavò, D. (2013). *Pramipexole induced psychosis in a patient with restless legs syndrome.* BMJ Case Reports.

Silva, J. A., Derecho, D. V., Leong, G. B., & Ferrari, M. M. (2000). Stalking behavior in delusional jealousy. *Journal of Forensic Sciences, 45*(1), 77–82.

Spencer, J., & Tie, A. (2013). Psychiatric symptoms associated with the mental health defence for serious violent offences in Queensland. *Australasian Psychiatry*, *21*(2), 147–152.

Strine, T. W., Dube, S. R., Edwards, V. J., Prehn, A. W., Rasmussen, S., Wagenfeld, M., Dhingra, S., & Croft, J. B. (2012). Associations between adverse childhood experiences, psychological distress, and adult alcohol problems. *American Journal of Health Behavior*, *36*(3), 408–423. http://dx.doi.org/10.5993/AJHB.36.3.11.

Sutherland, E. H. (1947). *Principles of criminology* (4th ed.). Chicago: J B Lippincott Co.

Teschler, S., Bartkuhn, M., Kunzel, N., Schmidt, C., Kiehl, S., Dammann, G., & Dammann, R. (2013). Aberrant methylation of gene associated CpG sites occurs in borderline personality disorder. *PLoS One*, *8*(12). e84180 http://dx.doi.org/10.1371/journal.pone.0084180.

Thompson, C. M., Dennison, S. M., & Stewart, A. L. (2013). Are different risk factors associated with moderate and severe stalking violence?: examining factors from the integrated theoretical model of stalking violence. *Criminal Justice and Behavior*, *40*(8), 850–880.

Tiihonen, J., Isohanni, M., Räsänen, P., Koiranen, M., & Moring, J. (1997). Specific major mental disorders and criminality: a 26-year prospective study of the 1966 Northern Finland birth cohort. *American Journal of Psychiatry*, *154*(6), 840–845.

Van Dongen, J. D. M., Buck, N. M. L., & Van Marle, H. J. C. (2012). The role of ideational distress in the relation between persecutory ideations and reactive aggression. *Criminal Behaviour and Mental Health*, *22*(5), 350–359.

Waddington, C. H. (1939). *Introduction to Modern Genetics*. London: Allen and Unwin.

Wertham, F. (1937). The catathymic crisis: a clinical entity. *Archives of Neurology and Psychiatry*, *37*(4), 974–978. http://dx.doi.org/10.1001/archneurpsyc.1937.022601602740.

Wertham, F. A. (1978). Catathymic crisis. In I. L. Kutash., S. B. Kutash., & L. B. Schlesinger (Eds.), *Violence: perspectives on murder and aggression* (pp. 165–170). San Franscisco: Jossey-Bass.

Whittington, R., Hockenhull, J. C., McGuire, J., Leitner, M., Barr, W., Cherry, M. G., et al. (2013). A systematic review of risk assessment strategies for populations at high risk of engaging in violent behaviour: Update 2002-8. *Health Technology Assessment*, *17*(50). i-xiv+1–128.

Wockner, L. F., Noble, E. P., Lawford, B. R., Young, R. M., Morris, C. P., Whitehall, V. L., & Voisey, J. (2014). Genome-wide DNA methylation analysis of human brain tissue from schizophrenia patients. *Translational psychiatry*, *4*, e339. http://dx.doi.org/10.1038/tp.2013.111.

Wong, C. C. Y., Mill, J., & Fernandes, C. (2011). Drugs and addiction: an introduction to epigenetics. *Addiction*, *106*(3), 480–489.

World Health Organisation. (1992). *International Statistical Classification of Diseases and Related Health Problems, 10th Revision, version: 2010 (ICD-10) [online edition]*. Geneva: World Health Organisation. Available at http://apps.who.int/classifications/icd10/browse/2010/en.

Zaidi, S. K., Grandy, R. A., Lopez-Camacho, C., Montecino, M., van Wijnen, A. J., Lian, J. B., Stein, J. L., & Stein, G. S. (2014). *Bookmarking target genes in mitosis: a shared epigenetic trait of phenotypic transcription factors and oncogenes? Cancer Research*. http://dx.doi.org/10.1158/0008-5472.CAN-13-2837.

Zugman, A., Pan, P. M., Gadelha, A., Mansur, R. B., Asevedo, E., Cunha, G. R., Silva, P. F. R., Brietzke, E., & Bressan, R. A. (2013). Brain tumor in a patient with attenuated psychosis syndrome. *Schizophrenia Research*, *144*(1–3), 151–152.

Report Writing, Style, and Components

Wayne Petherick
Bond University, Gold Coast, QLD, Australia

Key Terms

Basic components
Relevance
Definitions
Administrative inclusions
Materials examined
Examinations performed
Victimology
Investigative suggestions

INTRODUCTION

A well-written report is a valuable tool that can dictate or change the outcome of a case. The report must conform to good writing conventions; a poorly worded report can seriously undermine the opinions therein. A report in which there are numerous spelling and grammatical errors can confuse the reader as to tone and meaning such that it is rendered almost worthless. Worse still, a bad report for which one gets paid can seriously undermine professional credibility, directly impacting your ability to get more work from the same (or new) clients. Reports must also be technically accurate, representing the correct times, places, people, and any other relevant case details. Where numerous errors are present, the reader may also lose confidence in the analyst's ability to be thorough, or they may doubt the materials have undergone the appropriate level of scrutiny—which is exactly what they have been paid to do.

This chapter will present a suggested crime analysis report writing style and some of the various components that may go into it, followed by numerous examples. It will be based on sound logic and reasoning and adhere to modern writing conventions. It also covers in detail the appropriate language, definitions, relevance, fees (including contingency fees), and report structure, including the introduction, materials section, the main body of the report, and conclusions or further suggestions.

12.1 Basic components

12.1.1 Language

Although seemingly common sense, one of the most typical areas that reports fail in is the language used to convey the message. This could be because of spelling and grammar mistakes that detract from the meaning, either in part or in total, or it could be a result of the technical nature of the report. It is not uncommon to find professional reports written by subject matter experts that are full of technical jargon they may be intimately familiar with, but that leaves others floundering as to what is being said. To describe an injury to someone without a medico-legal background as a "large periorbital hematoma" may be less instructive than simply saying the victim had a black eye.

Despite the fact that much academic work is still written in lofty and inaccessible language, there is a general move back to "keeping it simple." This philosophy translates to the written report, which must be kept simple and accessible. Any chronicle that cannot be understood by the majority of people who access it, or need to use it to determine some material fact, is literally not worth the paper it is printed on.

12.1.2 Audience

Related to language discussed above, the audience of the report must be kept in mind. Because the range of people who may read your report is vast, it is a good idea to consider the lowest common denominator—that is, someone who may not be as familiar with the case as you are, or someone who may not be as familiar with the technical concepts discussed.

Considering the audience is about more than the use of appropriate language, however. Once a report is written, the author cannot guarantee what happens to the report after submission or who else may see or have access to it. Police officers retire, get seconded to other positions, go on leave, and are transferred to new duty stations. Lawyers may move firms, be allocated a new portfolio, or enter into other areas of law (from criminal to civil, for example). Should the report contain information previously unknown or undisclosed, it may be passed along the chain of command to escalate the matter further. The point is that you cannot assume the person who requested the report is the only one who will see it or use it.

In one case from the author's files, a report was requested on a homicide by a university innocence project. The project was shut down shortly after the report was tendered, and it was not passed along to the accused, their counsel, or their family. Some 10 years later, the mother of the accused found a comment about the report in the case files, but no actual report. They contacted the author and asked for a copy of the report in an attempt to overturn the original conviction. Elements of the report were also then used at the federal administrative level to rebut an attempt by the federal government to withdraw a granted immigration status and deport the individual concerned.

In a false allegation of stalking case, the original report was written on request by the father of the accused. After much frustration and seemingly little legal advance in the case (which had seen three court appearances), the father sought an appointment with a local politician, who then escalated the case by contacting a state minister. This resulted directly in the charges against the accused being dropped, and investigations launched into a number of individuals involved in the false allegation. At least one individual lost their job as a consequence.

12.1.3 **Definitions**

Whenever concepts or terms are used in a report, they must be fully defined and referenced as to source. This is more than an academic exercise to show people how smart and widely read you may be. Without an operational definition of terms, the reader cannot necessarily know how the conclusions were arrived at, and it becomes more difficult to demonstrate the features of the case without seeing a concise explanation of the meaning of a word or phrase. In addition, the theory of logic dictates that a term is first defined, then the evidence that supports or refutes this concept as defined is demonstrated.

12.1.4 **Relevance**

A report must be relevant. This means that it must contain information that applies to the current case and is material in answering probative (or courtworthy) questions. One of the pitfalls in using inductive logic without determining whether the aggregate information applies to the current case is that you are essentially describing an abstract. Put another way, the "average" or "common" findings presented may well represent the mean or typical group, but it may be a poor fit for the individual(s) under examination in the extant case. Sherlock's logic and the wisdom of Arthur Conan Doyle (1986, p. 60) provides the following on this very problem:

> *"Winwood Reade is good upon the subject," said Holmes. "He remarks that, while the individual man is an insoluble puzzle, in the aggregate he becomes a mathematical certainty. You can, for example, never foretell what any one man will do, but you can say with precision what an average number will be up to. Individuals vary, but percentages remain constant. So says the statistician."*

Simply restating the conclusions of a number of others may also undermine the relevance of the report to the case, should the client wish for more. Ideally, the reports and conclusions of others constitute a firm base on which to provide your own opinions; otherwise, the report's utility and relevance will be seriously undermined. It should be noted that this is a general rule. Involvement in some cases may simply be a matter of putting the facts together in a meaningful way, especially if this has not been done previously by anyone else.

In one case from the author's files, a woman was falsely charged with stalking her former fiancé. On the face of it, the complaint appeared legitimate. However, upon a detailed examination of the facts, it became apparent that her behavior was in context to a number of completely reasonable requests. Furthermore, when a timeline of events was compiled, it became evident that much of the reported offense-related behavior could not have occurred in the way the former fiancé had claimed in his complaint. In this instance, putting the timeline together and comparing this to the allegations was highly instructive and proved to be the key to resolution of the case.

12.1.5 **Fees**

Fees are an important consideration, especially in areas where fee schedules are dictated by professional or accrediting bodies, as is the case with psychology, for example. In areas where no guidelines exist, the analyst must determine if there is an industry standard or award, tempered with a consideration of his or her qualifications, associations, and affiliations. It would not be unreasonable to find that the more qualifications a professional has, the higher the fees. This may also be found in experts who are considered industry, national, or international subject matter experts.

The fees may also be dictated by the case at hand and the level of involvement. Some have a standard fee regardless of what is required, whereas other will charge one fee for report preparation and another fee for testifying in court, should that be required. Fees may also be dictated by the client agency or in some cases; organizations such as legal aid may pay a standard fee or be restricted by available budgets.

Fee structures must also factor in the capacity of the client to pay them. It would be pointless to state that a case requires 50 hours at $400/hour with an indigent client who simply could not bear this cost. Therefore, the client's ability to cover the costs may need to be a consideration.

If writing crime analyses, you may find that you receive more inquiries for reports than you actually write reports. This may be owing to unrealistic expectations the client has about what can be achieved, where the level of information on which to base a report is totally insufficient, or when the time required and the fees charged are too far outside of an individual's financial capacity. Whatever the reason, you may find yourself spending considerable time in the initial consultation process, or in examining the materials to determine whether a report is actually even a possibility. If consultation constitutes the entirety of your income, spending time in the initial phases only to determine that a report is not feasible will detract from other paid work. If consultation is a secondary income, it will detract from your primary employment. As a result, it may be prudent to charge a document examination fee. If the client follows through with a full report, it would be up to the individual analyst as to whether this was considered a separate fee or whether it was deducted from the final total cost of the report. Doing this can also stop those on a "fishing expedition" who may have no plans to retain your services or are looking to determine what your opinions are, which they may then use without paying for your services.

Along with considerations of charging for the initial examination of materials, it is worth considering what amounts to a deposit. Numerous anecdotes have been told and experienced firsthand where payment is never made after submission of a report. Claims of empty trust funds and clients not paying lawyers and other professionals may mean there is no remuneration for the work you have done. Whether to include a deposit and the proportion of split is up to the individual but it can be prudent. To put it simply: all others involved in the case have likely been paid or will expect to be. You should too, especially if that was what was initially agreed upon.

Considerations of fees go beyond ensuring you can meet your financial obligations, though. Charging for a service establishes a professional relationship between you and the client, setting up boundaries and clarifying expectations from all parties concerned. Regardless of who requests your services, you can and should expect treatment like a professional as this represents the services you are providing.

Some individuals and agencies will ask you to sign a contract outlining expectations, and it is not unreasonable to expect one in return. This would state any peculiar conditions of the relationship, an estimate of the total number of hours (including a statement that this may change with other fees being charged if the report should go over this estimate, within reason), an estimated date of completion, payment schedules or expectations including hourly rate, and any other necessary information. This contract should be signed prior to offering any advice or opinion and provides you with legal recourse should the client not be forthcoming with payment.

Another type of fee called a *contingency fee* is a fee paid based on the outcome of a case. Should the client win, you receive higher awards. This type of payment is perhaps more common in civil cases than in criminal ones, where financial reward can be directly mapped to outcome. Regardless of the crime type, contingency fees should be avoided at all costs. They make the analyst too invested in the outcome, potentially setting up a range of destructive cognitive biases that may render the report useless in the long term and establish a professional reputation as an advocate.

12.1.6 **Length**

The length of the overall report will be dictated by many factors. This includes but is not limited to: the amount of information or evidence available; the specific questions or issues you are asked to address; if used, the amount of research and literature available in the domain of the report; the knowledge base of the author; and legal or professional restrictions or requirements. There are no hard and fast rules here, however. A general crime analysis will likely be longer than one in which you are only asked to address one issue. A case with 30 images and 100 pages of documentation will not be as lengthy as one with several hundred photos and thousands of pages of documents of other information. One case from the author's files included 1,732 photographs, recorded interviews, and 212 individual documents, some of which were several hundred pages in length. Another included just over 30 photos and 600 pages of documentary evidence.

Discussions concerning report length produce mixed opinions. Some argue that a short report is better, while others (this author included) argue that there is no fixed length and that the report should be as inclusive as possible or as dictated by the nuances of the case. From the immortal words of a good friend and barrister in Victoria, Australia, *a report has to be as long as it has to be.* It has also been said that if the client gave you the information, then you do not need to list out the evidence or the case background in detail, which can expand length exponentially. This is not a safe assumption, because as the reader will recall, the person who commissioned the report may not be the only person who sees it or uses it, and there is no guarantee that other professionals in the future will be as familiar with the case as those in the past.

12.1.7 **Headings**

The use of appropriate headings throughout the entire document is strongly encouraged. This allows the reader to readily identify various parts of the document, and it saves them having to read it in its entirety should they already be familiar with certain facts. Reports without headings can be a cumbersome read, and these also help the author to stay on task and not wander off into issues other than those relevant to the current section.

Headings should follow the appropriate format, such as identifying a Level A heading (a main section) versus a Level B heading (a subheading or subsection). The use of bold, italics, and underlining or other formatting is strongly encouraged. A heading should reflect content and easily identify what is discussed and where one section ends and another begins. Generally, it is a good idea to keep your headings as brief as possible. For example, *An Examination of Potential Staging Indicators* may just be titled *An Examination of Staging* or simply *Staging* with further explanation in the opening paragraph that the section will be limited to those indicators of staging, if this is what is in fact being done.

12.1.8 **Signatures**

Once the report is finished, it is necessary at the very end to sign the document before submission (if it is more than a draft). This establishes ownership of the contents. If more than one person wrote the report, then it is necessary for both to sign. This refers strictly to authorship though, so if one only compiled background research they need not sign. If the report is to be tendered in court or used for some other legal purpose, a signature may be a legal requirement, where the report may not be accepted without it. With the proliferation of computing technology, electronic reports are commonplace and as such an electronic signature may be perfectly acceptable. Check with the jurisdiction in which the report will be submitted or with the client to comply with their particular demands.

12.2 Suggested report layout

For disciplines such as forensic psychology, clear guidelines exist as to what must be included in a forensic report. As previously stated, these guidelines would supersede some of what is discussed herein. Recall that the purpose of this book is to provide some structure for those disciplines in which clear guidelines are few or nonexistent.

The following has proven useful and instructive over the years and has been the subject of review and refinement based on "what works." Actual details from examples used such as phone numbers, names, and addresses have been changed for obvious privacy reasons.

12.2.1 Administrative inclusions

This is one section of the report that may be taken for granted, but over the years a number of analyses have been seen where the author has failed to provide any contact details at all. This makes it difficult to contact the analyst should further information or clarification be needed or should the author be required for further analysis, such as where they may be required to testify in court. This section is the same as that discussed by Petherick and Ferguson in Forensic Victimology (see Chapter 4).

Many word processing programs provide templates for this information, so its inclusion may be simple and largely automated. Should the analyst be working from a clean document, this information should be listed clearly at the top of the first page, and include who you are along with your qualifications listed in brief (usually the highest level of qualification achieved). It should also list basic contact details, such as telephone or mobile phone numbers, and contact addresses. Many professionals maintain post office boxes and separate contact numbers or email addresses so as to avoid giving out their personal details, such as their home address. This is always a wise decision. The date the report was finalized or submitted should be included at the end of this information.

Example
Wayne Petherick, PhD
Email: wayne@wayne.com
Phone: +617 1234-5678
Address: PO Box 111, Anytown, 4321, Australia
Date: 12 April, 2013.

The next brief section of the report details who wrote the report or assisted with its compilation. If the report was only prepared by one person, this would simply involve restating the name above. In some cases, however, reports may be prepared by more than one person, such as when working with an intern or a research assistant. The relative contribution of each person should be listed, such as when one person wrote the report with the other doing background research on the crime type or the crime itself.

Example
Report by Wayne Petherick and John Smith. Dr. Petherick examined the evidence for the case and prepared the report in its totality. Mr Smith prepared the background research on the crime type and visited the accused in custody. Mr Smith did not contribute to the authoring of the report.

The next section usually describes the individual or agency requesting the crime analysis. If the report is for an agency, you should identify here any individuals within that agency that you have had contact with. This should also include any changes in the contact person such as when a case gets handed over to a new law firm or to a new lawyer within the same firm. Also include how you were initially contacted, and how the person commissioning the report found or identified you as an expert, along with any relevant times or dates. If compiled for a legal matter, a brief statement identifying the matter should also be made.

Example

This analyst compiled the following report for Jane Jones of Jones, Smith and Johnson Lawyers. I was originally contacted on the 1st of March by Ms Jones and asked to prepare a report in the matter of R v Hammond[1]. *I subsequently met with Ms Jones on the 5th and 11th of March at her law offices to obtain the information and full brief. Ms Jones contacted me based on comments I had made in a media interview regarding homicide, and believed based on my comments that I may be able to answer questions related to the behavior in the current case.*

12.2.2 Caveats and disclaimers

Any caveats or disclaimers regarding the case or your involvement with it should be noted in this section. This should include whether information was missing or not disclosed to you, whether you visited the crime scene if this was necessary or possible, whether you interviewed the accused or other relevant parties if this was necessary or possible, and whether you charged fees for providing the report, with justifications for why no fees were charged, if that is the case.

These caveats and disclaimers are an important part of the report as they tell the reader what was not done or what could not be done and why. It documents potential limitations and provides some account of how this may impact the case or report. Doing this may also prevent some level of misfeasance, such as withholding information or evidence without sufficient moral or ethical reason. Even though it may be injurious to their case, it is not impossible for this to happen. In one case, omissions from the documentation were noted, and when the client was asked where this information was they replied "I just wanted to see what you could come up with without it." In short, it was a test to determine what the author could infer.

Example

It should be noted that while full access was provided to the documentary materials including autopsy report, witness statements, crime scene photos and videos, among others, this examiner was not able to visit the crime scene for contextual purposes. The property in which the crime occurred is a rental property that has since been relet, and the property manager was reluctant to provide access owing to disruption to the current tenants.

The following is from a report in which the client could not afford further expenses because of the amount of money spent on legal fees thus far. The report was compiled without financial remuneration in exchange for access to the case materials for teaching purposes:

In the interests of disclosure, no fees were charged for this report, but full access was granted to the material for teaching purposes in exchange for my time and opinion.

---

[1] Note: in Commonwealth jurisdictions *R* refers to *Regina*, meaning the Queen, and refers to a prosecution matter brought by the state (typically) against an individual. In the USA, this would be replaced by the state that has brought the charges, for example CA v Smith.

12.2.3 Qualifications

If you have been consulted, this would usually mean you are an expert in a particular field, which will mean you have some relevant education, training, and experience. Any relevant qualifications should be listed out in full in this section. This essentially informs the reader why you are the right person for the job. It should be noted that this should be limited to those things that relate directly to the case or your experience. You would not need to include the fact that you won a spelling competition in the fifth grade, or that you volunteer at a local animal shelter.

University qualifications would usually be listed first, from lowest to highest, along with any other achievements, such as whether you received an honors classification because of high grades. It is also usual for postnominals to include the institution through which the degree was obtained (not all institutions are created equally).

Other criteria by which expertise may be judged include honorary positions, society and professional body membership, classes taught, and works authored.

Example

I hold a Bachelor of Social Science (Psychology) from the Queensland University of Technology, a Master of Criminology from Bond University, and a Doctor of Philosophy (PhD) from Bond University, where I am currently Associate Professor and Coordinator of Criminology. I teach in the areas of criminal motivations, crime and deviance, alcohol, drugs and crime, criminal profiling, forensic victimology, forensic criminology, and applied crime analysis.

I am author or coauthor of over 50 scholarly articles and book chapters, including three textbooks related to my teaching matter. These include Profiling and Serial Crime, Forensic Victimology, *and* Forensic Criminology. *I have authored articles and chapters on stalking, risk assessment, crime analysis, profiling, logic and reasoning, personality disorder and crime, victimology, and various types of serial crime. I am currently a Member General Section and Secretary of the Australasian Association of Threat Assessment Professionals.*

12.2.4 Materials examined

The materials examined refers to any and all documentary, testimonial, physical, and other evidence on which you based your findings. This could include interviews with lawyers, witnesses, police, and others. These should be listed in sufficient detail so as to be able to identify individual documents available and reviewed but not in such detail that this section consumes page after page as this can make a report tedious to read.

Good Format:

Crime scene photographs of 11 Main Street, Anytown, numbered 1–75 and 80–100.

Bad Format:

Crime scene photograph of 11 Main Street, Anytown, number 1 of the front door of the residence.

Crime scene photograph of 11 Main Street, Anytown, number 2 of the entry hallway.

Crime scene photograph of 11 Main Street, Anytown, number 3 of the bloodstain in the entry hallway…

The latter example would substantially and unnecessarily increase the size of the report. In the first example, there are 95 photos identified, which occupies one line item. Consider how much space in the report this would occupy should you list all 95 out individually, or do so in the case mentioned earlier in which there were 1,732 photographs—and that is before you have even gotten to describe any other information you have.

Where possible, the original details of individual items should be listed as the source, such as when crime scene photographs are identified by number from the investigating agency. This will allow for uniformity between individuals involved in the case so that others know to which items you refer. Should you implement your own number or filing system for individual documents, for whatever reason, a key or a legend must be provided with reference back to the original classification used.

As a rule, the full title of the document should be provided if the item has one. For court transcripts, as one example, this should include the level of court system (Court of Appeal, Supreme Court, High Court, etc.), the location of the court (city, state, county, etc.), the date of the hearing(s), and the legal reference for the matter (*US v Smith, CA v Smith, R v Smith*, etc.).

> *Example*
>
> *CIB Brief of Evidence, Defense Copy*
> *Exhibit Transcripts and Records of Interview*
> *Joint Investigation Team re: alleged abuse of X*
> *Legal Documents re: Assault Occasioning Actual Bodily Harm of X*
> *Crime scene photographs 1–1732*
> *Police crime scene video of 123 Main Street, Anytown*
> *Transcript of proceedings in the matter of* R v V
> *Telephone discussion with Mr. B on the 20th of September, 2012*
> *Telephone discussion with Mr. D on the 20th of September, 2012*
> *Telephone discussion with Ms. F on the 21st of September, 2012*

12.2.5 Examinations

Following a detailed (but not too detailed) listing of the evidence, you should describe any particular examinations that you have undertaken with the evidence. Should an examination be requested by you of another professional or individual, this should be included here also, including details of the person or organization that performed the service, their qualifications, the nature of the examination, and any other relevant information such as why a new examination was necessary.

It is suggested that this section be reserved for examinations done by others and not include basic information, such as "this examiner read all of the available material," which would be implied. However, if part of the analysis was to compile a timeline of events, it would be perfectly acceptable to state that "this examiner read all of the available material for the purpose of constructing a timeline of events."

If nothing additional has been done with the case information, this section could be omitted but must be included if others have performed additional analyses. For example, if you had a firearm tested for malfunction, this would need to be cited. This would apply to bloodstain pattern analysis, crime reconstruction, additional DNA testing, and any other type of extra examination performed.

Example

  This examiner performed a thorough review of the information provided to determine a precise sequence of events with a view to determining which of the various statements are most in accord with the physical evidence. To assist with this determination, a full crime reconstruction was performed by Dr. David Allen of Forensic Services on the 5th of October 2012 to determine the exact nature and sequence of all acts evidenced in the crime.

12.2.6 Summary of opinions

There is some debate as to whether this is necessary or helpful given the opinions and their evidentiary basis will be provided and discussed in depth later in the report. It has been suggested that these be included in bullet form, with just the opinion and no supporting evidence so that the client can quickly assess what your opinions are without having to read the entire report.

  It is also possible to include these later in the report where you detail and describe not only your opinions but also outline the evidence on which they are based. If this is the preferred option, they should be clearly identified, such as formatted in bold or a larger font, so that they can be easily found should a summary of opinions section not be included.

12.2.7 Statement of conflicts of interest

A conflict of interest may occur where the analyst's involvement is clouded by prior knowledge, prior involvement, or by the expectations of interested parties, among others. Should the analyst be aware of these, they can be controlled for by adopting good ethical practices and adhering to sound principles of scientific analysis. The code of ethics of one's profession, should these exist, can also be a shield to rebuke for bias or conflict. Where this presents an insurmountable obstacle to involvement, the analyst must withdraw at the first possible instance noting the conflict and the reasoning behind it. One good example of this is a homicide where the analyst was to suggest to investigating police that the husband was a good and logical suspect, and later testified at a bail hearing suggesting that the husband be denied bail as he posed a risk of future violence (premised on the notion that the best indicator of future behavior is past behavior—the husband is being tried for a homicide; therefore, he presents a future risk of same).

12.2.8 Introduction or overview

The previous sections could be considered general or administrative in nature. The introduction or overview section then could be considered the first main or substantive section of the report. It is suggested that the title be one or the other but not both. This would be entirely up to the author of the report, and may be dictated by the actual content of this section.

  This section can be easily summarized by the following three components: main people, main times, and main places. This is obviously somewhat simplistic, as other information may be necessary here, but as a general rule limiting this section to these three things will keep it relatively brief, succinct, and on task. More detail on each follows.

  Main people: Depending on the type of incident being examined and the type of analysis requested this would include, but is not limited to, the accused, suspects (and the reason they are suspects), witnesses, victim(s), investigators, and others involved in the case (such as a pathologist who determined the cause of death, for example). The main thing to bear in mind with this section is that it should be limited to main people, and one good way to determine if someone meets the criteria is to determine if their inclusion is material to some fact or issue under consideration.

Main times: In general, timeframes can be somewhat arbitrary. It is often stated that the last 24 or 48 hours are important, but this tends to draw a bright yellow line between "important" and "not important." If we limit our investigation to the last 24 hours, do we exclude an event that happened 25, 26, or 27 hours prior to the event? A good general rule is that if it is important then include it, regardless of the timeframe set. One good example here is a murder where the wife of the deceased made an inquiry as to how much it costs to have someone killed some time prior to the actual death, but it is materially relevant.

Main places: Main places would include those locations that are important or relevant to the event. This would include the location of the crime scene, locations where other evidence was found, the location of witnesses, and other places where things may have occurred. It is good practice to identify these places by address, and to state the distance between related locations. For example, the distance between a witness who reports hearing a gun shot and the actual crime scene could be crucial in determining whether that is what they actually heard. The distance between the location someone claims to be at while a homicide occurs in another location could be pivotal if part of their alibi cannot be substantiated (put another way, is it possible they could have traveled to commit the crime and returned to their original location to continue their alibi?)

Example

   On the morning of the 17th of October, 2009, John Vidler invited a number of friends over to 38 Main Street, Anytown, for dinner and drinks. Paul Freeman and Jane Freeman arrived at approximately 5.00 pm. The reason given for the dinner was a welcome to his new house. Trish Tate and Tony Haddon also attended the dinner. According to Mr. Vidler, Ms Mullen became abusive towards him in front of their friends and she stormed off. He tried to get her to calm down, but she was crying and "going off." At about 6.30 pm, while playing darts, Ms Mullen had left to go to the toilet. When she did not return, Trish and Jane went to look for her and found her in her office.

   She appeared visibly upset and was crying and said words to the effect of "all I want him to do is include me in decision making and who was coming over." This was repeated half a dozen times or so, before she further stated "I always have to finish, like cooking the meal, I always have to finish what you start." Jane states she believed that Ms Mullen was referring to every, or every other, relationship.

   Upon rejoining the gathering, it was noted that Ms. Mullen was not listening to Mr. Vidler, and that he was trying to avoid confrontation. When asked to join them, Ms. Mullen said she did not want to come out and have dinner.

12.2.9 Physical evidence

While the materials examined section outlines what evidence you have in brief, and the examinations performed section outlines what you did with it (if anything), this section should detail the physical evidence and its meaning at some length. Again, there are no hard and fast rules with this section, and some may prefer to incorporate it into the introduction or overview section. Should this be the preferred option, each piece of evidence would have to be listed at that place in the timeline where it is important. For example, firearms evidence may be listed at the point in the timeline where the discovery of the deceased is noted and his gunshot injuries discussed. The cause of death would be listed at that point in the timeline where the autopsy is stated as having been done.

It should be noted, however, that incorporating this information into the introduction or overview is difficult because some of the evidence will be interpreted some time after the event, such as DNA, which may takes weeks or months to acquire. It will also substantially increase the length of the opening section. For these reasons, the author finds it prudent to include this in a separate section. Should the client not want this information included, it means it can also be easily removed without having to scour through the introduction and remove only certain parts.

This is another element that will be dictated almost entirely by the case at hand, but may include such things as wound patterns, bloodstain patterns, firearms and other weapons information, autopsy results, toxicology results, closed circuit television footage, and crime scene video as some examples. For all of this information you should again list the main features: the type of evidence, what it indicates, and who examined it. Based upon this information, you should also discuss what type of crime scene it is (indoor, outdoor, etc.), and other relevant information (such as whether it is the primary crime scene or simply where the body came to rest).

This section will likely just be a retelling of the opinions of other people, unless you have done other examinations yourself.

Example

In the entrance hallway, there is a considerable amount of drop-down bloodstains, what appears to be cast off, as well as blood swipe on the hall walls. There are trodden bloodstains throughout this area. Superior to Ms. Mullen's head is a large area of blood, where she was resting supine prior to being moved by the paramedics. A large bloodstained kitchen knife can be seen on the border between the carpet and tiled areas. Above the light switches immediately above the largest area of bloodstain are several areas of what looks like blood cast off. In certain places along the walls, there are also tiny droplets of blood, which are usually referred to as "high-velocity" blood splatter. In the absence of a firearm, these are most likely to result from aspirated blood, typically from blood appearing in the airway that is coughed or spat up, or from other wounds such as a punctured lung. The latter was posited during testimony as the likely cause.

12.2.10 Victimology

Victimology is the thorough study of all aspects of the victim's life, including their demographic information, hobbies, habits, routines, and other important factors as discussed in Chapter 4. The victimology should include all of the information discussed in the previous chapter, such as relationship history, drug and alcohol history, and prior victimization, along with any others for which there is evidence.

Should the crime include an initial aggressor who subsequently gets killed by the person they attack, a victimology of both parties should be included (as both are victims in the case under examination). By extension, any other parties involved who suffer harm or loss should also be included here. The source of any information should also be provided, along with the potential validity of that source. Should a witness state that the victim had a certain type of personality, unless this is corroborated by other evidence, the author of the report should be appropriately conservative in the way this is stated. For example, you would say that "the victim's son alleges his father was confrontational and violent" rather that "the victim was confrontational and violent toward his son" in the absence of other information.

This section should also include the risk assessment of all victims detailing not only the level of risk present, but where in the environment that risk is coming from. For children, as a rule, much of their victimology and the vast majority of their risk is drawn directly from the parents.

If property is the target rather than people, much of the same approach would apply. You would still need to describe the location, what and who it services (type of industry, for example), where it is located relative to other areas, and any past threats to the property or business. As a rule, property can be easier to establish because there are less "X factors" to consider, such as human emotion, among others. However, the more interaction between the people and place, the more these "X factors" can play a role.

Example

Ms. Mullen has had a variety of tumultuous past relationships, involving both psychological and physical abuse. As noted by Dr. James Gold in his report on Ms Mullen, in altercations with her former husband the fight became violent with possessions being broken. This was to become a relatively common theme running throughout other relationships. Ms Mullen is known to have a volatile, erratic, and unpredictable temper and has an extensive history of contact with the Department of Community Safety as well as numerous past partners taking out Domestic Violence Orders on her.

12.3 Conclusions

Next to victimology, this is perhaps one of the most substantial sections of the report. It is, in theory, what the entire report has been leading up to at this point. By the time you arrive at this section of the report, you will have a full and complete understanding of what has happened, where, and to whom. You should also have built up a good idea of the conclusions that you are going to argue, and the evidence that supports or refutes them. It could be said then, that the conclusions will virtually write themselves.

As with other parts, there is no rule about how many conclusions you will include. This will again be dictated by the complexity of the case and by directives given by the client. The comparison would again be the difference between a general crime analysis (lots of potential conclusions) versus a report directed only at the issue of staging (potentially a single conclusion—whether the crime was staged or not).

It is good practice to incorporate conclusions that have a sound theoretical basis. Indeed, this should be a consideration throughout every part of the report, but this tells the client that the report is based on sound theory that is generally accepted within the field in which it exists. This is accomplished by referencing literature, theory, and statistics where appropriate and demonstrating their relevance to the extant case.

The conclusion section can also be the place where you critique the conclusions of others for their validity. Despite the protestations of some, it is not rude or impolite to examine the work of others to determine whether it has been done properly and that the conclusions are fair and accurate. It is good scientific practice and an absolute necessity if we are being thorough and complete. If they have done their job properly, this finding will likely support whatever conclusions they have made about the case. If they have not, this could well be a point on which to hang doubt or suspicion about some aspect of the case.

The following example is drawn from a case involving the death of a 32-month-old male. The time at which the injury was sustained was material to the case with regards to who was likely responsible for the child's death. The child's father told police at the hospital that he must have caused the injury earlier that day when the child was misbehaving, which was later treated as a confession. Medical reports placing the injury leading to death prior to the day of death (and therefore not in accord with the father's "confession") were ignored.

Example

There are many estimates by medical professionals of the time at which the injury leading to the subdural haematoma (ruptured artery in the brain) occurred. These are made by a number of professionals ranging from neurosurgeons to ER doctors to pediatric specialists.

It is my opinion that the injuries sustained by Alexander that led to his death were sustained some days before the 24th of September. This opinion is based on the following examination of the statements by various medical practitioners, factoring in the pathologist's opinions and findings.

1. The Autopsy Report of Dr. Michael Johnson

Page 12 of the autopsy report reveals a softly clotted subdural hemorrhage over the right cerebral hemisphere. Page 25 of the autopsy report notes that the cause of death is head injuries arising from the subdural hemorrhage with associated brain swelling. It is also noted that the presence of bruises on the head resulted from hitting the head or the head hitting an object. This is in line with other medical examinations citing an acceleration/deceleration or rotational injury.

Dr. Johnson suggests histology of the hemorrhage is acute, indicating that the wound is "no more than a few days old," and also that the injuries "occurred within a few days of death."

2. Dr. Emma Cohen

Dr. Cohen mirrors conclusions and recommendations made by other medical professionals in their analysis of the case. Notable within Dr. Cohen's report are the following conclusions:

That subdural hematomas can occur by (1) direct impact to the overlying bone with trauma to the dura in relation to the skull fracture; (2) by disruption of the blood vessels underlying the dura with trauma which causes acceleration/deceleration with angular rotation and subsequent shearing of blood vessels. This latter scenario can occur following impact on one side of the head which causes subsequent trauma to the other side of the head (this is known as a contrecoup lesion, where the initial acceleration brings about injury from sudden cessation of movement).

Dr. Cohen's conclusion as to the possible age of the injury matches (the father's) statement given at hospital that

Alexander has been unwell for several days prior to his admission to hospital and subsequent death. Page 9 of this report states:

This may have resulted in altered neurological status and possible seizures or other compromise in respiration that could lead to progressive hypoxic/ischemic damage and increasing cerebral edema until cardiac-respiratory arrest had occurred. This would imply that there had been a period of up to 2–3 days where this child had been gravely unwell and his medical needs have not been met.

3. Dr. Karen McDonald, Pediatric Registrar

Dr. McDonald notes the presence of bruises on the ears with the left ear also evidencing a laceration. The bruising on the left ear is dark and does not appear diffuse suggesting it is a recent injury (Note: (the father) claims that (the mother) used to hit Alexander with a wooden spoon in this area if he dropped something off his plate).

4. Dr. David Nielson, Neurosurgeon

According to the report of Dr. Nielson, neurosurgeon, at #8 p. 11, the presence of hemosiderin at the site of the subdural hematoma indicates the wound was between 2 and 4 days old (citing Troncoso, Rubioa, & Fowler, 2009). Dr. Nielson also notes that in 24 hours there is intact red blood cells present at the site of injury, that in 24–48 h neutrophils form (but not invariably), and that by the third and fourth day, macrophages with and without hemosiderin appear. Dr. Nielson also notes that, based on his own visual inspection of the photographs, he would place the timeframe of the injury to be within 2–4 days of Joshua's death.

(The father) informed paramedics who responded to the emergency call that Alexander had been unwell for several days. This was confirmed by ambulance officer Casey Robinson. It is possible, perhaps even likely, that Alexander was symptomatic at this point in time. If this is true, this would place the initial injury days before the 24th of September.

12.3.1 Investigative suggestions

This could be considered the last part of the report, barring any closing statements made about contacting the report author for any further information or clarification. In each and every previous section of the report, questions may arise about missing information, or information that is contradicted by other information or witnesses. In one case, three different witnesses gave three different descriptions for the time for two gun shots. Any missing information or contradictions must be accounted for, and advice given as to how best to secure the information or resolve conflicting statements.

This is the purpose of the investigative suggestions section. This too will be limited only by the amount of information in a case. Evidence of good quality with little to none missing and cases where there are few inconsistencies will usually mean a shorter suggestions section. In cases where there are vast gaps in the knowledge base or where there are a large number of inconsistencies, this usually means more questions will be asked resulting in more suggestions being put forth.

A common pitfall here is just in asking questions without any guidance as to what should be done to answer the question. As a result, it is critical to keep the "problem-solution" approach in mind when crafting this section.

Example

Photographs 37, 38, and 41 of the lounge room show a number of cups and glasses at various locations. There is no discussion of whether any of these were collected and/or tested for DNA. Investigation needs to be undertaken as to whether any DNA analyses were done, and what the results of these were. If none were done, it needs to be determined whether the glasses and cups are still available, and if they are, they need to be tested to determine whether any other people were at the crime scene.

With the above example, the problem is that we have evidence, but we do not know whether the appropriate tests have been conducted. The solution is to see if it was done in the first place, and if not, whether the evidence still exists to be tested.

The author is often asked whether this is even necessary, as surely this would be done during any investigation. The short answer is yes it is necessary, and that there is no guarantee that even a basic level of investigation has been done, or done competently. Over the years I have seen even the most rudimentary inquiries not made, evidence overlooked, or in the worst cases, completely ignored. In one inquiry, the investigator's log had a red line drawn under the time at which a suspect was found with the words written all in capitals, "CEASE ALL FURTHER LINES OF INQUIRY SUSPECT IDENTIFIED."

As a final note, the purpose of this section is to advance the case to a point where some resolution can be achieved. Ensure this section does just that.

Example

According to Jane, when interviewed by police the following day, there seemed to be significant interest in the BBQ forks, such as whether they were used, what they looked like, or whether they were washed etc. Jane reports having the impression that police may have considered at least one of the forks as the murder weapon. In the crime scene photographs I have, there seems to only be two types of instrument/utensil in these images, and these are the knife used to kill Ms. Mullen, three BBQ forks, and a sharpening steel. There is even mention in the trial transcript that these were collected and sent off for DNA testing. However, there is no further mention of the results of any DNA test, or indeed why there would be so much investigative interest in these forks. Without being too specula- tive, it is my opinion that these may have been suspected of inflicting the wounds on either Ms. Mullen or Mr. Vidler early in the investigation. If so, and if his wounds were caused by such an instrument, it is imperative that the DNA results be obtained to see if his DNA is on any of that number collected. By extension, if a fork was used to inflict the wounds to his arm, this may indeed be suggestive that she was still armed during the confrontation. His level of intoxication would certainly explain the confusion surrounding this issue.

Conclusion

A well-written report can help the client fully understand the particulars of their case. These reports will vary in content and length, a result of the amount of information available in the case, the demands of the client, and the questions to be answered. The education, training, and experience of each report writer will also be different and this may also dictate length as well as quality.

The report itself may have different purposes and be seen during its life by any different number of people. As a result, you should ensure that they conform to the conventions of good spelling and gram- mar and be technically and factually accurate. Because you do not know who may end up using your report, it should be written in as accessible language as possible, but also be aware that your report may end up in court as part of the evidentiary package.

REVIEW QUESTIONS

1.   A report must be relevant. True or false?

2.   Why is it important to have an Administrative Inclusions section?

3.   Contingency fees are a good idea. True or false? Why or why not?

4.   What are the three elements or important considerations of the introduction or overview?

5.   What is the purpose of the Investigative Suggestions section?

6.   In general, it is necessary to provide the report author's name and contact details. Why might this be so?

Reference

Conan Doyle, A. (1986). *The sign of four: The complete works of Sherlock Holmes*. New York: Bantam Classics.

Appendix A

Threat assessment and management process—organizational settings

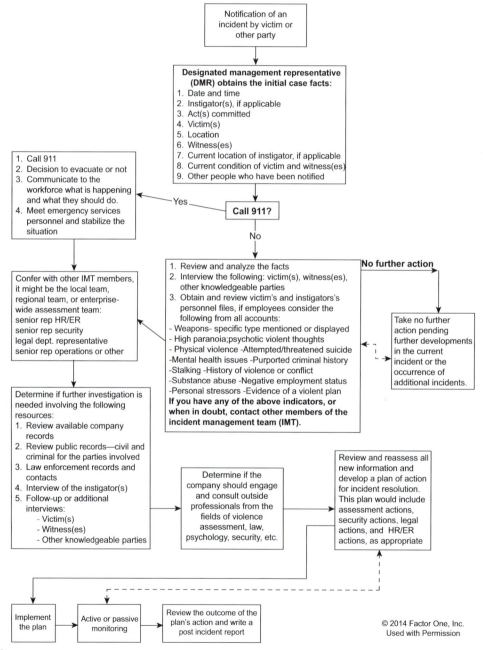

Notification of an incident by victim or other party

Designated management representative (DMR) obtains the initial case facts:
1. Date and time
2. Instigator(s), if applicable
3. Act(s) committed
4. Victim(s)
5. Location
6. Witness(es)
7. Current location of instigator, if applicable
8. Current condition of victim and witness(es)
9. Other people who have been notified

1. Call 911
2. Decision to evacuate or not
3. Communicate to the workforce what is happening and what they should do.
4. Meet emergency services personnel and stabilize the situation

Call 911?

Yes

No

No further action

1. Review and analyze the facts
2. Interview the following: victim(s), witness(es), other knowledgeable parties
3. Obtain and review victim's and instigators's personnel files, if employees consider the following from all accounts:
- Weapons- specific type mentioned or displayed
- High paranoia;psychotic violent thoughts
- Physical violence -Attempted/threatened suicide
-Mental health issues -Purported criminal history
-Stalking -History of violence or conflict
-Substance abuse -Negative employment status
-Personal stressors -Evidence of a violent plan
If you have any of the above indicators, or when in doubt, contact other members of the incident management team (IMT).

Take no further action pending further developments in the current incident or the occurrence of additional incidents.

Confer with other IMT members, it might be the local team, regional team, or enterprise-wide assessment team:
senior rep HR/ER
senior rep security
legal dept. representative
senior rep operations or other

Determine if further investigation is needed involving the following resources:
1. Review available company records
2. Review public records—civil and criminal for the parties involved
3. Law enforcement records and contacts
4. Interview of the instigator(s)
5. Follow-up or additional interviews:
   - Victim(s)
   - Witness(es)
   - Other knowledgeable parties

Determine if the company should engage and consult outside professionals from the fields of violence assessment, law, psychology, security, etc.

Review and reassess all new information and develop a plan of action for incident resolution. This plan would include assessment actions, security actions, legal actions, and HR/ER actions, as appropriate

Implement the plan

Active or passive monitoring

Review the outcome of the plan's action and write a post incident report

Index